Dictionary of
Agriculture

Specialist dictionaries

Dictionary of Accounting	0 7475 6991 6
Dictionary of Banking and Finance	0 7136 7739 2
Dictionary of Business	0 7136 7918 2
Dictionary of Computing	0 7475 6622 4
Dictionary of Economics	0 7136 8203 5
Dictionary of Environment and Ecology	0 7475 7201 1
Dictionary of Human Resources and Personnel Management	0 7136 8142 X
Dictionary of ICT	0 7475 6990 8
Dictionary of Information and Library Management	0 7136 7591 8
Dictionary of Law	0 7475 6636 4
Dictionary of Leisure, Travel and Tourism	0 7475 7222 4
Dictionary of Marketing	0 7475 6621 6
Dictionary of Media Studies	0 7136 7593 4
Dictionary of Medical Terms	0 7136 7603 5
Dictionary of Nursing	0 7475 6634 8
Dictionary of Politics and Government	0 7475 7220 8
Dictionary of Publishing and Printing	0 7136 7589 6
Dictionary of Science and Technology	0 7475 6620 8

Easier English™ titles

Easier English Basic Dictionary	0 7475 6644 5
Easier English Basic Synonyms	0 7475 6979 7
Easier English Dictionary: Handy Pocket Edition	0 7475 6625 9
Easier English Intermediate Dictionary	0 7475 6989 4
Easier English Student Dictionary	0 7475 6624 0
English Thesaurus for Students	1 9016 5931 3

Check Your English Vocabulary workbooks

Academic English	0 7475 6691 7
Business	0 7475 6626 7
Human Resources	0 7475 6997 5
Law	0 7136 7592 6
Medicine	0 7136 7590 X
FCE +	0 7475 6981 9
IELTS	0 7136 7604 3
PET	0 7475 6627 5
Phrasal Verbs and Idioms	0 7136 7805 4
TOEFL®	0 7475 6984 3
TOEIC®	0 7136 7508 X

Visit our website for full details of all our books: **www.acblack.com**

Dictionary of

Agriculture

third edition

A & C Black • London

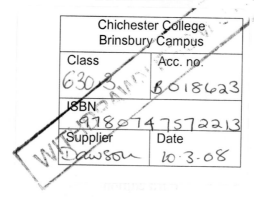
First published in Great Britain in 1990, reprinted 1994, 1995
Second edition published 1996, reprinted 1997, 1998
This third edition published 2006

A & C Black Publishers Ltd
38 Soho Square, London W1D 3HB

A CIP record for this book is available from the British Library

ISBN-10: 0 7136 7778 3
ISBN-13: 978 0 7136 7778 2

Text Production and Proofreading
Heather Bateman, Steve Curtis, Katy McAdam

This book is produced using paper that is made from wood grown in
managed, sustainable forests. It is natural, renewable and recyclable. The
logging and manufacturing processes conform to the environmental
regulations of the country of origin

Text typeset by A & C Black
Printed in Italy by Rotolito Lombarda

Preface

This dictionary provides a basic vocabulary of agricultural terms. It is ideal for students of land management, environmental and veterinary sciences, and is a handy reference for those working in the agriculture industries, especially those for whom English is not a first language.

Each headword is explained in clear, straightforward English and quotations from newspapers and specialist magazines show how the words are used in context. There are also supplements including conversion tables for weights and measures and a list of world commodity markets.

Thanks are due to Dr Mark Lyne, Department of Science, Agriculture and Technology, Writtle College and to Dr Stephen Chadd, Course Director at the Royal Agricultural College, for their help and advice during the production of this new edition.

Pronunciation Guide

The following symbols have been used to show the pronunciation of the main words in the dictionary.

Stress is indicated by a main stress mark (') and a secondary stress mark (,). Note that these are only guides, as the stress of the word changes according to its position in the sentence.

Vowels		*Consonants*	
æ	back	b	buck
ɑː	harm	d	dead
ɒ	stop	ð	other
aɪ	type	dʒ	jump
aʊ	how	f	fare
aɪə	hire	g	gold
aʊə	hour	h	head
ɔː	course	j	yellow
ɔɪ	annoy	k	cab
e	head	l	leave
eə	fair	m	mix
eɪ	make	n	nil
eʊ	go	ŋ	sing
ɜː	word	p	print
iː	keep	r	rest
i	happy	s	save
ə	about	ʃ	shop
ɪ	fit	t	take
ɪə	near	tʃ	change
u	annual	θ	theft
uː	pool	v	value
ʊ	book	w	work
ʊə	tour	x	loch
ʌ	shut	ʒ	measure
		z	zone

Pronunciation is given only for technical entries. For a full guide to pronunciation of all basic words and phrases, the A & C Black *Easier English Intermediate Dictionary* (0-7475-6989-4, £9.99) is recommended.

A

AA *abbreviation* Arboricultural Association

AAPP *abbreviation* Average All Pigs Price

AAPS *abbreviation* Arable Area Payments Scheme

Abandonment of Animals Act 1960 *noun* an Act of Parliament which made it an offence to leave an animal unattended if this was likely to cause distress or pain to the animal

abattoir *noun* a place where animals are slaughtered and prepared for sale to the public as meat

abdomen *noun* a space in the body situated below or behind the diaphragm and above or in front of the pelvis, containing the stomach, intestines, liver and other vital organs

Aberdeen Angus /ˌæbədiːn ˈæŋgəs/ *noun* an early maturing breed of beef cattle, which are naturally hornless and usually black all over. Aberdeen Angus cattle usually have a rather small headed and a long deep body. They are highly valued for quality beef.

abiotic /ˌeɪbaɪˈɒtɪk/ *adjective* not relating to a living organism ○ *abiotic factors*

abiotic stress *noun* stress caused by environmental factors such as drought or extreme heat or cold, not by biological factors

abiotic stress resistance *noun* resistance in organisms to stress arising from non-biological causes such as drought

abomasal ulcer /ˌæbəʊmeɪs(ə)l ˈʌlsə/ *noun* a disease common in both calves and adult cattle. Calves show poor growth and lose appetite. In rare cases, cows may bleed to death.

abomasum /ˌæbəʊˈmeɪsəm/ *noun* the fourth stomach of a ruminant. ◊ **omasum, reticulum, rumen**

abort *verb* **1.** to stop a process or the development of something before it is finished, or to stop developing ○ *The flowers abort and drop off in hot, dry conditions, with no fruit developing.* **2.** to end a pregnancy in an animal and prevent the birth of young **3.** to give birth before the usual end of a pregnancy (*technical*) Also called **miscarry**

abortion *noun* a situation when a pregnancy in a farm animal ends prematurely, generally as the result of a disease or infection (*technical*)

abreast parlour /əˈbrest ˌpɑːlə/ *noun* a type of milking parlour where the cows stand side by side with their heads facing away from the milker. ◊ **herringbone parlour, rotary parlour**

abscess *noun* a painful swollen area where pus forms

abscission /æbˈsɪʃ(ə)n/ *noun* the shedding of a leaf or fruit due to the formation of a layer of cells between the leaf or fruit and the rest of the plant (NOTE: It occurs naturally in autumn, e.g. leaf fall, or at any time of the year in response to stress.)

absorb *verb* to take something in ○ *Warm air absorbs moisture more easily than cold air.* ○ *Salt absorbs moisture from the air.*

absorption *noun* **1.** the process of taking in water, dissolved minerals and other nutrients across cell membranes **2.** the taking into the body of substances such as proteins or fats which have been digested from food and enter the bloodstream from the stomach and intestines **3.** the taking up of one type of substance by another, e.g. of a liquid by a solid or of a gas by a liquid

abstract /æb'strækt/ *verb* to remove water from a river so that it can be used by industry, farmers or gardeners

abstraction /æb'strækʃən/ *noun* **1.** the removal of water from a river or other source for use by industry, farmers or gardeners **2.** the removal of something such as gas, oil, mineral resources or gravel from the ground

Acacia /ə'keɪʃə/ *noun* a species of tree often grown for its pretty leaves and blossoms

ACAF *abbreviation* Advisory Committee on Animal Feedingstuffs

acariasis /ˌækə'raɪəsɪs/ *noun* a skin disease caused by ticks or mites

acaricide /ə'kærɪsaɪd/ *noun* a substance used to kill mites and ticks. Also called **acaridicide**

acarid /'ækərɪd/ *noun* a small animal which feeds on plants or other animals by piercing the outer skin and sucking juices, e.g. a mite or tick

Acarida /ə'kærɪdə/ *noun* the order of animals including mites and ticks. Also called **Acarina**

acaridicide /ˌækə'rɪdɪsaɪd/ *noun* same as **acaricide**

Acarina /ˌækə'riːnə/ *noun* same as **Acarida**

ACAS /'eɪkæs/ *abbreviation* Advisory, Conciliation and Arbitration Service

ACC *abbreviation* Agricultural Credit Corporation

acceptable daily intake *noun* the quantity of a substance such as a nutrient, vitamin, additive or pollutant which a person or animal can safely consume daily over their lifetime. Abbr **ADI**

'A UK wide consultation on the use of the colourant canthaxanthin in animal feed, used to give farmed salmon its pink colour, has been launched by the Food Standards Agency. Brussels' Scientific Committee on Animal Nutrition said: "Some consumers of high levels of produce from farmed fish were likely to exceed the acceptable daily intake for canthaxanthin".' [*The Grocer*]

access *noun* **1.** a place of entry, or the right of entry, to somewhere **2.** the right of the public to go onto uncultivated private land for recreation. ◊ **Countryside and Rights of Way Act**

acclimatisation /əˌklaɪmətaɪ'zeɪʃ(ə)n/, **acclimatization, acclimation** /ˌæklə'meɪʃ(ə)n/ *noun* the process of adapting to a different environment (NOTE: This process is known as **acclimatisation** if the changes occur naturally and **acclimation** if they are produced in laboratory conditions.)

COMMENT: When an organism such as a plant or animal is acclimatising, it is adapting physically to different environmental conditions, such as changes in food supply, temperature or altitude.

accommodation land *noun* land available for short-term tenancy

accredit /ə'kredɪt/ *verb* to recognise officially

accredited herd *noun* a herd of cattle registered under a scheme as being free from Brucellosis

accredited milk *noun* milk from a herd accredited as being free from Brucellosis

accumulated temperature *noun* the number of hours during which the temperature is above a particular point, taken as the minimum temperature necessary for growing a specific crop (NOTE: In the UK, this is usually taken to be the number of hours above 6°C.)

acer /'eɪsə/ *noun* a maple or sycamore tree. Genus: *Acer*.

acetonaemia /əˌsiːtəʊ'niːmiə/ *noun* a disease affecting cows, caused by ketone bodies accumulating. The animal loses appetite and the smell of acetone affects the breath, the urine and milk.

acetone *noun* a colourless liquid that has a sweetish smell and is flammable, used as a solvent and in the manufacture of organic chemicals. Formula: CH_3COCH_3.

achene /ə'kiːn/ *noun* a dry single-seeded fruit that does not split open (NOTE: Achenes are produced by plants such as dandelions and sunflowers.)

acid *noun* **1.** a chemical compound containing hydrogen which dissolves in water and forms hydrogen ions, or reacts with an alkali to form a salt and water, and turns litmus paper red **2.** any bitter juice

acid deposition /ˌæsɪd ˌdepə'zɪʃ(ə)n/ *noun* same as **acid rain**

acid grassland *noun* a type of vegetation that typically grows on soils that drain freely and are low in mineral nutrients, and may also occur on post-industrial sites. The range of plant species found is small. ◊ **calcareous grassland**

acidic *adjective* referring to acids ○ *acidic properties*

acidification /ə,sɪdɪfɪ'keɪʃ(ə)n/ *noun* the process of becoming acid or of making a substance more acid ○ *Acidification of the soil leads to the destruction of some living organisms.*

acidify /ə'sɪdɪfaɪ/ *verb* to make a substance more acid, or to become more acid ○ *Acid rain acidifies the soils and waters where it falls.* ○ *The sulphur released from wetlands as sulphate causes lakes to acidify.* (NOTE: Feed additives can be used to acidify animals' urine, which makes them less susceptible to infections of the urinary tract.)

acidity *noun* the proportion of acid in a substance ○ *The alkaline solution may help to reduce acidity.*

COMMENT: Acidity and alkalinity are measured according to the pH scale. pH7 is neutral. Numbers above pH7 show alkalinity, while those below show acidity.

acidophilus milk /,æsɪ'dɒfɪləs ,mɪlk/ *noun* a cultured milk made from fresh milk which is allowed to go sour in a controlled way. One of the most popular types of acidophilus milk in Europe is yoghurt.

acidosis /,æsɪ'dəusɪs/ *noun* an unusually high proportion of acid waste products such as urea in the blood, sometimes caused by a metabolic dysfunction (NOTE: As acidity increases the rumen wall becomes inflamed. The animal dehydrates progressively, the blood turns more acidic and in extreme cases the animal may die.)

acid rain *noun* precipitation such as rain or snow which contains a higher level of acid than normal. Also called **acid deposition, acid precipitation**

COMMENT: Acid rain is mainly caused by sulphur dioxide, nitrogen oxide and other pollutants being released into the atmosphere when fossil fuels such as oil or coal containing sulphur are burnt. Carbon combines with sulphur trioxide from sulphur-rich fuel to form particles of an acid substance. The effects of acid rain are primarily felt by wildlife. The water in lakes becomes very clear as fish and microscopic animal life are killed. It is believed acid rain kills trees, especially conifers, making them gradually lose their leaves and die. Acid rain can also damage surfaces such as the stone surfaces of buildings when it falls on them.

acid soil *noun* soil which has a pH value of 6 or less (NOTE: Farming tends to make the soil more acid, but most farm crops will not grow well if the soil is very acid. This can be cured by applying one of the materials commonly used for adding lime, such as ground chalk or limestone.)

ACOS *abbreviation* Advisory Committee on Organic Standards

ACP *abbreviation* Advisory Committee on Pesticides

ACPAT *abbreviation* Association of Chartered Physiotherapists in Animal Therapy

ACP states /,eɪ siː 'piː ,steɪts/ ♦ **Lomé Convention**

ACR *abbreviation* automatic cluster removal

acre *noun* a unit of measurement of land area, equal to 4840 square yards or 0.4047 hectares

ACRE /'eɪkə/ *abbreviation* Advisory Committee on Releases to the Environment

acreage /'eɪkərɪdʒ/ *noun* the area of a piece of land measured in acres

acreage allotment *noun* a quota system operated in the USA, which limits the area of land which can be planted with a certain type of crop

acreage reduction programme *noun* an American federal programme under which farmers are only eligible for subsidies if they reduce the acreage of certain crops planted. Abbr **ARP** (NOTE: The British equivalent is **set-aside**.)

actinobacillosis /,æktɪnəubæsɪ'ləusɪs/ *noun* a disease of cattle affecting the tongue and throat. It also occurs in sheep as swellings on the lips, cheeks and jaws. Also called **cruels, wooden tongue**

actinomycete /,æktɪnəu'maɪsiːt/ *noun* a bacterium shaped like a rod or filament. Order: Actinomycetales. (NOTE: Some actinomycetes cause diseases while others are sources of antibiotics.)

actinomycosis /,æktɪnəumaɪ'kəusɪs/ *noun* a disease of cattle and pigs, where the animal is infected with bacteria which form abscesses in the mouth and lungs. Also called **lumpy jaw**

activate *verb* to start a process or to make something start working ○ *Pressing this switch activates the pump.*

activated sludge *noun* solid sewage containing active microorganisms and air, mixed with untreated sewage to speed up the purification process

activator /ˈæktɪveɪtə/ *noun* a substance which activates a process ○ *a compost activator*

active ingredient *noun* the main effective ingredient of something such as an ointment or agrochemical, as opposed to the base substance. Abbr **AI**

actuals /ˈæktʃuəlz/ *plural noun* stocks of commodities such as cotton or rice which are available for shipping. Compare **futures**

acute *adjective* **1.** referring to a disease which comes on rapidly and can be dangerous ○ *acute mastitis* **2.** referring to a pain which is sharp and intense ○ *acute stomach pain* ▶ compare **chronic**

ADAS /ˈeɪdæs/ *noun* a commercial research-based organisation that offers technical advice on agricultural, food and environmental matters to rural industries. Former name **Agricultural Development and Advisory Service**

additive *noun* **1.** a chemical which is added to food to improve its appearance or to keep it fresh ○ *The tin of beans contains a number of additives.* ○ *These animal foodstuffs are free from all additives.* **2.** a chemical which is added to something to improve it ○ *A new fuel additive made from plants could help reduce energy costs.* **3.** a substance which is added to animal feedingstuffs to provide antibiotics, mineral supplements, vitamins or hormones

addled egg /ˈæd(ə)ld eg/ *noun* a rotten egg, an egg which produces no chick

ADHAC *abbreviation* Agricultural Dwelling House Advisory Committee

ADI *abbreviation* acceptable daily intake

adipose /ˈædɪpəʊs/ *adjective* containing or made of fat

adipose tissue *noun* a type of tissue where the fibrous parts of cells are replaced by fat when too much food is eaten

Adjusted Eurospec Average /ə ˌdʒʌstɪd ˈjʊərəʊspek ˌæv(ə)rɪdʒ/ *noun* formerly, the average price for pigs. It was replaced in 2003 by the Deadweight Average Pig Price. Abbr **AESA**

ad lib feeding /ˌæd ˈlɪb ˌfiːdɪŋ/ *noun* the unrestricted supply of feed, day and night

admixture /ˈædmɪktʃə/ *noun* the proportion of a seed crop which is made up of weed seeds or other crop species

ADP *abbreviation* Agricultural Development Programme

ADRA *abbreviation* Animal Diseases Research Association

adrenal gland /əˈdriːn(ə)l glænd/ *noun* one of two endocrine glands at the top of the kidneys which produce adrenaline and other hormones

adrenaline /əˈdrenəlɪn/ *noun* a hormone secreted by the medulla of the adrenal glands which has an effect similar to stimulation of the sympathetic nervous system (NOTE: The US term is **epinephrine**.)

ADS *abbreviation* Agriculture Development Scheme

adsorb /ædˈzɔːb/ *verb* (*of a solid*) to bond with a gas or vapour which touches its surface

adsorbable /ədˈzɔːbəb(ə)l/ *adjective* referring to a gas or vapour which is able to bond with a solid when it touches its surface

adsorbent /ədˈzɔːbənt/ *adjective* able to adsorb something such as a gas or vapour

adulterate /əˈdʌltəreɪt/ *verb* to reduce the quality of something, such as by adding water to milk

advanced register *noun* a book which records breeding performance of outstanding livestock

adventitious /ˌædvənˈtɪʃəs/ *adjective* referring to a root which develops from a node on a plant stem and not from another root

Advisory, Conciliation and Arbitration Service *noun* an organisation which advises on employment disputes and rights in the workplace. Abbr **ACAS**

Advisory Committee on Animal Feedingstuffs *noun* a committee set up by the Food Standards Agency in 1999 to advise on health and safety in animal feeds and feeding practices. Abbr **ACAF**

Advisory Committee on Organic Standards *noun* a non-departmental public body set up by the Government to advise ministers on organic standards. Abbr **ACOS**

Advisory Committee on Pesticides *noun* a statutory body set up under the UK Food and Environment Protection Act 1985 to advise on all matters relating to the control of pesticides. Abbr **ACP**

Advisory Committee on Releases to the Environment noun an independent advisory committee giving statutory advice to UK government ministers on the risks to human health and the environment from the release and marketing of genetically modified organisms (GMOs). It also advises on the release of some non-GM species of plants and animals that are not native to Great Britain. Abbr **ACRE**

AEA abbreviation Agricultural Engineers Association

AEBC abbreviation Agriculture, Environment and Biotechnology Commission

aerate /'eəreɪt/ verb to allow air to enter a substance, especially soil or water ○ Worms play a useful role in aerating the soil.

aeration /eə'reɪʃ(ə)n/ noun the replacement of stagnant soil air with fresh air

'Soil nutrient availability can also be influenced by compaction as one of the main effects of compaction is on soil aeration, which can lead to de-nitrification (loss of nitrogen into the atmosphere).' [Arable Farming]

COMMENT: The process of aeration of soil is mainly brought about by the movement of water into and out of the soil. Rainwater drives out the air and then, as the water drains away or is used by plants, fresh air is drawn into the soil to fill the spaces. The aeration process is also assisted by changes in temperature, good drainage, cultivation and open soil structure. Sandy soils are usually well aerated. Clay soils are poorly aerated.

aerial adjective referring to something which exists in the air

aerial root noun a root of some plants, which hangs above the ground or clings to other plants so that it can take up moisture from the air

aerobic adjective needing oxygen for its existence or for a biochemical reaction to occur. Compare **anaerobic**

AESA abbreviation Adjusted Eurospec Average

AFB abbreviation American foul brood

afforest /ə'fɒrɪst/ verb to plant an area with trees

afforestation /ə,fɒrɪ'steɪʃ(ə)n/ noun 1. the planting of trees in an area or as a crop ○ There is likely to be an increase in afforestation of upland areas if the scheme is introduced. 2. the planting of trees on land previously used for other purposes

aflatoxin /,æflə'tɒksɪn/ noun a toxin produced by species of the fungus Aspergillus, especially Aspergillus flavus, which grows on seeds and nuts and affects stored grain

African swine fever /,æfrɪkən 'swaɪn ,fiːvə/ noun a virus disease which is highly contagious among pigs. Animals suffer fever and high temperature followed by death. In Europe, it occurs in parts of Spain.

AFS abbreviation Assured Food Standards

afterbirth /'ɑːftəbɜːθ/ noun the remains of the placenta pushed out of the uterus of the dam at the birth of a young animal. Also called **cleansing**

aftermath noun grass which grows quickly after cutting for hay, and which will provide a second cut

agalactia /,eɪɡə'læktiə/ noun a disease of pigs, a form of post-farrowing shock. The sow does not secrete milk.

agbiotech /'æɡ,baɪəʊtek/ noun biotechnology applied to agriculture or an agricultural industry

age noun the number of years during which a person or thing has existed ○ The size varies according to age. ■ verb to treat flour to make the dough more elastic and whiter

agglutination /ə,ɡluːtɪ'neɪʃ(ə)n/ noun a process in which cells come together to form clumps. For example, agglutination takes place when bacterial cells are in the presence of serum or affects blood cells when blood of different types is mixed.

agglutination tests plural noun 1. tests to identify bacteria 2. tests used to detect Brucellosis in cattle

aggregate noun a mass of soil and rock particles stuck together

aggregate measure of support noun an index which shows the actual monetary value of the support given by the Government to a sector such as agriculture. Abbr **AMS**

agist /ə'dʒɪst/ verb to take another person's livestock to feed on your land

agistment /ə'dʒɪstmənt/ noun money paid for grazing stock on land owned by another person. The owner of the land is responsible for the feeding and care of the livestock.

agitator /ˈædʒɪteɪtə/ *noun* the part of a machine for harvesting root crops, such as potatoes, which shakes the earth off the crop after it has been lifted

agrarian /əˈgreəriən/ *adjective* referring to matters of land tenure and problems arising from land ownership

agri- /ægri/ *prefix* referring to agriculture or to the cultivation or management of land. ◊ **agro-**

agri-biotechnology *noun* biotechnology as applied to agriculture

agribusiness /ˈægrɪbɪznəs/ *noun* a large-scale farming business run along the lines of a conventional company, often involving the processing, packaging and sale of farm products

'The Committee of Public Accounts has told Defra that farm business support should be targeted at those enterprises which need it most. This means smaller and intermediate sized farming businesses run by families and individuals, rather than large agribusinesses.' [*Farmers Guardian*]

agricultural *adjective* referring to farming

Agricultural and Food Research Council *noun* formerly, a council established to organise and provide funds for agricultural and food research. It was replaced by the Biotechnology and Biological Sciences Research Council in 1993. Abbr **AFRC**

agricultural burning *noun* the burning of agricultural waste as part of farming practice, e.g. stubble burning

Agricultural Chemicals Approval Scheme *noun* a scheme which gave advice to farmers on the use and efficiency of chemicals and which tested chemicals before use by farmers. It was operated by the Agricultural Chemicals Approved Organisation and was a voluntary scheme which has now been replaced by the FEPA legislation. Abbr **ACAS**

agricultural depopulation /ˌægrɪkʌtʃ(ə)rəl ˌdiːpɒpjʊˈleɪʃ(ə)n/ *noun* the fact of people leaving farms to live and work elsewhere

Agricultural Development and Advisory Service *noun* former name for **ADAS**

Agricultural Development Programme *noun* a plan to improve the agricultural productivity of a community

through training and modernisation of equipment. Abbr **ADP**

agricultural economist *noun* a person who studies the economics of the agricultural industry

agricultural engineer *noun* **1.** a person trained in applying the principles of science to farming **2.** a person who designs, manufactures or repairs farm machinery and equipment

agricultural engineering *noun* the applying of the principles of science to farming

Agricultural Engineers Association *noun* an organisation which protects the interests of manufacturers and suppliers of agricultural machinery in the UK. Abbr **AEA**

agricultural holding *noun* a basic unit for agricultural production, consisting of all the land and livestock under the management of one particular person or group of people

Agricultural Holdings Act 1984 *noun* an Act of Parliament which gives protection to tenants in questions of the fixing of rent and security of tenure. It makes provision for tenancies for a life time and for short-term lettings.

Agricultural Industries Confederation *noun* a trade association for suppliers of feed, fertilisers, seeds and grain to the agricultural sector. Abbr **AIC**

agriculturalist /ˌægrɪˈkʌltʃ(ə)rəlɪst/ *noun* a person trained in applying the principles of science to farming

agricultural labourer *noun* a person who does heavy work on a farm, formerly a rural worker with no land, and sometimes still a worker with a special skill, such as ditching or hedging

Agricultural Land Tribunal *noun* a court established in 1947 to hear appeals against decisions affecting owners or tenants of agricultural land

Agricultural Mortgage Corporation *noun* a corporation which makes loans available to borrowers on the security of agricultural land and buildings in England and Wales

agricultural policy *noun* the decisions and commitments that make up a government's attitude to and programme for agriculture

Agricultural Revolution *noun* the changes in agriculture which transformed

Britain's countryside in the 18th and 19th centuries

Agricultural Stabilization and Conservation Service *noun* formerly, a service of the federal Department of Agriculture which operated the department's various schemes throughout the USA. It was incorporated into the Farm Service Agency in 1993. Abbr **ASCS**

Agricultural Wages Board *noun* a board which fixes minimum wages and holiday entitlements for agricultural workers, and deals with terms and conditions of their employment. Abbr **AWB**

agricultural waste *noun* waste matter produced on a farm, e.g. plastic containers for pesticides

Agricultural Waste Stakeholders' Forum *noun* a group that includes representatives of government, farming organisations, waste companies and farm suppliers with the aim of identifying and dealing with issues of waste management in agriculture

agriculture *noun* the cultivation of land, including horticulture, fruit growing, crop and seed growing, dairy farming and livestock breeding

COMMENT: The use of land to raise crops for eating first started about 10,000 years ago. All plants grown for food have been developed over many centuries from wild plants, which have been progressively bred to give the best yields in different types of environment. Genes from wild plants are likely to be more hardy and resistant to disease, and are still kept in gene banks to strengthen new cultivated varieties.

Agriculture, Environment and Biotechnology Commission *noun* the UK government advisory body on biotechnology issues affecting agriculture and the environment

Agriculture (Miscellaneous Provisions) Act 1968 *noun* an Act of Parliament which defines what constitutes livestock and makes it an offence to cause unnecessary pain or distress to a farm animal

Agriculture Acts *plural noun* Acts of Parliament, introduced to update legislation affecting agricultural policy

Agriculture and Rural Affairs Department *noun* the department of the devolved Welsh Assembly government which deals with farming, the environment, animal welfare and rural development in Wales. Abbr **ARAD**

Agriculture Industry Advisory Committee *noun* the committee that advises the Health and Safety Commission on the protection of people at work and others from hazards to health and safety arising within the agricultural and related industries

'Agriculture in the United Kingdom' *noun* a review undertaken each year by the British government, reporting on the state of the agricultural industry (NOTE: Formerly called the **Annual Review of Agriculture**.)

agri-environmental indicator *noun* an indicator designed to provide information on the various ways in which agriculture affects the environment

agri-environment scheme *noun* a scheme to give money to farmers to persuade them to adopt land management practices that benefit the environment, e.g. the Environmental Stewardship Scheme

agri-food *adjective* relating to industries which are involved in the mass production, processing and inspection of food products made from agricultural commodities

'The agri-food industry in Wales is already growing at a faster rate than in any other part of the UK with employment in the Welsh food sector since 1998 showing a 1.7 per cent increase compared to a 5.2 per cent fall in the rest of Great Britain.' [*Farmers Guardian*]

agri-tourism *noun* a type of tourism where visitors can help out on a working farm, buy produce from a farm shop or be involved in other leisure activities on the farm's land

agro- /'ægrəʊ/ *prefix* referring to agriculture or to the cultivation or management of land. ◊ **agri-**

agrobiodiversity /ˌægrəʊbaɪəʊdaɪ'vɜːsɪti/ *noun* the aspects of biodiversity that affect agriculture and food production, including within-species, species and ecosystem diversity

agrochemical industry /'ægrəʊkemɪk(ə)l ˌɪndəstri/ *noun* the branch of industry which produces pesticides and fertilisers used on farms

agrochemicals /'ægrəʊˌkemɪk(ə)lz/ *plural noun* pesticides and fertilisers developed for agricultural use

agroclimatology /ˌægrəʊklaɪmə'tɒlədʒi/ noun the study of climate and its effect on agriculture

agroecology /'ægrəʊˌkɒlədʒi/ noun the ecology of a crop-producing area

agroecosystem /'ægrəʊˌiːkəʊsɪstəm/ noun a community of organisms in a crop-producing area

agroforestry /'ægrəʊˌfɒrɪstri/ noun the growing of farm crops and trees together as a farming unit

agroindustry /'ægrəʊˌɪndəstri/ noun an industry dealing with the supply, processing and distribution of farm products

agronomist /ə'grɒnəmɪst/ noun a person who studies the cultivation of crops and provides advice to farmers

agronomy /ə'grɒnəmi/ noun the scientific study of the cultivation of crops

AHDO abbreviation Animal Health Divisional Office

AHO abbreviation 1. Animal Health Office 2. Animal Health Officer

A horizon noun topsoil. ◊ horizon

AI[1] abbreviation 1. active ingredient 2. artificial insemination

AI[2] /ˌeɪ 'aɪ/ verb to inseminate an animal artificially ○ Twenty ewes were AI'd.

AIAC abbreviation Agriculture Industry Advisory Committee

AIC abbreviation Agricultural Industries Confederation

AI centre noun a centre which keeps breeding bulls, boars and rams, and quantities of their semen for use in artificial insemination

air layering noun a method of propagation where a stem is partially cut, then surrounded with damp moss, which is tied securely to the stem. Roots will grow from the cut at the point where it is in contact with the moss.

air pollution noun the contamination of the air by substances such as gas or smoke. Also called **atmospheric pollution** (NOTE: Odour nuisance from livestock units and other farming activities is governed by the Environmental Protection Act. The burning of agricultural crop residues is now banned.)

'Several countries, including the US, Brazil and Denmark, have plants up and running already and reports suggest the blended fuel produces a higher performance than pure petrol as well as significantly reducing air pollution.' [Farmers Weekly]

albinism /'ælbɪnɪz(ə)m/ noun an inherited lack of pigmentation in an organism (NOTE: A person or animal with albinism has unusually white skin and hair.)

albino /æl'biːnəʊ/ noun an organism that is unusually white, having little or no pigmentation in its skin, hair or eyes because it is deficient in the colouring pigment melanin

albumen /'ælbjʊmɪn/ noun the white of an egg, containing albumin

albumin /'ælbjʊmɪn/ noun a common protein, soluble in water and found in plant and animal tissue and digested in the intestine

albumose /'ælbjʊməʊz/ noun an intermediate product in the digestion of protein

ALC abbreviation agricultural land classification

alder /'ɔːldə/ noun a hardwood tree in the birch family. Genus: Alnus. (NOTE: The wood is resistant to decay in wet conditions.)

aldosterone /æl'dɒstərəʊn/ noun a hormone secreted by the adrenal gland which regulates the balance of sodium and potassium in the body and the amount of body fluid

aldrin /'ɔːldrɪn/ noun an organochlorine insecticide that is banned in the European Union

aleurone /'æluːrəʊn/ noun a protein found in the outer skin of seeds

alfalfa /æl'fælfə/ noun same as **lucerne**

algae plural noun tiny plants living in water or in moist conditions, which contain chlorophyll and have no stems or roots or leaves

COMMENT: Algae grow rapidly in water which is rich in phosphates. When the phosphate level increases, as when fertiliser runoff enters the water, the algae multiply greatly to form enormous floating mats (or blooms), blocking out the light and inhibiting the growth of other organisms. When the algae die, they combine with all the oxygen in the water so that other organisms suffocate.

algaecide /'ældʒɪsaɪd/ noun same as **algicide**

algae poisoning noun poisoning caused by toxic substances released when algae decompose

algicide /'ældʒɪsaɪd/ noun a substance used to kill algae

alien *adjective* same as **exotic** ○ *A fifth of the area of the national park is under alien conifers.* ○ *Alien species, introduced by settlers as domestic animals, have brought about the extinction of some endemic species.* ■ *noun* same as **exotic**

alimentary canal /ˌælɪˌment(ə)ri kə 'næl/ *noun* a tube in the body going from the mouth to the anus, including the throat, stomach and intestines, through which food passes and is digested

alkali *noun* a substance which reacts with an acid to form a salt and water. It may be either a soluble base or a solution of a base that has a pH value of more than 7. (NOTE: The plural is **alkalis**; an alternative US plural is **alkalies**.)

alkaline *adjective* containing more alkali than acid and having a pH value of more than 7

alkalinity /ˌælkə'lɪnɪti/ *noun* the amount of alkali in something such as soil, water or a body ○ *Hyperventilation causes fluctuating carbon dioxide levels in the blood, resulting in an increase of blood alkalinity.*

COMMENT: Alkalinity and acidity are measured according to the pH scale. pH7 is neutral, and pH8 and upwards are alkaline. Alkaline solutions are used to counteract the effects of acid poisoning and also of bee stings.

alkaloid /'ælkələɪd/ *adjective* similar to an alkali ■ *noun* one of many poisonous substances found in plants, which use them as a defence against herbivores (NOTE: Many alkaloids such as atropine, morphine or quinine are also useful as medicines.)

allele /ə'liːl/ *noun* one of two or more alternative forms of a gene, situated in the same area (**locus**) on paired chromosomes and controlling the inheritance of the same characteristic

allelopathy /ˌælɪ'lɒpəθi/ *noun* the release by one plant of a chemical substance that restricts the germination or growth of another plant

allergen *noun* a substance which produces a hypersensitive reaction in someone. Allergens are usually proteins, and include foods, the hair of animals and pollen from flowers, as well as dust.

allergy *noun* a sensitivity to substances such as pollen or dust, which cause a physical reaction ○ *She has an allergy to household dust.* ○ *He has a penicillin allergy.*

alley cropping *noun* the planting of crops such as maize or sorghum between

trees (NOTE: The trees help to prevent soil erosion, especially on slopes, and may benefit soil fertility if the leaves are used as mulch or if the trees are legumes.)

Allium /'æliəm/ *noun* the Latin name for a family of plants including the onion, leek, garlic and chives

allo- /æləʊ/ *prefix* different

allogamy /ə'lɒgəmi/ *noun* fertilisation by pollen from different flowers or from flowers of genetically different plants of the same species

COMMENT: Some fruit trees are self-fertile, that is, they fertilise themselves with their own pollen. Others need pollinators that are usually different cultivars of the same species.

allograft /'æləʊgrɑːft/ *noun* a graft of tissue from one individual to another of the same species. Also called **homograft**

allopatric /ˌælə'pætrɪk/ *adjective* referring to plants of the same species which grow in different parts of the world and so do not cross-breed

allotment *noun* a small area of land, owned by a municipality, which is let to a person called an allotment-holder for the cultivation and production of vegetables and fruit for the consumption of the holder and his or her family

all-terrain vehicle *noun* a vehicle which can be driven over all types of land surface. Abbr **ATV**

alluvial /ə'luːviəl/ *adjective* referring to alluvium

alluvium /ə'luːviəm/ *noun* the silt deposited by a river or a lake

alm /ɑːm/ *noun* an alpine pasture normally only grazed in summer

almond *noun* a small tree (*Prunus dulcis*) grown for its edible nuts, or an edible nut produced by this tree

almond oil *noun* an oil from almond seed used for toilet preparations and for flavouring

alp /ælp/ *noun* a high mountain pasture, above the treeline

alpaca /æl'pækə/ *noun* an animal which is similar to the llama. A native of the Andes, it is domesticated and reared for its very soft and elastic wool.

alpha acids /'ælfə ˌæsɪdz/ *noun* a number of related compounds found in hops, which give hops their bitter taste

alpha amylase *noun* an enzyme present in wheat seed, which changes some starch to sugar. Excessive amounts can result in loaves of bread with sticky texture.

alpine pastures /ˌælpaɪn ˈpɑːstʃəz/ *plural noun* grass fields in high mountains which are used by cattle farmers in the summer

alpine plants *plural noun* plants which grow on high mountains ○ *alpine vegetation grows above the treeline*

alternate husbandry *noun* husbandry in which arable and grassland cultivation are alternated every few years

alternative technology *noun* the use of traditional techniques and equipment and materials that are available locally for agriculture, manufacturing and other processes

alveolus /ˌælvɪˈəʊləs, ælˈviːələs/ *noun* a thin-walled air sac that occurs in large numbers in each lung and allows oxygen to enter and carbon dioxide to leave the blood

AMC *abbreviation* Agricultural Mortgage Corporation

American bison *noun* ♦ **bison**

American foul brood *noun* a disease affecting bees that is caused by a bacterial parasite of the Bacillaceae family that infests the larvae. Abbr **AFB**

amino acid *noun* a chemical compound which is a component of proteins ○ *Proteins are first broken down into amino acids.* ◊ **essential amino acid**

COMMENT: Amino acids all contain carbon, hydrogen, nitrogen and oxygen, as well as other elements. Some amino acids are produced in the body itself, but others have to be absorbed from food.

ammonia *noun* a gas with an unpleasant smell that is easily soluble in water. Formula: NH_3.

COMMENT: Ammonia is released into the atmosphere from animal dung. It has the effect of neutralising acid rain, but in combination with sulphur dioxide it forms ammonium sulphate which damages the green leaves of plants.

ammoniacal /ˌæməˈnaɪək(ə)l/ *adjective* referring to ammonia

ammoniacal nitrogen *noun* nitrogen derived from ammonia

ammonia treatment *noun* a method of treating straw, using ammonia to make it more palatable and nutritious

ammonium /əˈməʊniəm/ *noun* an ion formed from ammonia

ammonium nitrate *noun* a popular fertiliser used as top dressing (NOTE: It is available in a special prilled or granular form, and can be used both as a straight fertiliser and in compounds.)

ammonium phosphate *noun* a fertiliser which can be used straight, but is more often used in compounds (NOTE: Applications may increase the acidity of the soil.)

ammonium sulphate *noun* a colourless crystalline solid that is soluble in water, used as a fertiliser. Formula: $(NH_4)_2SO_4$. Also called **sulphate of ammonia**

amniotic fluid *noun* the fluid that surrounds and protects a foetus

amoeba *noun* a single-celled organism found in water, wet soil, or as a parasite of other organisms (NOTE: The plural is **amoebae.**)

amoebic *adjective* referring to an amoeba or amoebae

AMS *abbreviation* aggregate measure of support

amylase /ˈæmɪleɪz/ *noun* an enzyme which converts starch into maltose

anabolic steroids /ˌænəbɒlɪk ˈsterɔɪdz/ *plural noun* hormones which encourage growth and muscle building

anabolism /æˈnæbəlɪz(ə)m/ *noun* the process of building up complex chemical substances on the basis of simpler ones

anaemia *noun* a condition where the level of red blood cells is less than normal or where the haemoglobin is reduced, making it more difficult for the blood to carry oxygen (NOTE: The US spelling is **anemia.**)

anaemic /əˈniːmɪk/ *adjective* affected by anaemia (NOTE: The US spelling is **anemic.**)

anaerobic /ˌænəˈrəʊbɪk/ *adjective* not needing oxygen for existence. Compare **aerobic**

anaerobically /ˌænəˈrəʊbɪkli/ *adverb* without using oxygen ○ *Slurry is digested anaerobically by bacteria.*

anaerobic decomposition *noun* the breaking down of organic material by microorganisms without the presence of oxygen

anaerobic digester *noun* a digester that operates without oxygen ○ *Anaerobic digesters can be used to convert cattle manure into gas.*

COMMENT: Anaerobic digesters for pig, cattle and poultry waste feed the waste into a tank where it breaks down biologically without the presence of oxygen to give off large amounts of methane. This gas is then used to generate electricity. The remaining slurry can be applied directly to the land.

anaerobic digestion *noun* the breakdown of organic material without the presence of oxygen, a process which permanently removes the unpleasant smell of many organic wastes so that they can be used on agricultural land

anaerobism /ˌænəˈrəʊbɪz(ə)m/ *noun* a lack of oxygen such as is found in gley soils

anaesthesia /ˌænəsˈθiːziə/ *noun* **1.** the loss of the feeling of pain **2.** a process that prevents a person or animal from feeling pain, usually by the use of drugs (NOTE: The US spelling is **anesthesia**.)

analyse *verb* **1.** to examine something in detail ○ *We analysed the milk yields from different breeds of cow.* **2.** to separate a substance into its parts ○ *The laboratory is analysing the soil samples.* ○ *When the water sample was analysed it was found to contain traces of bacteria.*

analysis *noun* **1.** the process of examining something in detail **2.** the process of breaking down a substance into its parts in order to study them closely ○ *Samples of material were removed for analysis.*

COMMENT: Chemical and electrical methods are used in soil analysis to determine the pH and lime requirements of a soil. Portable testing equipment, using colour charts, is sometimes used to test for pH.

analyst *noun* **1.** a person who examines samples of substances to find out what they are made of **2.** a person who carries out a study of a problem ○ *a health and safety analyst*

anaplasmosis /ˌænəplæzˈməʊsɪs/ *noun* an infectious disease of cattle, characterised by anaemia

anchorage /ˈæŋkərɪdʒ/ *noun* the ability of plant roots to hold firm in the soil

Ancona /ænˈkəʊnə/ *noun* a laying breed of chicken, with white tips on black feathers

Andalusian /ˌændəluːˈsiːən/ *noun* **1.** a dark red breed of cattle, used both as draught animals and for beef **2.** a laying breed of chicken with blue feathers

angelica /ænˈdʒelɪkə/ *noun* a plant with dark green stems, which are crystallised with sugar and used in confectionery

Angeln /ˈæŋgeln/ *noun* a German dual-purpose breed of cattle, red or brown in colour, with black hooves

angiosperm /ˈændʒiəʊspɜːm/ *noun* a plant in which the sex organs are carried within flowers and seeds are enclosed in a fruit. Compare **gymnosperm**

Anglo-Nubian /ˌæŋgləʊ ˈnjuːbiən/ *noun* a hardy breed of goat with high milk yields. It has a brown coat with white patches.

angora *noun* **1.** a breed of rabbit, bred mainly for its fur **2.** a breed of goat, important as a source of mohair

COMMENT: The original colour was white, but there are now grey, pale brown and other shades. The wool is extremely fine.

Angus /ˈæŋgəs/ *noun* ♦ **Aberdeen Angus**

Animal Diseases Research Association *noun* former name for **Moredun**

Animal Health Act 1981 *noun* an Act of Parliament which aimed to control the spread of diseases in farm animals and to set up rules protecting the welfare of animals on the farm, in transit and at market

Animal Health Act 2002 *noun* an Act of Parliament which put into place controls to deal with outbreaks of diseases such as foot-and-mouth disease and scrapie

Animal Health Divisional Office *noun* one of 24 regional branches of the State Veterinary Service. Abbr **AHDO**

Animal Health Officer *noun* an employee of an Animal Health Divisional Office, with veterinary training. Abbr **AHO**

animal health planning *noun* an official set of guidelines for controlling and treating diseases in farm animals

animal husbandry *noun* the process of breeding and looking after farm animals

animal inspector, animals inspector *noun* an official whose job is to inspect animals to see if they have notifiable diseases and are being kept in acceptable conditions

animal welfare *noun* the idea that animals raised on farms should be treated humanely and protected from unnecessary pain or distress

'In his case, the messages about traceability and confidence in animal welfare and food safety are simple and positive. Meat for the business comes from stock contract reared on a network of local farms, fed using mainly homegrown ingredients which are then mill-and-mixed from a mobile unit visiting the farms.' [*Farmers Guardian*]

Animal Welfare Bill *noun* an Act of Parliament due to come into force in 2006 to overhaul laws and regulations on animal welfare

animal welfare code *noun* ♦ **welfare code**

annual *adjective* **1.** happening or done once a year **2.** over a period of one year ■ *noun* a plant whose life cycle of germination, flowering and fruiting takes place within the period of a year. ◊ **biennial, perennial**

COMMENT: Typical examples of annuals are wheat and barley, which complete their life history in one growing season. So, starting from seed, they develop roots, stem and leaves, then produce flowers and seed before dying. Plants which develop over two years are called 'biennials' and those which do not die at the end of the fruiting period are called 'perennials'.

annual meadowgrass *noun* a widespread weed *(Poa annua)* found in all arable and grass crops

annual nettle *noun* same as **small nettle**

Annual Review of Agriculture *noun* ♦ **'Agriculture in the United Kingdom'**

annual ring *noun* a ring of new wood formed each year in the trunk of a tree which can easily be seen when the tree is cut down. ◊ **dendrochronology**. Also called **tree ring** (NOTE: As a tree grows, the wood formed in the spring has more open cells than that formed in later summer. The difference in texture forms the visible rings. In tropical countries, trees grow all the year round and so do not form rings.)

anoestrus /æn'iːstrəs/ *noun* a situation where a female animal does not come on heat at the usual time

antenna *noun* one of a pair of long thin sensors on the heads of insects, crustaceans and some other arthropods

anthelmintic /ˌænθel'mɪntɪk/ *noun* a substance such as thiabendazole which is used as a treatment against parasites, as in worming livestock

anther /'ænθə/ *noun* the part of the stamen of a flower that produces pollen

anthesis /æn'θiːsɪs/ *noun* the action of flowering, when the anthers emerge

anthocyanin /ˌænθəʊ'saɪənɪn/ *noun* a water-soluble plant pigment responsible for blue, violet and red colours

anthrax *noun* a highly infectious, often fatal, bacterial disease of mammals, especially cattle and sheep, that is transmissible to humans and causes skin ulcers (**cutaneous anthrax**) or a form of pneumonia when inhaled (**pulmonary anthrax**)

COMMENT: Anthrax is caused by the bacterium *Bacillus anthracis*, which is difficult to destroy and can stay in the soil and infect animals. Anthrax can be transmitted to humans by touching infected skin, meat or other parts of an animal (including bone meal used as fertiliser).

antibiotic *noun* a drug such as penicillin which was originally developed from fungi and which stops the spread of bacteria or fungi ○ *The vet prescribed a course of antibiotics for the infected animal.*

COMMENT: Penicillin is one of the commonest antibiotics, together with streptomycin, tetracycline and erythromycin. Although antibiotics are widely and successfully used, new forms of bacteria have developed which are resistant to them. Antibiotics were formerly used as feed additives to promote growth, but this practice has been banned in the EU since 2006.

antibody *noun* a protein which is produced in the body in response to foreign substances such as bacteria or viruses

anticaking additive /ˌæntɪ'keɪkɪŋ ˌædɪtɪv/, **anticaking agent** *noun* an additive added to food to prevent it becoming solid (NOTE: Anticaking additives have the E numbers E530–578.)

antifungal /ˌænti'fʌŋɡəl/ *adjective* referring to a substance which kills or controls fungi

antigen /'æntɪdʒən/ *noun* a substance in the body which makes the body produce antibodies to attack it, e.g. a virus or germ

antimicrobial /ˌæntimaɪ'krəʊbiəl/ *adjective* referring to something which is capable of killing or inhibiting the growth of microorganisms, especially bacteria, fungi or viruses

antioxidant /ˌænti'ɒksɪd(ə)nt/ *noun* a substance which prevents oxidation, used to prevent materials such as rubber from deteriorating and added to processed food to prevent oil going bad (NOTE: In the EU, antioxidant food additives have numbers E300–321.)

antiresistance strategy /ˌæntirə'zɪstəns ˌstrætədʒi/ *noun* a strategy designed to prevent pests and weeds from developing resistance to the chemicals used to control them, or the loss of disease resistance in crop plants

antiseptic *adjective* preventing or reducing the growth of harmful microorganisms ○ *Wash the wheels of the vehicle with antiseptic spray to reduce the risk of infection.* ■ *noun* a substance which prevents germs growing or spreading ○ *The nurse painted the wound with antiseptic.*

antiserum /ˌænti'sɪərəm/ *noun* a serum taken from an animal which has developed antibodies to bacteria and formerly used to give temporary immunity to a disease (NOTE: The plural is **antisera**.)

antitoxin /ˌænti'tɒksɪn/ *noun* an antibody produced by the body to counteract a poison in the body

antivenin /ˌænti'venəm/, **antivenene, antivenom** *noun* a serum which is used to counteract the poison from snake or insect bites

antlers *noun* the branched horns of male deer

anus *noun* the opening in the alimentary canal through which faeces leave the body

anvil /'ænvɪl/ *noun* a metal block which ends in a point, has a rounded bottom and a flat top, and on which metal objects such as horseshoes are made

AONB *abbreviation* Area of Outstanding Natural Beauty

apex *noun* the main growing shoot of a plant

aphicide /'eɪfɪsaɪd/ *noun* a pesticide designed to kill aphids on plants

aphid /'eɪfɪd/ *noun* an insect that sucks sap from plants and can multiply very rapidly, e.g. a greenfly ○ *The aphid population showed a 19% increase.* (NOTE: Aphids are pests of some garden plants such as roses and may transmit virus diseases in crops such as potatoes and sugarbeet.)

COMMENT: Cereal aphids are various species of greenfly. Winged females feed on cereal crops in May and June. The grain aphid causes empty or small grain by puncturing the grain as it develops, letting the grain contents seep out.

Aphis /'eɪfɪs/ *noun* the genus of insects which comprises aphids

apiarist /'eɪpɪərɪst/ *noun* a person who keeps bees

apiary /'eɪpɪəri/ *noun* a place where bees are kept

apiculture /'eɪpɪkʌltʃə/ *noun* the husbandry of bees for honey production

apple *noun* an edible fruit of the apple tree *(Malus domestica)*

COMMENT: The apple is the most important UK fruit crop, growing mainly in Kent and Worcestershire. In 1957, England had about 26,000 hectares growing dessert apples, but by 1987, there were less than 14,000 hectares. Ninety percent of all cooking apples grown are Bramleys. In the USA and Canada apples are mainly grown in Washington, Oregon, British Columbia and Nova Scotia. The apple is also grown extensively in temperate regions of Australia, South Africa and South America. Six thousand apple varieties once grew in Britain, and all of them are recorded in the UK's National Apple Register. Around 2,300 of these are growing at the National Fruit Trials at Brogdale in Kent. Of the recognised apple varieties the most important are Cox's Orange Pippin and Golden Delicious (dessert varieties) and Bramley's Seedling (cooking apple). Cider apples are grown mainly in Herefordshire and Somerset in the UK, and in Normandy in France.

Apple and Pear Development Council *noun* a former body providing research for apple and pear growers in the UK, part of the Horticultural Development Council from 2003. Abbr **APDC**

apple blossom weevil *noun* an insect which attacks apple flower buds, causing no fruit to develop

application *noun* **1.** a formal request ○ *an application for research funds* **2.** the act of putting a substance on a surface ○ *The crop received two applications of fungicide.* **3.** the act of using something that you already have, such as an ability or knowledge, in order to do something ○ *the application of knowledge and skills* **4.** a particular use ○ *This new technology has many applications.*

apportionment /ə'pɔːʃ(ə)nmənt/ *noun* an enclosure of part of common land for private farm use

appropriate technology *noun* a technology that is suited to the local environment, usually involving skills or materials that are available locally ○ *Biomethanation is an appropriate technology for use in rural areas.*

approved products *plural noun* chemicals used in agriculture and approved for use by the government

APRC *abbreviation* Apple and Pear Research Council

apricot *noun* a deciduous tree *(Prunus armeniaca)* bearing soft yellow fruit, similar to a small peach, but not as juicy

APRS *abbreviation* Association for the Protection of Rural Scotland

aqua- /ˈækwə/ *prefix* water. ◊ **aqui-**

aquaculture /ˈækwəkʌltʃə/, **aquafarming** /ˈækwəfɑːmɪŋ/ *noun* same as **fish farming**

aquatic *adjective* referring to water

aqueous /ˈeɪkwiəs, ˈækwiəs/ *adjective* referring to a solution made with water

aqueous ammonia *noun* ammonia in solution, obtained from gas works and as a by-product; sometimes used as a fertiliser

aqueous solution *noun* a solution of a substance in water

aqui- *prefix* water. ◊ **aqua-**

aquiculture /ˈækwɪkʌltʃə/ *noun* same as **fish farming**

aquifer /ˈækwɪfə/ *noun* a mass of porous rock or soil through which water passes and in which water gathers

Arab /ˈærəb/, **Arabian horse** *noun* a breed of horse, especially used for racing (NOTE: Arabian horses are those in whose pedigree there is no other blood than Arabian. Of great antiquity, they are the purest of the equine races. Nearly every breed has had an infusion of Arab blood.)

Arabis Mosaic /ˈærəbɪs məʊˌzeɪɪk/ *noun* a viral disease which infects a wide range of horticultural crops. Leaves become stunted and plants die back.

arable /ˈærəb(ə)l/ *adjective* referring to land on which crops are grown

COMMENT: In 2005 the UK was the 4th largest producer of cereal and oilseed crops in the EU, with arable crops representing about 16% of the total agricultural output for the country.

Arable Area Payments Scheme *noun* until 2005, a subsidy for growers of arable crops. Abbr **AAPS** (NOTE: Now superseded by the Single Payment Scheme.)

arable crops *plural noun* crops which are cultivated on ploughed land. They are

annual crops and include cereals, root crops and potatoes.

arable farming *noun* the growing of crops, as opposed to dairy farming, cattle farming, etc.

arable land *noun* **1.** land suitable for ploughing **2.** land which has been ploughed and cropped

Arable Research Institute Association *noun* a body set up by the Institute of Arable Crops Research to oversee the distribution of relevant research to farmers. Abbr **ARIA**

arable soil *noun* soil which is able to be used for the cultivation of crops

arable weed *noun* a weed that grows on or near land that has been cultivated for crops, e.g. poppy

arachidonic acid /əˌrækɪdɒnɪk ˈæsɪd/ *noun* an essential fatty acid

Arachnida /əˈræknɪdə/ *noun* a class of animals that have eight legs, e.g. spiders and mites

COMMENT: Arachnids have pincers on the first pair of legs. Their bodies are divided into two parts and they have no antennae.

ARAD *abbreviation* Agriculture and Rural Affairs Department (Welsh Assembly)

arbor- /ɑːbɔː/ *prefix* tree

arboretum /ˌɑːbəˈriːtəm/ *noun* a collection of trees from different parts of the world, grown for scientific study (NOTE: The plural is **arboreta.**)

arboricide /ɑːˈbɔːrɪsaɪd/ *noun* a chemical substance which kills trees

Arboricultural Association /ˌɑːbərɪ ˈkʌltʃ(ə)rəl əˌsəʊsieɪʃ(ə)n/ *noun* a society which promotes the study of arboriculture and represents the interests of its members. Abbr **AA**

arboriculture /ˈɑːbərɪkʌltʃə/ *noun* the study of the cultivation of trees

arborist /ˈɑːbərɪst/ *noun* a person who studies the cultivation of trees

arbovirus /ˈɑːbəvaɪrəs/ *noun* a virus transmitted by blood-sucking insects

Ardennes /ɑːˈden/ *noun* a short, hardy French breed of horse found in Britain mainly on hill farms

are /ɑː/ *noun* a unit of metric land measurement which equals 100 square metres

area *noun* a region of land ○ *The whole area has been contaminated by waste from the power station.*

Area of Outstanding Natural Beauty *noun* in England and Wales, a region which is not a National Park but which is considered sufficiently attractive to be preserved from unsympathetic development. Abbr **AONB**

area payment *noun* ♦ **Arable Area Payments Scheme**

arenaceous /ˌærɪ'neɪʃəs/ *adjective* sandy

arenaceous soils *plural noun* sandy soils, soils which have a high amount of sand particles

argente /ɑː'ʒɒnt/ *noun* French breed of rabbit

argillaceous /ˌɑːdʒɪ'leɪʃəs/ *adjective* clayey

argillaceous soils *plural noun* soils with a high amount of clay particles

arginine /'ɑːdʒɪniːn/ *noun* an essential amino acid and a constituent of proteins

ARIA /'ɑːriə/ *abbreviation* Arable Research Institute Association (NOTE: Now called the Rothamsted Research Association (RRA).)

arid *adjective* referring to soil which is very dry, or an area of land which has very little rain

aridity /ə'rɪdɪti/ *noun* the state of being extremely dry

aril /'ærɪl/ *noun* an extra covering to the seed, found in certain plants

ark /ɑːk/ *noun* **1.** a small hut made of wood or corrugated metal, used to shelter sows kept outdoors in fields **2.** a mobile poultry house with wire or slatted floors, formerly used to house chickens at night and in bad weather

aromatic *adjective* having a pleasant smell ■ *noun* a substance, plant or chemical which has a pleasant smell

aromatic herbs *plural noun* herbs, such as rosemary or thyme, which are used to give a particular taste to food

ARP *abbreviation US* acreage reduction programme

arrowroot /'ærəʊruːt/ *noun* a tropical plant *(Maranta arundinacea)* whose tubers yield starch. Used particularly in the preparation of invalid foods, since this form of starch is easily digested.

arsenic *noun* a grey semimetallic chemical element that forms poisonous compounds such as arsenic trioxide, which was formerly used in some medicines

arsenical /ɑː'senɪk(ə)l/ *noun* a drug or insecticide which is one of the group of poisonous oxides of arsenic

arsenide /'ɑːsənaɪd/ *noun* a compound of arsenic and a metallic element

arterial drain /ɑː'tɪəriəl dreɪn/ *noun* a main watercourse which carries water from smaller drains or ditches

artesian well /ɑːˌtiːziən 'wel/ *noun* a well which has been bored into a confined aquifer, the hydrostatic pressure usually being strong enough to force the water to the surface

arthropod /'ɑːθrəpɒd/ *noun* an invertebrate with jointed limbs, a segmented body, and a chitin exoskeleton. Phylum: Arthropoda. (NOTE: Insects, arachnids, centipedes and crustaceans are arthropods.)

artichoke *noun* a vegetable grown as a specialised crop

COMMENT: There are two types of artichoke: the globe artichoke, a tall thistle-like plant (the leaves of the head are cooked and eaten), and the Jerusalem artichoke, a tall plant which develops tubers like a potato.

artificial *adjective* made by humans and not existing naturally

artificial fertiliser *noun* a fertiliser manufactured from chemicals. Also called **chemical fertiliser**

artificial insemination *noun* a method of breeding livestock by injecting sperm from specially selected males into females. Abbr **AI**. Compare **multiple ovulation and embryo transfer**

artificial manure *noun* a manufactured chemical substance used to increase the nutrient level of the soil

artificial selection *noun* the selection by people of individual animals or plants from which to breed further generations because the animals or plants have useful characteristics

arvensis /ɑː'vensɪs/ *adjective* a Latin word meaning 'of the fields', used in many genetic names of plants

asbestos *noun* a fibrous mineral substance that causes lung disease if inhaled. It was formerly used as a shield against fire and as an insulating material in many industrial and construction processes.

Ascaris /'æskərɪs/ *noun* a nematode worm which infects the small intestine, causing the disease (**ascariasis**)

Ascochyta /ˌæskəʊˈʃiːtə/ *noun* a disease affecting pulses such as peas and beans

ascorbic acid /əˌskɔːbɪk ˈæsɪd/ *noun* vitamin C

asexual *adjective* not involving sexual reproduction

asexually /eɪˈsekʃjuəli/ *adverb* not involving sexual intercourse ○ *By taking cuttings it is possible to reproduce plants asexually.*

ash *noun* **1.** a hardwood tree. Genus: *Fraxinus.* **2.** a grey or black powder formed of minerals left after an organic substance has been burnt

asparagus *noun* a native European plant *(Asparagus officinalis)* of which the young shoots (called 'spears') are cut when they are about 25cm long and are eaten as a vegetable

asparagus pea *noun* a plant *(Lotus tetragonolobus)* grown in southern Europe, the pods of which are edible

aspect *noun* a direction in which something faces ○ *a site with a northern aspect* (NOTE: It can affect the amount of sunshine and heat absorbed by the soil. In the UK, the temperature of north-facing slopes may be 1°C lower than on similar slopes facing south.)

aspen /ˈæspən/ *noun* a hardwood tree with leaves that tremble in the wind. Genus: *Populus.*

aspergillosis /ˌæspɜːdʒɪˈləʊsɪs/ *noun* an infection of the lungs with *Aspergillus,* a type of fungus which affects parts of the respiratory system

ass *noun* a long-eared animal of the horse family. Used as a draught animal and still important in Mediterranean countries, where it thrives better than the horse on scanty herbage.

association *noun* **1.** a link between two things ○ *They were looking for an association between specific chemicals and the disease.* **2.** same as **biological association**

Association for the Protection of Rural Scotland *noun* an independent charity which promotes the preservation of Scotland's natural landscapes. Abbr **APRS**

Association of Chartered Physiotherapists in Animal Therapy *noun* a professional body in the UK representing animal physiotherapists. Abbr **ACPAT**

assortive mating /əˈsɔːtɪv ˌmeɪtɪŋ/ *noun* the practice of mating animals that have a similar appearance. Also called **mating likes**

assurance scheme *noun* a set of voluntary guidelines that farmers can sign up to in order to reassure customers about the quality of their produce

Assured Food Standards *noun* an organisation which is responsible for managing the 'Little Red Tractor' assurance scheme. Abbr **AFS**

Asteraceae /ˌæstəˈræsiaɪ/ *plural noun* a common and very large family of plants with flat flowers that consist of many florets arranged around a central structure. Former name **Compositae**

asthma *noun* a lung condition characterised by narrowing of the bronchial tubes, in which the muscles go into spasm and the person has difficulty breathing. ◊ **occupational asthma**

asulam /ˈæsjuləm/ *noun* a herbicide used around trees and fruit bushes

atavism /ˈætəvɪz(ə)m/ *noun* the reappearance of a genetically controlled feature in an organism after it has been absent for several generations, usually as a result of an accidental recombination of genes

atmospheric lifetime *noun* same as **lifetime**

atmospheric pollution *noun* same as **air pollution**

atrazine /ˈætrəziːn/ *noun* a herbicide that kills germinating seedlings, used especially on maize crops

atrium /ˈeɪtriəm/ *noun* one of the two upper chambers in the heart. Compare **ventricle**

atrophic rhinitis /æˌtrɒfɪk raɪˈnaɪtɪs/ *noun* a bacterial disease of young pigs causing inflammation of the nasal passages, which can cause deformity of the snout

atropine /ˈætrəpiːn/ *noun* a poisonous plant compound (**alkaloid**), found in *Atropa belladonna,* that affects heart rate and is used medically to relax muscles

attachment *noun* a device which can be attached to a machine, e.g. a straw chopper which can be attached to a combine harvester

attenuated strains /əˈtenjueɪtɪd ˌstreɪnz/ *plural noun* pathogenic microorganisms, mainly bacteria and viruses, which have lost their virulence

attested area /ə'testɪd ˌeəriə/ *noun* an area declared to be free from a specific animal disease

COMMENT: In 1960, the whole of the UK was declared an attested area, since bovine tuberculosis had by then been almost completely eradicated from all herds of cattle.

attested herd *noun* a herd tested and found to be free of bovine tuberculosis

attractant /ə'træktənt/ *noun* a chemical that attracts an organism. ◊ **pheromone**

ATV *abbreviation* all-terrain vehicle

aubergine *noun* a purple fruit of the eggplant (*Solanum melongena*), used as a vegetable. A native of tropical Asia, it is sometimes called by its Indian name 'brinjal'.

Aubrac /'ɔːbræk/ *noun* a rare breed of cattle from southern France. The animals are light yellow or brown in colour and are used for draught, milk and meat.

auction *noun* the selling of goods where people offer bids, and the item is sold to the person who makes the highest offer ■ *verb* to sell at an auction

auctioneer *noun* a person who conducts an auction

auger /'ɔːgə/ *noun* **1.** a tool for boring holes, made up of a long shank with a cutting edge and a screw point. Soil augers are used to obtain samples of soil for analysis. **2.** a device on a combine harvester (shaped like a large screw) which carries the grain up into the grain tank

Aujeszky's disease /aʊ'jeskiːz dɪ ˌziːz/ *noun* a virus disease of animals, characterised by intense itching. The disease is most serious in young pigs, and often leads to death. A notifiable disease.

auricle /'ɔːrɪk(ə)l/ *noun* **1.** an ear-shaped part in each upper chamber of the heart **2.** a part of a grass plant found at the base of a leaf

Australorp /'ɒstrəlɔːp/ *noun* an Australian breed of chicken; the birds are black with red combs

auto- /ɔːtəʊ/ *prefix* **1.** automatic or auto-mated **2.** self

autogamy /ɔː'tɒgəmi/ *noun* pollination with pollen from the same flower

automatic cluster removal *noun* the automatic removal of the cluster of the milking machine from the udder at the end of milking. Abbr **ACR**

automatic pick-up hitch *noun* a mechanism on a tractor, operated by the hydraulic system, which allows the driver to hitch up to a trailer without leaving the driving seat

automation *noun* the use of machinery to save manual labour

COMMENT: Automation has considerably reduced the labour involved in dairying. It has now been adopted almost every-where for the milking process (removing the cluster of the milking machine at the end of milking is done automatically) and for the making of conserved fodder.

autotroph /'ɔːtətrɒf/, **autotrophic organism** /ˌɔːtətrɒfɪk 'ɔːgənɪz(ə)m/ *noun* an organism which manufactures its own organic constituents from inorganic materials, e.g. a bacterium or green plant. Compare **chemotroph, heterotroph**

autumn fly *noun* an irritating non-biting fly (*Musca autumnalis*)

auxin /'ɔːkzɪn/ *noun* a plant hormone that encourages or suppresses tissue growth (NOTE: Some herbicides act as synthetic auxins by upsetting the balance of the plant's growth.)

available water *noun* water which can be taken up by the plant roots

available water capacity *noun* the amount of water held by a soil between the amounts at field capacity and wilting point. Abbr **AWC**

Avena /ə'viːnə/ *noun* the Latin name for the oat family

Average All Pigs Price *noun* formerly, the average price for pigs, calculated each week and used in contracts for payment. It has now been replaced by the Deadweight Adjusted Pig Price. Abbr **AAPP**

Aves /'eɪviːz/ *noun* the class that comprises birds

avian /'eɪviən/ *adjective* relating to birds

avian flu *noun* a type of influenza that affects birds and can infect humans. Also called **bird flu**

aviary /'eɪviəri/ *noun* a cage or large enclosure for birds

aviary system *noun* a poultry housing system using a deep litter house with extra raised areas inside it for the birds to feed, roost and exercise in

'Improving welfare was the key reason why Robert and Ethel Chapman became the first Scottish egg producer to invest in the new enriched aviary system. It consists of birds housed in groups of 60 in large cages containing a nesting area, scratching area and perches in addition to drinkers and feeders.' [*Farmers Weekly*]

avocado *noun* a pear-shaped green fruit of a tree *(Persea americana)* which is native of South and Central America and is cultivated in Israel, Spain, the USA and elsewhere. The fruit has a high protein and fat content, making it very nutritious and an important food crop.

away-going crop *noun* a crop sown by a tenant farmer before leaving the farm at the end of his tenancy. He is permitted to return and harvest the crop and remove it.

AWB *abbreviation* Agricultural Wages Board

AWC *abbreviation* available water capacity

awn /ɔːn/ *noun* the tip of a leaf which ends in a spine; in cereals awns are attached to the grains, and in barley each grain has a particularly long awn

axil /'æksɪl/ *noun* the angle between a leaf or branch and the stem from which it grows

axillary /æk'sɪləri/ *adjective* referring to an axil

axillary bud *noun* a bud in the angle between a leaf and the main stem, producing a side shoot, as in the case of tomatoes

Aylesbury /'eɪlzbəri/ *noun* a heavy table breed of duck, with white feathers

Ayrshire /'eəʃə/ *noun* a breed of dairy cow, originating in South-West Scotland. It is deep and broad at the hips and narrow at the shoulders. It was the main rival to the Friesian as a milk producer. It is hardy, and white and brown in colour.

Azarole /ˌæzə'rəʊl/ *noun* a small tree *(Crataegus azarolus)* bearing fruit which are used for making jellies. It is a native of the Mediterranean area.

azotobacter /ə'zəʊtəʊbæktə/ *noun* a nitrogen-fixing bacterium belonging to a group found in soil

B

B *symbol* boron

BAA *abbreviation* British Agrochemicals Association

Baars irrigator /ˈbɑːz ˌɪrɪɡeɪtə/ *noun* a flexible hose with sixteen folding sprinklers, used for irrigating

baby beef *noun* meat from cattle slaughtered around 12 months of age

bacillary /bəˈsɪləri/ *adjective* referring to a bacillus

bacillary white diarrhoea *noun* an acute, infectious and highly fatal disease of chicks, caused by *Salmonella pullorum*

bacillus /bəˈsɪləs/ *noun* a bacterium shaped like a rod (NOTE: The plural is **bacilli**.)

back *verb* (*of the wind*) to change direction, anticlockwise in the northern hemisphere and clockwise in the southern hemisphere. Opposite **veer**

back band *noun* a band of rope or chain over a cart saddle, to keep the shafts up

backboard /ˈbækbɔːd/ *noun* a board at the back of a cart

backcrossing /ˈbækkrɒsɪŋ/ *noun* breeding a crossbred offspring back to one of its parents, usually a purebred

back end *noun* late autumn

background *noun* a set of conditions which are always present in the environment, but are less obvious or less important than others

bacon *noun* the cured back and sides of a pig; bacon may be green or smoked

COMMENT: Bacon is cured in brine for several days. Some bacon is also smoked by hanging in smoke, which improves its taste. Unsmoked bacon is also known as 'green' bacon.

baconer /ˈbeɪk(ə)nə/ *noun* a pig bred and reared for bacon

bacteria *plural noun* very small organisms, invisible except through a microscope, belonging to a large group, some of which help in the decomposition of organic matter, some of which are permanently in the intestines of animals and can break down food tissue and some of which cause disease (NOTE: The singular is **bacterium**.)

COMMENT: Bacteria can be shaped like rods (bacilli), like balls (cocci) or have a spiral form (such as spirochaetes). Bacteria, especially bacilli and spirochaetes, can move and reproduce very rapidly.

bacteria bed *noun* a filter bed of rough stone, forming the last stage in the treatment of sewage

bacterial *adjective* referring to or caused by bacteria ○ *a bacterial disease*

bacterial canker *noun* a disease affecting the plant genus *Prunus*. Stems of the plant swell and exude a light brown gum and can cause browning of foliage (*Pseudomonas morsprunorum*).

bacterial digestion *noun* the process by which bacteria break down organic matter such as sewage and slurry

bacterial pea blight *noun* a fungal disease attacking peas. It is a notifiable disease.

bacterial pneumonia *noun* a form of pneumonia caused by pneumococcus

bacteriophage /bækˈtɪərɪəfeɪdʒ/ *noun* a virus that affects bacteria

badger *noun* a white-faced grey-coated carnivore (NOTE: Badgers are harmless as far as crops are concerned and useful as they kill insects, slugs and mice, but there have been calls for a cull of badgers as they are thought to be responsible for the spread of bovine tuberculosis.)

badlands /ˈbædlændz/ *plural noun* areas of land which are or have become unsuitable for agriculture

bag *noun* the udder of a cow ∎ *verb* to cut wheat with a sickle

bagasse /bæ'gæs/ *noun* fibres left after sugar cane has been crushed

BAGMA /'bægmə/ *abbreviation* British Agricultural and Garden Machinery Association

Bagot /'bægət/ *noun* a rare breed of goat with long hair, usually with black head, neck and shoulders and the rest of the body white. Both sexes have horns.

bail *noun* **1.** a bar separating horses in a shed **2.** a milking shed which can be moved from place to place

bailiff *noun* a person employed to manage a farm on behalf of the owner

bait *noun* feed of hay and oats for horses

bait plant *noun* a plant which is grown near to or among a crop in order to attract pests or infections to it rather than having them affect the crop itself

'The advantage of using weeds such as fat hen and chickweed is that they are in situ for a number of months acting as bait plants compared with soil testing, which relies on the nematodes being in the soil profile at the time of sampling.' [*Farmers Weekly*]

bake *verb* to cook food such as bread or cakes in an oven

bakers /'beɪkəz/ *plural noun* large potatoes selected to be cooked in an oven as baked potatoes

Bakewell, Sir Robert (1735–95) /'beɪkwel/ an important Leicestershire farmer whose experiments on longhorn cattle showed that the way to achieve rapid improvement in stock was by carefully controlled inbreeding. He also originated the improved Leicester sheep.

baking powder /'beɪkɪŋ ˌpaʊdə/ *noun* powder formed of dry acid which reacts with water to produce carbon dioxide, and so aerate dough

baking quality *noun* the ability of flour to retain carbon dioxide bubbles until it has become hard

balance *noun* **1.** a state in which two sides are equal or in proportion **2.** a state in which proportions of substances are correct ○ *to maintain a healthy balance of vitamins in the diet*

balanced diet *noun* a diet which provides all the nutrients needed in the correct proportions

balance of nature *noun* a popular concept that relative numbers of different organisms living in the same ecosystem may remain more or less constant without human interference □ **to disturb the balance of nature** to make a change to the environment which has the effect of putting some organisms at a disadvantage compared with others

balancer *noun* a food supplement which is given to livestock to counteract the lack of any nutrients in their normal diet

bald faced *adjective* referring to an animal with white on the face

bale *noun* **1.** a package of hay or straw, square, round or rectangular in shape, usually tied with twine **2.** a standard pack for cotton ∎ *verb* to take loose straw, grass or hay and press it into bales

bale buster *noun* a machine for unrolling big bales of straw, hay and silage, for bedding and feeding

bale loader *noun* an attachment to the front loader arms of a tractor, which allows it to lift bales onto a trailer or into a barn

baler /'beɪlə/ *noun* a machine that picks up loose hay or straw and compresses it into bales of even size and weight, and then ties them with twine. The completed bale is dropped onto the ground at the back of the baler.

baler twine *noun* a cord used to tie bales, now often made of rot-proof polypropylene

bale sledge *noun* a sledge pulled behind a baler to collect the finished bales

bale wrapper *noun* a machine mounted on a tractor used to wrap film around a bale

balling /'bɔːlɪŋ/ *noun* the dosing of larger animals, by rolling the drug into a ball which is then swallowed

balling gun *noun* an instrument used to force balls of drugs down the throats of animals

bamboo *noun* a plant (*Bambusa vulgaris*) growing mainly in temperate and subtropical regions of China, Japan and Korea

Bampton Nott /ˌbæmptən 'nɒt/ *noun* an ancient breed of sheep, the ancestor of the Devon Longwool

BANC *abbreviation* British Association of Nature Conservationists

band *noun* **1.** a strip or loop of fabric, metal or plastic **2.** a narrow area that is different in colour from other areas

banding /'bændɪŋ/ noun the placing of greasebands round the trunks of fruit trees to trap insects

band sprayer noun a crop sprayer that applies chemicals in narrow strips, mostly used with precision seeders

bane /beɪn/ noun same as **liver fluke**

bantam /'bæntəm/ noun a small breed of domestic fowl, most kinds being about half the size and weight of the common fowl

BAP abbreviation Biodiversity Action Plan

barban /'bɑːbən/ noun a herbicide formerly used on many varieties of winter wheat and barley to control wild oats but no longer approved in the UK

bar code noun data represented as a series of printed stripes of varying widths

COMMENT: Bar codes are found on most goods and their packages. The width and position of the stripes can be recognised by the reader and give information about the goods, such as price, stock quantities, etc. Many packaged foods, even fresh foods, are bar-coded to allow quicker data capture in the supermarket.

bare adjective referring to land which has no plants growing on it

bareback /'beəbæk/ adjective riding an unsaddled horse

bare fallow noun land left fallow and ploughed several times during the course of a year, the aim being to get rid of weeds

bark noun the outer layer of a tree trunk or branch

barking /'bɑːkɪŋ/ noun the process of removing the bark of a tree, such as when cork is harvested from cork oaks

barley noun a cereal crop used as animal feed and for making malt for beer or whisky. Latin name: *Hordeum sativum*.

COMMENT: Barley is widely grown in northern temperate countries, with the largest production in Germany and France. It is an important arable crop in the UK. The grain is mainly used for live-stock feeding and for malting. It is rarely used for making flour.

barley powdery mildew noun ⚬ **powdery mildew**

barley yellow dwarf virus noun a virus spread by aphids, which causes poor root development in barley plants. Red and yellow colour changes occur in leaves and yields are much reduced. Abbr **BYDV**

barn noun a farm building used for storing hay or grain. In the USA, barns are also used for keeping animals.

barn-dried hay noun hay dried indoors by blowing air through it. It is usually more nutritious than field-dried hay.

Barnevelder /'bɑːnveldə/ noun a breed of fowl

bar pig noun same as **barrow**

bar point noun a spring-loaded type of a plough body, useful when boulders occur below the surface of the land being ploughed

barren /'bærən/ adjective 1. unable to support plant or animal life ⚬ *a sparse and barren landscape at high altitude* 2. unable to reproduce

barren brome /'bærən brəʊm/ noun a widespread weed (*Bromus sterilis*) which affects winter cereals

barrener /'bærənə/ noun a female animal unable to produce young

Barrosa /bə'rəʊsə/ noun a breed of cattle found in northern Portugal and used both for draught and for meat. The animals are reddish-brown with large curved horns.

barrow /'bærəʊ/ noun a male pig after castration, while a suckler or weaner

basal area /'beɪs(ə)l ,eəriə/ noun the area covered by the trunks of trees or stems of plants

basal metabolic rate noun the amount of energy used by a body in exchanging oxygen and carbon dioxide when at rest. Abbr **BMR** (NOTE: It is a measure of the energy needed to keep the body functioning and the temperature normal.)

basal metabolism noun energy used by a body at rest, i.e. energy needed to keep the body functioning and the temperature normal. This can be calculated while an animal is in a state of complete rest by observing the amount of heat given out or the amount of oxygen taken in and retained.

BASC abbreviation British Association for Shooting and Conservation

basic adjective relating to a chemical which reacts with an acid to form a salt

basic price noun a support price fixed each year by the EU for certain fruit and vegetables

basic slag noun calcium phosphate, produced as waste from blast furnaces and formerly used as a fertiliser because of its phosphate content

basidiomycotes /bəˌsɪdiəʊˈmaɪkəʊts/ *plural noun* a large group of fungi, including mushrooms and toadstools

basil *noun* an aromatic herb *(Ocimum basilicum)*, used in cooking, particularly with fish

basin irrigation *noun* a form of irrigation where the water is trapped in basins surrounded by low mud walls, as in rice paddies

BASIS /ˈbeɪsɪs/ *noun* a scheme for registering distributors of chemicals used to protect crops, so that dangerous chemicals are stored and used correctly. Full form **British Agrochemicals Standards Inspection Scheme**

basmati /bæzˈmɑːti/ *noun* a common variety of long-grain rice

bastard fallow *noun* land left fallow for the time between harvesting and sowing, usually ploughed to control weeds

batch drying *noun* a process for drying bales of hay in batches. Bales are placed on a platform and heated air is blown up between the bales.

battery *noun* a large number of small cages, usually arranged in rows one above the other, in which many birds or animals, especially chickens, are kept

battery farming *noun* a system of farming where many birds, especially chickens, or animals are kept in small cages

battery hen *noun* a hen that is reared and kept in a battery, along with many others, for egg production

baulk *noun* a narrow strip of land left unploughed to mark the boundary between fields (NOTE: also written **balk**)

bay *noun* **1.** a stall in a stable **2.** a horse of a reddish-brown colour

BB *abbreviation* big bale

BBS *abbreviation* big bale silage

BBSRC *noun* an organisation which provides funding for research into biotechnology and areas such as livestock breeding, crop productivity and agricultural sustainability. Full form **Biotechnology and Biological Sciences Research Council**

BCMS *abbreviation* British Cattle Movement Service

BCPC *noun* an organisation which promotes the development, use and understanding of sustainable crop production practices. Former name **British Crop Protection Council**

BDFA *abbreviation* British Deer Farmers Association

BDM *abbreviation* Bleu du Maine

beak *noun* hard parts forming the mouth of a bird. Also called **bill**

beak trimming *noun* cutting off a bird's beak to prevent injury to other birds by pecking, especially when they are kept in close conditions (NOTE: Some beak trimming may be required to prevent birds being injured by aggressive pecking, but there are strict guidelines set out in the animal welfare code to ensure that the process is undertaken in a humane way.)

'Speaking in London last week at the launch of FAWC's annual review, Dr Potter said a multi-disciplinary approach is required for further research to prevent mutilations. A priority issue is beak trimming in laying hens, turkeys and broiler breeder males.' [*Farmers Guardian*]

beam *noun* the main frame of a plough, to which the parts that cut into the soil are attached

bean *noun* one of various varieties of legumes with edible seeds

COMMENT: Bacteria on the roots of bean crops can fix nitrogen and crops following will benefit from the nitrogen left in the soil by the beans. Beans normally have a single square hollow stem bearing a large number of compound bluish-green leaves. Black and white flowers appear in spring, developing into green pods. The main types of beans are field beans *(Vicia faba)*, used for stock feeding, or for producing broad beans, which are the immature seeds used for human consumption. Field beans are usually grown as a break crop. Winter beans are sown in October and spring beans in February/March. Spring beans include tick beans (with small seeds), minor beans (very small seeds, used to feed racing pigeons) and horse beans (large seeds).

bean aphid *noun* a very small, oval-bodied, black or dark green fly *(Aphis fabae)* that colonises plants in summer and makes them wilt. Also called **blackfly**

bean stem rot *noun* a fungus disease of beans

beastings /ˈbiːstɪŋz/ *noun* same as **colostrum**

beaters /'biːtəz/ *plural noun* **1.** steel bars on the drum of a combine harvester **2.** people who rouse game in hunting or shooting

Beaumont period /'bəʊmɒnt ˌpɪəriəd/ *noun* a period of 48 hours during which temperatures do not fall below 10°C (50°F) and relative humidity remains above 75%. Potato blight is likely to occur within 21 days of this, though spraying may prevent the blight from appearing.

beck *noun* a mountain stream

bed *noun* a specially planted area of land, e.g. an asparagus bed, flower bed or a strawberry bed

bedded set *noun* a young hop plant rooted from a cutting

bedding *noun* **1.** materials such as straw, shavings or sand, used as litter for animals to lie on **2.** the act of planting out small flower plants into a flower bed

bedding plants *plural noun* small annual flower plants which are used for bedding out

bed out *verb* to plant a flower bed with plants, especially in such a way as to give a decorative effect

bee *noun* a flying insect with a hairy body (NOTE: Bees pollinate some types of plant.)

beech *noun* a common temperate hardwood tree. Genus: *Fagus*.

beef *noun* meat of bull or cow

beefalo /'biːfələʊ/ *noun* an American breed of cattle, brown in colour. It is derived from crossing bison, Charolais and Hereford breeds.

Beef Assurance Scheme *noun* a scheme allowing farmers to register established herds of cattle which have never had a case of BSE, so that older cattle from this herd are exempted from the 30 month rule and can be slaughtered and sold for food. ◊ **OCDS**

beef bull *noun* a bull reared to produce beef

beef cow *noun* a cow kept for rearing calves for beef production

beef shorthorn *noun* a compact short-legged breed of beef cattle; the colour may be red, red and white, white or roan

Beef Special Premium Scheme *noun* until 2005, a subsidy for producers of male cattle (NOTE: Now superseded by the Single Payments Scheme.)

beef value *noun* the value of an animal, calculated by assessing its genetic background for a history of good growth and carcass conformation

beehive *noun* a structure for housing a colony of bees, containing the frames in which bees store honey

beekeeper /'biːkiːpə/ *noun* a person who keeps bees for honey

beer *noun* an alcoholic drink made by fermenting malted barley with large quantities of water

beet *noun* **1.** a plant with a succulent root, a source of sugar **2.** *US* same as **beetroot**

beet cyst nematode *noun* a type of nematode which causes white cysts on the roots of beets

beet flea beetle *noun* a sugar beet pest (*Chaetocnema concinna*)

beetle *noun* an insect with hard covers on its wings. Order: Coleoptera.

beetle bank *noun* an uncultivated ridge left in the middle of large fields for insects and spiders which survive there during winter and then spread rapidly into the crop in the following spring to eat pests such as aphids

'Plan and establish your beetle banks at the same time and in the same way as buffer strips and field corners, setting-up the ground in the autumn but leaving sowing until the spring for the best possible seedbed cleaning and grass establishment.' [*Arable Farming*]

beetroot *noun* a salad vegetable with a bright red root

beggary /'begəri/ *noun* same as **common fumitory**

BEIC *abbreviation* British Egg Industry Council

Belgian Blue /ˌbeldʒən 'bluː/ *noun* a dual-purpose breed of cattle, the result of crossing Friesians and Shorthorns with native Belgian stock. The animals are coloured blue and white.

Belgian hare *noun* a breed of rabbit

Belgian Heavy Draught *noun* a breed of horse of great weight and traction power

belladonna /ˌbelə'dɒnə/ *noun* same as **deadly nightshade**

bellwether /'belweðə/ *noun* a sheep with a bell hung round its neck, which leads a flock

belly *noun* the underside of an animal

belt *verb* to clean out a sheep's fleece with shears

belt drive *noun* the transmission of power from an engine or electric motor to another machine, by means of a belt

Belted Galloway /ˌbeltɪd 'gæləweɪ/ *noun* a breed of beef cattle, coloured black with a white belt round the body. The animals are long-haired with a dense undercoat. It is a medium-sized polled breed used for cross-breeding with White Shorthorn bulls to produce blue-grey suckler cows.

Beltsville /'beltsvɪl/ *noun* a breed of large turkey, with white feathers

benazolin /bə'næzəliːn/ *noun* a herbicide formerly used for controlling broad-leaved weeds but no longer approved in the UK

beneficial insect *noun* an insect which is encouraged to live on or near farmland to help control pests or for pollination

bennet /'benɪt/ *noun* same as **bent**

benomyl /'benəmiːl/ *noun* a fungicide formerly used against eyespot in cereals but no longer approved for use in the UK

bent *noun* **1.** stiff-stemmed grasses of the genus *Agrostis,* found in hill pastures and tolerant of poor conditions. Also called **bennet 2.** old dry stalks of dead grass

benzene hexachloride /ˌbenziːn ˌheksə'klɔːraɪd/ *noun* the active ingredient of the pesticide lindane, which was banned from use in the UK from 2001

Berkankamp scale /'beəkənkæmp skeɪl/ *noun* a scale used to describe the growth stages in oilseed rape crop

Berkshire /'baːkʃə/ *noun* a breed of small pig, dark coloured with a compact body, short head and prick ears. The snout, feet and tail are white.

Berkshire Knot *noun* a local breed of sheep crossed with Southdown to develop the Hampshire Down breed

Berrichon du Cher /ˌberiʃɒn du 'ʃeə/ *noun* a French breed of sheep, now imported into the UK

berry *noun* a small fleshy fruit with several seeds, e.g. a tomato or a grape

best-before date *noun* a date stamped on foodstuffs sold in supermarkets, which is the last date when the food is guaranteed to be in good condition. Compare **sell-by date**

best linear unbiased prediction *noun* full form of **BLUP**

best practice *noun* the most effective or efficient method of achieving an objective or completing a task

'The National Register of Sprayer Operators (NRoSO) was established for those committed to adopting best practice in both handling and applying pesticides. Some 21,500 members account for 80% of the arable area.' [*Arable Farming*]

Beulah speckle face /ˌbjuːlə 'spek(ə)l ˌfeɪs/ *noun* a breed of sheep which has a black and white speckled face and is native to the hills of Wales

BFREPA *abbreviation* British Free Range Egg Producers Association

BFSS *abbreviation* British Field Sports Society

BGH *abbreviation* bovine growth hormone

BGS *abbreviation* British Grassland Society

BHC *abbreviation* benzene hexachloride

B horizon *noun* the subsoil. ◊ **horizon**

bhp *abbreviation* brake horsepower

bible *noun* a name given to the omasum, the third stomach of ruminants

bid *noun* an offer to buy something at a certain price

bidder /'bɪdə/ *noun* a person who makes a bid at an auction

biddy /'bɪdi/ *noun* a popular name for a hen

bident /'baɪdənt/ *noun* a two-year-old sheep

biennial *adjective* happening every two years ■ *noun* a plant that completes its life cycle over a period of two years. ◊ **annual, perennial**

COMMENT: Biennial plants spend the first year producing roots, stem and leaves. In the following year the flowering stem and seeds are produced, after which the plant dies. Sugar beet and swedes are typical biennials, although they are grown as annuals, and are harvested at the end of their first year when the roots are fully developed.

biffin /'bɪfɪn/ *noun* a variety of cooking apple

bifoliate /baɪ'fəʊlɪət/ *adjective* referring to a plant which has two leaves only

big bale *noun* a large bale, weighing from 600 to 900 kilos. Abbr **BB**

COMMENT: There are two types of big bale, the round and the rectangular. The baling is slower with the round bale, but the completed bale is more weatherproof and can be safely left out in the field for longer periods of time. One of the main advantages of the big baler is that the number of bales per hectare is very low. This makes handling from the field to the barn very much simpler, provided the right equipment is used.

big bale silage *noun* silage stored in big bales. Abbr **BBS**

big bud *noun* the swelling of blackcurrant buds caused by gall mites

bigg /bɪg/ *noun* four-rowed barley

bilharzia /bɪlˈhɑːtsiə/ *noun* **1.** a flatworm which enters the bloodstream from infected water and causes schistosomiasis. Genus: *Schistosoma*. **2.** same as **schistosomiasis**

bilharziasis /ˌbɪlhɑːˈtsaɪəsɪs/ *noun* same as **schistosomiasis**

bill *noun* same as **beak**

billhook /ˈbɪlhʊk/ *noun* a cutting implement with a curved blade, used for trimming hedges

billy goat /ˈbɪli ɡəʊt/ *noun* a male goat

bin *noun* a large container for storage, e.g. a maize bin or a bin for holding wool in a shearing shed ○ *The grain is kept in large storage bins.*

bind *verb* to cut corn and tie it together in sheaves

binder *noun* a machine formerly used to cut and bind corn, drawn by a tractor. It has now been replaced by the combine harvester.

bindweed /ˈbaɪndwiːd/ *noun* a perennial weed with creeping roots, which climbs by twisting round other plants. ⋄ **black bindweed, field bindweed**

bine /baɪn/ *noun* **1.** a new shoot of a hop plant which is twisted round the strings up which it will start to grow **2.** the stem of a climbing plant such as the runner bean

bing /bɪŋ/ *noun* a feed passage

binomial classification *noun* the scientific system of naming organisms devised by the Swedish scientist Carolus Linnaeus (1707–78)

COMMENT: The Linnaean system of binomial classification gives each organism a name made up of two Latin words. The first is a generic name referring to the genus to which the organism belongs, and the second is a specific name referring to the particular species. Organisms are usually identified by using both their generic and specific names, e.g. *Homo*

sapiens (human), *Felis catus* (domestic cat) and *Sequoia sempervirens* (redwood). A third name can be added to give a subspecies. The generic name is written or printed with a capital letter and the specific name with a lowercase letter. Both names are usually given in italics or are underlined if written or typed.

bio- /baɪəʊ/ *prefix* referring to living organisms

bioaccumulation /ˌbaɪəʊəkjuːmjuːˈleɪʃ(ə)n/ *noun* the accumulation of substances such as toxic chemicals in increasing amounts up the food chain

biochemical oxygen demand *noun* same as **biological oxygen demand**

biocide /ˈbaɪəʊsaɪd/ *noun* a substance that kills living organisms ○ *Biocides used in agriculture run off into lakes and rivers.* ○ *The biological effect of biocides in surface waters can be very harmful.*

biocide pollution *noun* pollution of lakes and rivers caused by the runoff from fields of herbicides and other biocides used in agriculture

biocontrol /ˈbaɪəʊkənˌtrəʊl/ *noun* same as **biological control**

biodegradable *adjective* referring to something which is easily decomposed by organisms such as bacteria or by natural processes such as the effect of sunlight or the sea ○ *Organochlorines are not biodegradable and enter the food chain easily.*

biodiversity *noun* the range of species, subspecies or communities in a specific habitat such as a rainforest or a meadow. Also called **biological diversity**

'Modern intensive farming is probably the main cause of declining biodiversity in the countryside.'
[*Delivering the evidence. Defra's Science and Innovation Strategy 2003–06*]

Biodiversity Action Plan *noun* a detailed scheme to maintain the biological diversity of a specific area. Abbr **BAP**

biodiversity indicator *noun* a factor that allows change in the environment over time to be assessed. Also called **bioindicator**

biodynamic agriculture /ˌbaɪəʊdaɪnæmɪk ˈæɡrɪkʌltʃə/ *noun* a view of agriculture based on a holistic and spiritual understanding of nature and humans' role in it, which considers a farm as a self-contained evolving organism, relying on home-produced feeds and manures with external inputs kept to a minimum

bioenergy /ˈbaɪəʊˌenədʒi/ *noun* energy produced from biomass

bioethanol /ˈbaɪəʊˌeθənɒl/ *noun* a fuel for internal-combustion engines that is made by fermenting biological material to produce alcohol (NOTE: Typically 5–10% bioethanol is added to petrol.)

'In less than a lifetime we are returning to growing crops for energy. Crops grown for the production of biodiesel and bioethanol have a particularly important role in the Government's pledge to reduce carbon emissions.' [*Arable Farming*]

biofuel /ˈbaɪəʊfjuːəl/ *noun* a fuel produced from organic domestic waste or other sources such as plants (NOTE: Coppiced willow is sometimes grown for biofuel.)

biogas /ˈbaɪəʊgæs/ *noun* a mixture of methane and carbon dioxide produced from fermenting waste such as animal dung ○ *Farm biogas systems may be uneconomic unless there is a constant demand for heat.* ○ *The use of biogas systems in rural areas of developing countries is increasing.*

'The Green Fuels Challenge was launched by the Government in 2000 to stimulate the development of low-emission fuels. It offers duty concessions to the most promising alternative fuels such as LPG, bio-diesel and compressed natural gas. Further research on methanol, biogas and hydrogen under a pilot project scheme would qualify for either a reduction or exemption from tax.' [*UK: Environment News*]

bioindicator /ˈbaɪəʊˌɪndɪkeɪtə/ *noun* same as **biodiversity indicator**

bioinsecticide /ˌbaɪəʊɪnˈsektɪsaɪd/ *noun* an insecticide developed from natural plant toxins, e.g. pyrethrum. ◊ **microbial insecticide**

biological association *noun* a group of organisms living together in a large area, forming a stable community

biological control *noun* the control of pests by using predators and natural processes to remove them. Also called **biocontrol**

COMMENT: Biological control of insects involves using bacteria, viruses, parasites and predators to destroy the insects. Plants can be controlled by herbivorous animals such as cattle.

biological diversity *noun* same as **biodiversity**

biological magnification *noun* same as **bioaccumulation**

biological mass *noun* same as **biomass**

biological oxygen demand *noun* a measure of the amount of pollution in water, shown by the amount of oxygen needed to oxidise the polluting substances. Abbr **BOD**

COMMENT: Diluted sewage passed into rivers contains dissolved oxygen, which is utilised by bacteria as they oxidise the pollutants in the sewage. The oxygen is replaced by oxygen from the air. Diluted sewage should not absorb more than 20 ppm of dissolved oxygen.

biological pesticide *noun* same as **biopesticide**

biology *noun* the study of living organisms

biomagnification /ˌbaɪəʊmægnɪfɪˈkeɪʃ(ə)n/ *noun* same as **bioaccumulation**

biomarker /ˈbaɪəʊˌmɑːkə/ *noun* a distinctive indicator of a biological or biochemical process, e.g. a chemical whose occurrence shows the presence of a disease

biomass /ˈbaɪəʊmæs/ *noun* **1.** the sum of all living organisms in a given area or at a given trophic level, usually expressed in terms of living or dry mass **2.** organic matter used to produce energy (NOTE: Willow and miscanthus are grown as biomass for fuel.)

COMMENT: About 2,500 million people, half the world's population, rely on biomass for virtually all their cooking, heating and lighting. Most of these people live in rural areas of developing countries. There are many environmental benefits from using biomass for energy. They include: less climate change, lower levels of acid rain, less pressure on landfill sites, new wildlife habitats and enhanced biodiversity, and improved woodland management.

biomethanation /ˌbaɪəʊmeθəˈneɪʃ(ə)n/ *noun* a system of producing biogas for use as fuel

'Biomethanation is attractive for use in rural areas for several reasons: it is an anaerobic digestion process, which is the simplest, safest way that has been found for treating human excreta and animal manure.' [*Appropriate Technology*]

biopesticide /'baɪəʊˌpestɪsaɪd/ *noun* a pesticide produced from biological sources such as plant toxins that occur naturally. ◊ **microbial insecticide**

COMMENT: Biopesticides have the advantage that they do not harm the environment as they are easily inactivated and broken down by sunlight. This is, however, a practical disadvantage for a farmer who uses them, since they may not be as efficient in controlling pests as artificial chemical pesticides, which are persistent but difficult to control.

bioremediation /ˌbaɪəʊrɪmiːdi'eɪʃ(ə)n/ *noun* the use of organisms such as bacteria to remove environmental pollutants from soil, water or gases (NOTE: Bioremediation is used to clean up contaminated land and oil spills.)

biosecurity /'baɪəʊsɪˌkjʊərɪti/ *noun* the management of the risks to animal, plant and human health posed by pests and diseases

'Given the difficulties with a badger cull, the scientists agreed that efforts should focus instead on reducing cattle-to-cattle transmission through pre-movement testing and improving biosecurity on farms to prevent infection by badgers.' [*Farmers Weekly*]

biosolids /'baɪəʊˌsɒlɪdz/ *plural noun* a nutrient-rich organic material, solid or semi-solid before processing, that is derived from sewage as a product of wastewater treatment and used as a fertiliser

biotechnology *noun* the use of biological processes in industrial production, e.g. the use of yeasts in making beer, bread or yoghurt. ◊ **genetic modification**

COMMENT: Biotechnology offers great potential to increase farm production and food processing efficiency, to lower food costs, to enhance food quality and safety and to increase international competitiveness.

Biotechnology and Biological Sciences Research Council *noun* full form of **BBSRC**

birch *noun* a common hardwood tree found in northern temperate zones. Genus: *Betula*.

bird *noun* a warm-blooded animal that has wings, feathers and a beak and lays eggs

'Birds are sensitive to even small climate changes. For instance the North Atlantic Oscillation – the Atlantic's version of El Niño – has subtle effects on weather. These go unnoticed by most humans, but research shows that as the oscillation waxes and wanes, populations of larks and sandpipers rise and fall. (Climate Change, UK farmland birds and the global greenhouse. RSPB 2001)'

COMMENT: All birds are members of the class Aves. They have feathers and their forelimbs have developed into wings, though not all birds are able to fly. Birds are closely related to reptiles, and have scales on their legs. Some birds, for example rooks, pigeons and pheasants, can cause very serious damage to crops: various controls can be used such as shooting, scarecrows and destruction of nests. Birds also destroy many pests, for example wireworms, leatherjackets and caterpillars. Some birds such as chickens are farmed for food.

bird flu *noun* same as **avian flu**

bird haven *noun* same as **bird sanctuary**

bird reserve *noun* same as **bird sanctuary**

bird sanctuary *noun* a place where birds can breed and live in a protected environment

Birds Directive *noun* a European Union directive relating to the conservation of all species of naturally occurring wild birds

birdseed /'bɜːdsiːd/ *noun* same as **groundsel**

bird's-eye *noun* same as **ivy-leaved speedwell**

birdsfoot trefoil /ˌbɜːdzfʊt 'triːfɔɪl/ *noun* a plant with small yellow flowers, grown for fodder

biscuit-making quality *noun* a quality in certain types of grain, especially hardwheat grain, which produces a weak flour, with the effect that the dough does not rise when cooked

bison *noun* US a type of large cattle found in North America, now largely restricted to protected areas because of hunting. Also called **buffalo**

COMMENT: The European bison *(Bos bonasus)* is now only found in zoos and some reserves. The American bison *(Bos bison)* is still found in reserves in the USA and Canada. Both are very large animals with massive heads and shoulders, and coats of short curly brown hair.

bit *noun* the metal part of a bridle, placed in a horse's mouth to give the rider control over the animal

bite *noun* grazing

bitter *adjective* referring to something which has a sharp taste, and is not sweet

bitter pit *noun* a disease of apples

black bent *noun* a grassweed plant *(Agrostis gigantea)* which affects cereals

blackberry *noun* a soft black fruit of the bramble

black bindweed *noun* a common weed *(Bilderdykia convolvulus)* which is widespread in spring arable crops

blackcurrant *noun* a soft fruit *(Ribes)* grown for its small black berries, used also in making soft drinks

black disease *noun* a liver disease of sheep and cattle, rarely found in pigs and horses

black dot *noun* a fungal disease of potatoes that covers the skin with tiny black spots, of importance only in washed ware tubers

black earth *noun* same as **chernozem**

black eye bean, blackeyed bean *noun* a common name for the cow pea *(Vigna unguiculata)* in the USA and West Indies

blackface /'blækfeɪs/, **blackfaced sheep** *noun* one of several breeds of sheep with black faces, e.g. the Scottish Blackface or the Suffolk

Blackface mountain *noun* a common breed of sheep found in Ireland

blackfly /'blækflaɪ/ *noun* **1.** same as **bean aphid 2.** a very small, oval-bodied, black or dark green fly

blackgrass /'blækgrɑːs/ *noun* a weed *(Alopecurus myosuroides)* which is widespread among winter cereals, especially on heavy soils. In some cases it can be resistant to a wide range of herbicides. Also called **slender foxtail**

black grouse *noun* ♦ **grouse**

blackhead /'blækhed/ *noun* a common and fatal disease of young turkeys

blackleg /'blækleg/ *noun* **1.** a bacterial disease of potatoes, affecting the base of the stem which turns black and rots **2.** a bacterial disease of sheep and cattle. It causes swellings containing gas on the shoulders, neck and thigh, and can cause death within 24 hours of the appearance of the symptoms.

Black Leghorn *noun* one of the several varieties of leghorn chicken. The breed is black all over, with a large red single comb. It has yellow legs and flesh and lays white eggs.

black mould *noun* a fungus growth on cereals *(Cladosporium herbarum)*

black mustard *noun* a variety of rape *(Brassica nigra)* sown in spring and harvested by combine in August. The oil and powder from the seeds are important ingredients of table mustard.

black rust *noun* a disease of cereals *(Puccinia graminis)*

black scurf *noun* a disease of potatoes, caused by the *Rhizoctonia solani* fungus. Symptoms include raised black patches on the surface of tubers.

Blacksided Trondheim /ˌblæksaɪdɪd 'trɒndhaɪm/ *noun* a Norwegian breed of polled cattle. The animals are small and white, with black patches.

blacksmith *noun* a person who makes things, especially horseshoes, from wrought iron

black spot *noun* a fungal disease that attacks plants, causing black spots to appear on the leaves

black tea *noun* tea where the leaves wither and are then allowed to ferment in a dry place before drying and crushing

blackthorn /'blækθɔːn/ *noun* same as **sloe**

Black Welsh mountain *noun* a breed of sheep with dark brown wool. A completely black strain has been isolated in separate flocks to supply the specialist demand for black wool.

BLAD *abbreviation* bovine leucocyte adhesion deficiency

blade *noun* **1.** a thin flat leaf, e.g. a leaf of a grass, iris or daffodil **2.** the sharp flat cutting part of a knife, mowing machine, etc

blanch /blɑːntʃ/ *verb* **1.** to make plants white by covering them up (NOTE: Celery is blanched either by putting dark tubes round the stems or by earthing the plants up.) **2.** to partly cook vegetables by putting them in boiling water for a short time. This is done before preserving.

blast *noun* same as **bloat**

blast freezing *noun* a method of quick-freezing oddly-shaped food, by subjecting it to a blast of freezing air

bleach *verb* to make something whiter or lighter in colour, or to become whiter or lighter in colour

bleat *noun* a sound made by a sheep or goat ○ *The ewes seem to be able to recognise the bleat of their own lambs.* ■ *verb* (of *a sheep or goat*) to make the typical sound of a sheep or goat. Compare **grunt, low, neigh**

Blenheim orange /ˌblenəm ˈɒrɪndʒ/ *noun* an apple which ripens late in season; the skin is golden in colour

Bleu du Maine /ˌblɜː du ˈmeɪn/ *noun* a breed of sheep, originating in France and introduced into the UK to produce cross-bred ewes with good conformation. Abbr **BDM**

blight *noun* a disease caused by different fungi, that rapidly destroys a plant or plant part ■ *verb* to ruin or spoil the environment ○ *The landscape was blighted by open-cast mining.*

blind gut *noun* same as **caecum**

blind quarter *noun* a nonfunctional udder, usually of sheep

blinkers /ˈblɪŋkəz/ *noun* a head-covering with leather eyeshields which allows a horse to see only what is in front of it

bloat /bləʊt/ *noun* the swelling of a cow's rumen due to a buildup of gas from lucerne and clovers which is unable to escape, or due to an obstruction in the oesophagus. The gas presses on the diaphragm and the animal may die of asphyxiation.

Blonde /blɒnd/, **Blonde d'Aquitaine** /ˌblɒnd ˈdækiten/ *noun* a breed of cattle, originating in Southwest France and now established in Britain, which produces large calves which develop into good beef animals. The colour varies from off-white to light brown.

blood *noun* the red liquid that is pumped by the heart around an animal's body

COMMENT: Blood is formed of red and white corpuscles, platelets and plasma. It circulates round the body, going from the heart and lungs along arteries and returning to the heart through veins. As it moves round the body it takes oxygen to the tissues and removes waste material from them. Blood also carries hormones produced by glands to the various organs that need them.

bloodline /ˈblʌdlaɪn/ *noun* a general term used to describe the relationship between animals and their ancestors, such as the pedigree line in a flock or herd

'While pedigree Holstein breeders will continue to utilise superior proven bloodlines, commercial milk producers now have a unique opportunity to assess their own breeding goals following the results of several cross-breeding trials.' [*Dairy Farmer*]

bloodmeal /ˈblʌdmiːl/ *noun* a protein-rich feedstuff

blood sports *plural noun* sport in the countryside involving the killing of animals such as foxes and hares

bloodstock /ˈblʌdstɒk/ *noun* a collective term for thoroughbred horses

bloodstream *noun* blood as it passes round the body

blood sucker *noun* an insect or parasite which sucks blood from an animal

blood typing *noun* a method of classifying the blood group of an animal and establishing parentage from this

bloom *noun* 1. a flower ○ *The blooms on the orchids have been ruined by frost.* □ **tree in bloom, field in bloom** a tree or field covered with flowers 2. a powdery substance on the surface of a fruit such as grapes. Bloom is in fact a form of yeast. 3. a fine hairy covering on some fruit such as peaches ■ *verb* to flower ○ *The plant blooms at night.* ○ *Some cacti only bloom once every seven years.*

blossom *noun* a flower or flowers which come before an edible fruit, e.g. apple blossom or cherry blossom □ **trees in blossom** trees covered with flowers ■ *verb* to open into flower

blotch *noun* ♦ **leaf blotch, net blotch**

blower unit *noun* the part of an agricultural machine which blows out waste

blowfly /ˈbləʊflaɪ/ *noun* a name for a number of species of fly such as *Lucila cuprina*, which deposit their eggs in flesh. Also called **meat fly** (NOTE: A sheep which has been attacked by blowfly is said to be **blown**.)

blueberry *noun* a wild plant (*Vaccinium*) now cultivated for its blue berries

bluebottle /ˈbluːˌbɒt(ə)l/ *noun* a two-winged fly, whose maggots live in decomposing flesh, but are sometimes also found on living sheep

blue cross *noun* a term used for blue and white pigs, usually produced by crossing saddleback and landrace breeds

blue-ear *noun* same as **porcine reproductive respiratory syndrome**

Blue-faced Leicester /ˌbluː feɪst ˈlestə/ *noun* a middle-sized longwool breed of sheep, with dense curly fleece. The ram is excellent for crossing with smaller types of ewe. It is the sire of the 'mule' and the so-called 'Welsh mule'.

Blue grey *noun* a crossbred type of cattle, resulting from mating a white shorthorn bull with a Galloway cow

blue nose *noun* a disease affecting horses

blue tongue *noun* a viral disease of cattle, sheep and goats. It is a notifiable disease.

BLUP /blʌp/ *noun* a system under which pedigree information about animals is used to assess how successful a breeding programme involving them might be. Full form **best linear unbiased prediction**

'Since 1986, the majority of modern breeding programmes have adopted BLUP to estimate breeding value. More recent alternative methods which select for or against particular sequences of DNA, using markers, may or may not be used in conjunction with estimates of breeding value.' [*Pig Farming*]

BMR *abbreviation* basal metabolic rate

boar *noun* a male uncastrated pig

board *noun* the floor of a shearing shed

boarding /ˈbɔːdɪŋ/ *noun* the practice of tilting a plough towards the ploughed land so as to increase the pressure on the mould-boards

BOAT *abbreviation* byway open to all traffic

bobby calf /ˈbɒbi kɑːf/ *noun* an unwanted male calf in extreme dairy breeds, such as the Channel Island breeds

bocage /bɒˈkɑːʒ/ *noun* a type of rural landscape with small fields, hedges and trees

BOD *abbreviation* biological oxygen demand ○ *The main aim of sewage treatment is to reduce the BOD of the liquid.*

bog *noun* **1.** soft wet land, usually with moss growing on it, which does not decompose, but forms a thick layer of acid peat (NOTE: The mosses that grow on bogs live on the nutrients that fall in rain.) **2.** an area of bog

boggy /ˈbɒgi/ *adjective* soft and wet like a bog

bogland /ˈbɒglænd/ *noun* an area of bog

boil *verb* **1.** to heat a liquid until it reaches a temperature at which it changes into gas **2.** to reach boiling point ○ *Water boils at 100°C.*

boiling fowl /ˈbɔɪlɪŋ faʊl/, **boiler** /ˈbɔɪlə/ *noun* a hen sold for its meat, when no longer used for laying

bole /bəʊl/ *noun* the wide base of a tree trunk

boll /bəʊl/ *noun* a seed pod of the cotton or flax plant

boll number *noun* the number of bolls per cotton plant

bolt *verb* **1.** (*of a vegetable*) to produce flowers and seeds too early, as in the case of beetroot or lettuce **2.** (*of biennial plants*) to behave as an annual

COMMENT: Bolting occurs as a response to low temperatures and results in the plant failing to build up food reserves in the root, so producing seed in its first year. Bolting is highly undesirable in many plants, especially in the sugar beet, where the roots become woody and have a low sugar content.

bolter /ˈbəʊltə/ *noun* a plant which grows too fast, producing a low seed yield

bolting *noun* the production of seed too early, especially in dry conditions

bolus /ˈbəʊləs/ *noun* a ball of partly digested food regurgitated by ruminants

Bon Chretien /ˌbɒn ˈkretien/ *noun* a common type of pear

bonding *noun* an important time when a newly-born young animal becomes 'tied' to its mother in the period just after birth. The young animal then begins to imitate its mother and learn from her.

bone *noun* one of the calcified pieces of connective tissue which make up the skeleton, e.g. a leg bone

COMMENT: Bones are formed of a hard outer layer (compact bone) which is made up of a series of layers of tissue (Haversian systems) and a softer inner part (cancellous bone or spongy bone) which contains bone marrow.

bonemeal /ˈbəʊnmiːl/ *noun* a fertiliser made of ground bones or horns, reduced to a fine powder

boost *noun* an instrument for marking sheep, usually with hot tar, often bearing the owner's initials

boot *noun* a characteristic swelling in the stem of a cereal plant, produced as the developing ear moves up the stem

borage /ˈbɒrɪdʒ/ *noun* a herb used as an oilseed break crop

Bordeaux mixture /bɔːˈdəʊ ˌmɪkstʃə/ *noun* a mixture of copper sulphate, lime and water, used to spray on plants to prevent infection by fungi

border disease *noun* a disease affecting sheep, caused by a virus. Ewes may abort, or lambs may be born weak and may show muscle tremors. Infection enters the flock when diseased sheep are brought in and an infected ewe excretes the virus.

Border Leicester /ˈbɔːdə ˌlestə/ *noun* a breed of longwool sheep, derived from the English Leicester; its head is bare of wool and has a pronounced 'Roman' nose. The breed is used to produce ram lambs for breeding.

Boreray /ˈbɔːreɪ/ *noun* a rare breed of horned sheep, with grey or cream coloured fleece

-borne *suffix* carried by ◦ *wind-borne pollen of grasses*

borogluconate /ˌbɔːrəʊˈɡluːkəneɪt/ *noun* ♦ **calcium borogluconate**

boron /ˈbɔːrɒn/ *noun* a chemical element. It is essential as a trace element for healthy plant growth.

Bos /bəʊs/ *noun* the genus, part of the family Bovidae, to which cattle, buffalo, bison, yaks and gaur belong

bosk /bɒsk/ *noun* a small wood

botanical /bəˈtænɪk(ə)l/ *adjective* referring to botany

botanical horticulture *noun* the activity of growing different species of plants to study and maintain them

botanical insecticide *noun* an insecticide made from a substance extracted from plants, e.g. pyrethrum, derived from chrysanthemums, or nicotine, derived from tobacco plants

botanist *noun* someone who studies plants as a scientific or leisure activity

botany *noun* the study of plants as a scientific or leisure activity

bot fly /ˈbɒt flaɪ/ *noun* a two-winged fly of the *Oestrus* family whose maggots are parasitic on sheep and horses

bothy /ˈbɒθi/ *noun* a sparsely furnished farm worker's cottage in Scotland

Botrytis /bɒˈtraɪtɪs/ *noun* a fungal disease affecting plants, especially young seedlings. ◊ **chocolate spot**

bots /bɒts/ *noun* the larval stage of the bot fly which lays its eggs on horses' forelegs in summer. It is a cause of restlessness.

bottle *noun* a glass container for liquids or preserves ■ *verb* to preserve food by

heating it inside a glass jar with a suction cap

bottle feeding *noun* a process by which young animals are fed with liquids from a bottle with a teat

botulism /ˈbɒtʃʊlɪz(ə)m/ *noun* a type of food poisoning, caused by a toxin of *Clostridium botulinum* in badly canned or preserved food

boulder *noun* a large rounded piece of rock

boundary *noun* a line that separates one area from another

boundary stone *noun* a stone used to mark the edge of a property

Bovidae /ˈbɒvɪdiː/ *plural noun* the largest class of even-toed ungulates, including cattle, antelopes, gazelles, sheep and goats

bovine *adjective* referring to cattle

bovine growth hormone *noun* a hormone supplement for cows which increases their production of milk. Abbr **BGH**

bovine immunodeficiency virus /ˌbəʊvaɪn ˌɪmjʊnəʊdɪˈfɪʃ(ə)nsi ˌvaɪrəs/ *noun* a disease of cattle which affects the immune system

bovine leucocyte adhesion deficiency *noun* a hereditary condition which makes cattle less able to resist disease. Abbr **BLAD**

bovine leucosis /ˌbəʊvaɪn luːˈkəʊsɪs/ *noun* a cancerous disease in cattle

bovine somatotrophin /ˌbəʊvaɪn ˌsəʊmətəˈtrəʊfɪn/ *noun* full form of **BST**

bovine spongiform encephalopathy /ˌbəʊvaɪn ˌspʌndʒɪfɔːm enˌkefə ˈlɒpəθi/ *noun* full form of **BSE**

bovine tuberculosis *noun* a bacterial disease of cattle which has been almost completely eliminated from the UK by the attested herds scheme. It is a notifiable disease.

'The government's policy to tackle bovine tuberculosis will continue to focus on minimising the disease's spread from infected areas to clean ones until a point when new technologies become available to eradicate it.' [*Farmers Weekly*]

bovine viral diarrhoea *noun* a serious infection of cattle which causes ulcers to the mucus membranes, severe diarrhoea and high rates of abortion. Abbr **BVD**

bowel /ˈbaʊəl/ *noun* an animal's large intestine

bowel oedema /ˈbaʊəl ɪˌdiːmə/ *noun* a bacterial infection of pigs associated with sudden changes of diet and management. Pigs may throw fits, stagger and become paralysed. In acute cases piglets are found dead.

box *verb* to mix different flocks of sheep

BPA *abbreviation* British Pig Association

BPC *abbreviation* British Potato Council

BPEX /ˈbiːpeks/ *noun* a small board of members which determines the Pig Strategy of the MLC and ensures that levies paid by pig farmers are used appropriately. Full form **British Pig Executive**

brace *noun* a pair of dead game birds

bracken /ˈbrækən/ *noun* a tall fern with large triangular fronds that grows on hillsides and in woodland. Latin name: *Pteridium aquilinum.* (NOTE: The word **bracken** has no plural form.)

COMMENT: Bracken is hard to eradicate as it sprouts easily after being burnt. The spores of bracken contain a carcinogenic substance.

bracken poisoning *noun* poisoning of livestock from eating bracken. It can cause serious illness or death.

bract /brækt/ *noun* a small green leaf at the base of a flower or flowering stem

Bradford worsted count /ˌbrædfəd ˈwɜːstɪd ˌkaʊnt/ *noun* a method of measuring wool quality

brahman /ˈbrɑːmən/ *noun* a breed of cattle of Indian origin. Brahmans have a large hump over the neck and shoulders, tolerate high temperatures, are resistant to insects and tropical diseases, and are important for beef production in hot and humid regions.

braird /breɪrd/ *noun* fresh shoots of corn or other crops

brake harrow *noun* a harrow used for breaking up large clods of earth

brake horsepower *noun* the power developed by a tractor's engine, calculated as the force applied to the brakes. Abbr **bhp**

bramble /ˈbræmb(ə)l/ *noun* a wild blackberry bush *(Rubus fruticosus),* with edible black fruits

Bramley's seedling /ˌbræmliːz ˈsiːdlɪŋ/ *noun* a common variety of cooking apple

bran *noun* **1.** the outside covering of a cereal grain (NOTE: It is removed from wheat in making white flour, but is an important source of roughage in the human diet and is used in muesli and other breakfast cereals.) **2.** a feedingstuff, used especially for cattle and poultry

branch *noun* **1.** a woody stem growing out from the main trunk of a tree **2.** a smaller stream separating from but still forming part of a river

branched /brɑːnʃt/ *adjective* with branches

brand *noun* a mark burnt with a hot iron on an animal's hide, to show ownership

Brangus /ˈbræŋɡəs/ *a trademark for a* crossbreed of cattle, obtained by crossing a Brahman with an Aberdeen Angus. It is a black polled breed, and is found in Central and South America, the USA, Canada, Australia and Zimbabwe.

brank /bræŋk/ *noun* same as **buckwheat**

brash *noun* soil containing many stone particles or fragments of rock

brasics /ˈbræsɪks/ *noun* a common name for the charlock *(Sinapsis arvensis)*

brassica /ˈbræsɪkə/ *noun* a generic term for members of the cabbage family, e.g. broccoli, Brussels sprouts, cauliflowers, kales, savoys, swedes and turnips

Brassicaceae /ˌbræsɪˈkæsiaɪ/ *noun* a family of common plants, including cabbage, whose flowers have four petals. Former name **Cruciferae**

bratting /ˈbrætɪŋ/ *noun* the act of putting a jacket on sheep, usually hoggs, to protect them from cold in winter

brawn /brɔːn/ *noun* **1.** meat dish prepared from cut, boiled and pickled pig's head **2.** an obsolete term for a stag boar. Also called **brawner**

braxy /ˈbræksi/ *noun* a bacterial disease of sheep. The affected animals show loss of appetite, dullness, difficult breathing, although these signs of the disease are rarely seen since death occurs rapidly, after five or six hours.

breadcorn /ˈbredkɔːn/ *noun* corn from which the flour for bread is obtained

break *verb* **1.** to train a horse to wear a saddle **2.** to mill flour, by passing it through break rolls

break crop *noun* a crop grown between periods of continuous cultivation of a main crop. On a cereal-growing farm, crops such as sugar beet or potatoes may be introduced to give a 'break' from continuous cereal growing.

break feeding *noun* a method of rationing animal feed by moving animals from one grazing area to another

break in *verb* **1.** to train a wild horse **2.** to bring uncultivated land into cultivation

break rolls *plural noun* heavy rollers with grooves in them, through which the grain passes when being milled. As the grain goes through it is broken into pieces and finally reduced to flour.

breast *noun* **1.** the front part of the body of a bird **2.** the lower part of the body of a sheep, similar to the belly on a pig

Brecknock Hill Cheviot /ˌbreknək ˈhɪl ˌtʃiːviət/ *noun* a breed of sheep found in the Welsh border counties for producing ideal halfbreds, and also in lowlands for prime lamb production

breeching /ˈbriːtʃɪŋ/ *noun* a leather strap round the hindquarters of a shaft horse, which allows the horse to push backwards and so make the cart go backwards

breed *noun* a group of animals of a specific species which have been developed by people over a period of time so that they have desirable characteristics ○ *a hardy breed of sheep* ○ *Two new breeds of rice have been developed.* ■ *verb* **1.** (*of organisms*) to produce young ○ *Rabbits breed very rapidly.* **2.** to encourage something to develop ○ *Insanitary conditions help to breed disease.* **3.** to produce an improved animal or plant by crossing two parent animals or plants showing the desired characteristics ○ *Farmers have bred new hardy forms of sheep.*

breeder *noun* a person who breeds new forms of animals or plants ○ *a cat breeder* ○ *a cattle breeder* ○ *a rose breeder* ○ *a plant breeder*

breed in *verb* to introduce a characteristic into an animal breed or plant variety, by breeding until it is a permanent characteristic of the breed

breeding *noun* the crossing of different plants or animals to produce offspring with desirable characteristics

breeding chicken *noun* a chicken which is raised to produce chicks rather than for its meat or eggs

breeding crate *noun* a box construction designed to take the weight of a bull, when mating with a heifer or cow

breeding ground *noun* an area where birds or animals come each year to breed

breeding herd *noun* **1.** a group of animals kept for breeding purposes **2.** the total number of animals kept in a country for breeding purposes

'Brazil has over 158 million beef cattle and the UK beef breeding herd is estimated at 1.7 million head. When you hear the word globalisation, these figures show we compete in a very large market.' [*Western Mail*]

breeding stock *noun* prime animals kept for breeding purposes to maintain and improve quality of stock

breeze fly *noun* a large two-winged blood-sucking fly, similar to a gadfly

brew *verb* to make beer by fermenting water and malted barley

brewers' grain *noun* a by-product of brewing, the residue of barley after malting. It is a valuable cattle food, fed wet or dried, and is a source of fibre and protein.

brickearth /ˈbrɪkɜːθ/ *noun* a common name for loess. Brickearth soils are fertile and well drained and are used for intensive crop production.

bridle *noun* the part of a horse's harness which goes over the head, including the bit and rein

bridle path, bridle way *noun* a track along which a horse can be ridden

Brie /briː/ *noun* a soft French cheese, made in large flat round shapes

brindled /ˈbrɪndəld/ *adjective* brownish coloured, with spots or marks of another colour

brine /braɪn/ *noun* a solution of salt in water, used for preserving food

COMMENT: Some meat, such as bacon, is cured by soaking in brine. Some types of pickles are preserved by cooking in brine. Some foodstuffs are preserved in brine in jars.

brisket /ˈbrɪskɪt/ *noun* **1.** the lower front part of the body of a cow, between the front legs ○ *When a cow is down for a length of time it should be supported on the brisket.* **2.** the meat from this part of the cow

brisket board *noun* a board across the front of a cubicle to prevent a cow from lying down too near the front

British Agricultural and Garden Machinery Association *noun* a trade association which represents manufacturers and suppliers of agricultural machinery. Abbr **BAGMA**

British Agrochemicals Association *noun* the former name for the Crop Protection Association. Abbr **BAA**

British Agrochemicals Standards Inspection Scheme *noun* full form of **BASIS**

British alpine *noun* a large goat of the Swiss type, which is black with white markings on the face and legs. Males have a large beard. It is a good milking breed.

British Association for Shooting and Conservation *noun* an organisation which promotes shooting for sport in the UK. Abbr **BASC**

British Association of Nature Conservationists *noun* a charity which functions as a think tank on the Government's conservation policies. Abbr **BANC**

British Belgian Blue *noun* a breed of cattle in the UK, originating from the Belgian Blue and registered by the breed society

British Cattle Movement Service *noun* the organisation that registers births, deaths, imports and exports of cattle

British Crop Protection Council *noun* former name for **BCPC**

British Dane /ˌbrɪtɪʃ ˈdeɪn/ *noun* a dairy breed of cattle

British Egg Industry Council *noun* a member association which promotes the use and sale of eggs, and allows its subscribers to use the Lion Quality mark on their produce. Abbr **BEIC**

British Field Sports Society *noun* former name for **Countryside Alliance**

British Free Range Egg Producers Association *noun* an organisation which provides industry news for suppliers and producers of free-range eggs

British Holstein *noun* a breed of dairy cattle developed in the UK from the Holstein

British Lop *noun* a rare breed of pig, which is very large, white, has lop ears and is found mainly in southwest England

British Milk Sheep *noun* a medium polled sheep, with white face and legs. Mainly used as a dairying ewe.

British Oldenburg /ˌbrɪtɪʃ ˈəʊldənbɜːɡ/ *noun* a breed of sheep giving a high yield of wool of good quality

British Pig Association *noun* a society which maintains herd records for native and rare pig breeds in Britain. Abbr **BPA**

British Pig Executive *noun* full form of **BPEX**

British Potato Council *noun* a non-departmental government body which works to promote British potatoes. Abbr **BPC**

British Romagnola *noun* a breed of beef cattle developed in the UK from stock originating in north Italy

British Saanen /ˌbrɪtɪʃ ˈsɑːnən/ *noun* a large white breed of goat with a high milk yield. The breed originated in Switzerland.

British Saddleback *noun* a breed of pig, derived from the Wessex, Hampshire and Essex breeds. It is coloured black with a white band round the front of the body and is used for breeding hardy crosses.

British Society of Plant Breeders *noun* a member association which represents the interests of plant breeders in the UK. Abbr **BSPB**

British Standards Institute *noun* an organisation that monitors design and safety standards in the UK. Abbr **BSI**

British Sugar *noun* a company which buys sugar beet each year from farmers and is responsible for processing and marketing the sugar

British Toggenburg *noun* a dark-brown coloured goat, developed from the Swiss Toggenburg breed

British Veterinary Association *noun* a professional body which represents veterinary surgeons

British White *noun* a rare breed of cattle, white with coloured points. The breed is hornless, short-legged and medium-sized. It is still used as a suckling breed for beef production.

broadacre agriculture /ˈbrɔːdeɪkə ˌæɡrɪkʌltʃə/ *noun* the large-scale cultivation of field crops

broad bean *noun* a common bean (*Vicia faba*) which has large flattish pale green seeds and is grown throughout the world. Also called **field bean**

broadcast *verb* to scatter seeds freely over an area of ground, as opposed to sowing in drills

broadcaster *noun* a machine for sowing seeds broadcast

broadleaf /ˈbrɔːdliːf/, **broadleaf tree** *noun* a deciduous tree that has wide leaves, e.g. beech or oak. Compare **conifer**

broadleaved /'brɔːdliːvd/ *adjective (of a tree)* having wide leaves rather than needles. Compare **coniferous**

broadleaved evergreens *plural noun* evergreen trees with large leaves, e.g. rhododendrons or tulip trees

broadshare /'brɔːdʃeə/ *noun* a cultivator used on heavy ground for stubble clearing and land reclamation work

broad spectrum antibiotic *plural noun* an antibiotic used to control many types of bacteria

'Dairy producers and their vets can now treat milking cows against bovine respiratory disease, bacterial pneumonia and foul-in-the-foot with a broad spectrum antibiotic which has no requirement to discard milk.' [*Farmers Weekly*]

broccoli *noun* a plant of the brassica type that exists in several varieties. One is the winter-grown type of cauliflower. Other varieties include sprouting broccoli, of which there are purple and white kinds, the plants making their curds on numbers of side shoots. Perennial broccoli grows into large plants and forms a number of quite large cauliflower-like heads each season for two or three years. ◊ **calabrese**

broiler *noun* a chicken raised for the table, usually under intensive conditions. It is usually less than 3 months old.

broken mouthed *adjective* referring to a sheep which has lost some of its teeth

broken-winded *adjective* referring to a horse which has ruptured cells in the lungs

brome grass /'brəʊm grɑːs/ *noun* a weed grass found in arable fields and short leys. There are many species (*Bromus*) and none are of any value. ◊ **barren brome**

bronchi /'brɒŋkaɪ/ *plural noun* air passages leading from the throat into the lungs

bronchitis *noun* the inflammation of the membranes in the bronchi, often caused by viruses. ◊ **husk**

Bronze *noun* a breed of large table turkey

brood *noun* a group of offspring produced at the same time, especially a group of young birds ○ *The territory provides enough food for two adults and a brood of six or eight young.* ■ *verb* to raise chicks after they have hatched

brooder /'bruːdə/ *noun* a container with a source of heat, used for housing newly hatched chicks

brooding time *noun* the length of time a bird sits on its eggs to hatch them out

brood mare *noun* a female horse kept for breeding

broody hen /'bruːdi hen/ *noun* **1.** a hen which persists in sitting on its eggs to hatch them **2.** the stage in a fowl's productive life when it has finished laying and sits on the eggs

brownfield site /'braʊnfiːld saɪt/ *noun* a development site that is in a town and formerly had buildings on it, preferred for building development to open fields. Compare **greenfield site**

'The conversion of redundant farm buildings does seem to score highly with European funding because it is focusing on brownfield sites, while it also creates jobs in the rural community and it improves the visual amenity aspect of the countryside.' [*Farmers Guardian*]

brown foot rot *noun* a disease affecting wheat seedlings (*Fusarium avenaceum*)

brown rat *noun* a very destructive type of rat that eats and damages growing and stored crops and can also carry infection to cattle and pigs. Also called **Norway rat**

brown rice *noun* a rice grain that has had the husk removed, but has not been milled and polished to remove the bran

brown rot *noun* a disease of ripe tree fruits such as apples and peaches, caused by fungi. Genus: *Rhizoctonia*.

brown rust *noun* a fungal disease (*Puccinia*) of barley and wheat, causing the grain to shrivel

brown scale *noun* a disease (*Parthenolecanium corni*) causing stunted growth and leaf defoliation. It attacks vines, currants and, in greenhouses, peaches.

Brown Swiss /ˌbraʊn 'swɪs/ *noun* a medium-sized Swiss dual-purpose breed of grey-brown cattle, now found in many parts of the world including Canada, Mexico and Angola

browse *verb* to feed on plant material, especially the leaves of woody plants, which is not growing close to the ground. Compare **graze**

browser *noun* an animal which browses

Brucella /bruː'selə/ *noun* a type of rod-shaped bacterium

brucellosis /ˌbruːsɪ'ləʊsɪs/ *noun* a disease which can be caught from cattle or goats or from drinking infected milk, spread by a species of the bacterium

Brucella. It is a notifiable disease. Also called **undulant fever**

bruise *verb* to harm the flesh under the skin, usually by hitting ○ *Ripe pears have to be handled carefully to avoid bruising.*

brush *noun* same as **brushwood**

brush drill *noun* an old type of seed drill, in which a rotating brush moves the seed into the drill tube

brush killer *noun* a powerful herbicide which destroys the undergrowth

brushwood /'brʌʃwʊd/ *noun* undergrowth with twigs and small branches

Brussels sprouts /ˌbrʌs(ə)lz 'spraʊts/ *noun* a variety of cabbage with a tall main stem, which develops buds which are picked and eaten as fresh vegetables or may be kept frozen

BSE *noun* a fatal brain disease of cattle. Also called **mad cow disease**. Full form **bovine spongiform encephalopathy**

COMMENT: BSE is caused by the use of ruminant-based additives in cattle feed, by which 'scrapie' (the disease of sheep) infects cattle. BSE-infected meat is believed to be the cause of a new strain of Creutzfeldt-Jacob disease in humans.

BSI *abbreviation* British Standards Institute

BSPB *abbreviation* British Society of Plant Breeders

BST *noun* a growth hormone of cattle, formerly added to feed to improve milk production but banned in the EU since 2000. Full form **bovine somatotrophin**

buck *noun* a male goat or rabbit

bucket feeding *noun* a process by which a young animal is fed with milk from a bucket

buckrake /'bʌkreɪk/ *noun* a machine for collecting green crops cut for silage. The machine has a number of steel tines, usually rear-mounted in a tractor linkage system. A buckrake is used by going backwards along a swath with the tines pointing downwards.

buckwheat /'bʌkwiːt/ *noun* a grain crop (*Fagopyrum esculentum*) that is not a member of the grass family like other cereals. It can be grown on the poorest of soils.

bud *noun* a young shoot on a plant, which may later become a leaf or flower □ **tree in bud** a tree which has buds which are swelling and about to produce leaves or flowers ■ *verb* to propagate plants by

grafting a bud from one plant in place of a bud on the stock of another plant

budding *noun* a way of propagating plants in which a bud from one plant is grafted onto another plant

buffalo *noun* **1.** a common domestic animal in tropical countries, used for milk and also as a draught animal **2.** *US* same as **bison**

buffer area *noun* an area of land that separates other areas of land used for different purposes

buffer grazing *noun* a form of set stocking, where a site is divided into three areas, a grazing area, a silage area and a buffer area, which change as the silage is cut during the season

buffer stock *noun* stock of a commodity such as coffee held by an international organisation and used to control movements of the price on international commodity markets

Buff Orpington /ˌbʊf 'ɔːpɪŋgtən/ *noun* a dual-purpose breed of chicken, brown in colour

built environment *noun* the buildings, roads and other structures made by people and in which they live, work or travel. Compare **natural environment**

buisted /'bjuːstɪd/ *noun* a sheep marked after shearing

bulb *noun* an underground plant organ of fleshy scale leaves and buds. It can be planted and will produce flowers and seed.

bulk milk collection *noun* the collection of large quantities of milk by tankers from farms

bulk storage *noun* the storing of fertiliser or grain in dry covered barns, rather than in bags

bull *noun* an uncastrated adult male ox

bull beef *noun* beef from a bull, which is leaner than meat from a steer

bull calf *noun* a male calf

bullets *plural noun* doses of mineral given to cattle and sheep by means of a dosing gun. Bullets provide the animal with a long-lasting supply of minerals to overcome deficiency diseases.

bulling /'bʊlɪŋ/ *noun* a heifer cow of right inclination and body condition to be served

bulling hormone *noun* ♦ **oestrogen**

bullock /'bʊlək/ *noun* **1.** a young male ox **2.** a castrated bull. Bullocks are more docile than bulls and 'finish' quicker than entire animals. They are used as draught animals in many parts of the world. Also called **steer**

bumble foot /ˈbʌmb(ə)l fʊt/ *noun* a condition of the feet of poultry, characterised by abscesses, causing lameness

bund /bʌnd/ *noun* a soil wall built across a slope to retain water or to hold waste in a sloping landfill site

bunding /ˈbʌndɪŋ/ *noun* the formation of bunds

'Farmers have traditionally relied on a system of bunding to grow sorghum in the area's semi-arid soil.' [*New Scientist*]

bundle *noun* a number of twigs or stems of plants, tied together

bunt /bʌnt/ *noun* a disease of wheat caused by the smut fungus

bunt order *noun* the order of social dominance established by cattle and pigs which is the order in which the animals feed and drink (NOTE: The equivalent in birds is called the **pecking order**.)

bur /bɜː/ *noun* a seed case of the cleavers plants, with tiny hooked hairs, which catch on the coats of animals, and help to disperse the seed. Also called **burr**

burdizzo /bɜːˈdiːtsəʊ/ *noun* an implement used in 'bloodless' castration by crushing the spermatic cord

burial site *noun* a place where animals that have died from an infectious disease such as foot and mouth disease are buried

burn *noun* **1.** an injury to skin and tissue caused by light, heat, radiation, electricity or a chemical **2.** in Scotland, a small stream ■ *verb* **1.** to destroy or damage something by fire ○ *Several hundred hectares of forest were burnt in the fire.* ◊ **slash and burn agriculture 2.** to use fuel or food to produce energy

burning nettle *noun* same as **small nettle**

burr /bɜː/ *noun* another spelling of **bur**

burrow *noun* a hole made in the earth by an animal such as a rabbit or fox

bush *noun* **1.** a plant with many woody stems ○ *a coffee bush* ○ *a rose bush* **2.** in semi-arid regions, natural land covered with bushes and small trees

bushel *noun* a measure of capacity for corn and fruit, equivalent to eight gallons

COMMENT: The amount contained in a bushel varies: in the UK it is equivalent to 36.4 litres, but in the USA it is equivalent to 35.3 litres. The weight of a bushel varies with the crop: a bushel of wheat weighs 60lb; of barley 48lb; and of rice 45lb.

bushel weights *plural noun* the weight of average bushels of a product

bush-fallow *noun* a form of subsistence agriculture in which land is cultivated for a few years until its natural fertility is exhausted, then allowed to rest for a long period during which the natural vegetation regrows, after which the land is cleared and cultivated again

bush fruit *noun* fruit from bushes, e.g. gooseberries and red currants, as opposed to fruit from trees

butter *noun* a fatty substance made from cream by churning

butter bean *noun* a leguminous plant (*Phaseolus lunatus*) with large white seeds. Also called **Lima bean**

buttercup *noun* a common weed with yellow flowers

butterfat /ˈbʌtəfæt/ *noun* a fatty substance contained in milk

butterfat content *noun* the amount of butterfat in milk, which will vary with the breed of cow. Jersey cow milk contains 5.14% butterfat, and is much used for butter-making.

butterfly *noun* a flying insect with large, often colourful wings

buttermilk /ˈbʌtəmɪlk/ *noun* the liquid which remains after the churning of cream, when making butter

butter mountain *noun* a popular term for vast quantities of dairy produce in the form of butter, which has been paid for by EU governments and put into cold store

BVA *abbreviation* British Veterinary Association

BVD *abbreviation* bovine viral diarrhoea

BYDV *abbreviation* barley yellow dwarf virus

byre /baɪə/ *noun* a cow house

byway *noun* a small country road or track

byway open to all traffic *noun* a road mainly used by the public as a footpath or bridleway on which vehicles are allowed. Abbr **BOAT**

C

C *symbol* carbon

C₃ *noun* a metabolic pathway in plants, which uses three-carbon compounds to fix CO_2 from the atmosphere

C₄ *noun* a metabolic pathway for CO_2 fixation, which uses four-carbon compounds. Plants with this mechanism, e.g. maize, are adapted to high sunlight and arid conditions. They are called C_4 plants and have low photorespiration.

Ca *symbol* calcium

CA *abbreviation* Countryside Agency

cab *noun* housing for the driver on a tractor, usually protected by anti-roll bars

cabbage *noun* a cultivated vegetable (*Brassica oleracea*) with a round heart or head, a useful food for stock. Other varieties are grown for human consumption.

cabbage root fly *noun* a fly whose larvae attack the roots of Brassica seedlings, causing the plants to turn bluish in colour and wilt and die

cabbage white butterfly *noun* a common white butterfly (*Pieris brassicae*) which lays eggs on the leaves of plants of the cabbage family. The caterpillars cause much damage to the plants.

cactus *noun* a succulent plant with a fleshy stem often protected by spines, found in the deserts of North and Central America (NOTE: The plural is **cacti** or **cactuses**.)

cade lamb /'keɪd læm/ *noun* a lamb reared from a bottle because of the death of its mother

cadmium /'kædmiəm/ *noun* a metallic element naturally present in soil and rock in association with zinc

CAE *abbreviation* caprine arthritis-encephalitis

caecum /'siːkəm/ *noun* a wide part of the intestine leading to the colon

Caerphilly /kə'fɪli/ *noun* a hard white cheese, originally made in South Wales

caesar /'siːzə/ *verb* to perform a surgical intervention to enable an animal to give birth, as is necessary with Belgian Blue cattle

caesium /'siːziəm/ *noun* a metallic alkali element which is one of the main radioactive pollutants taken up by fish

caffeine *noun* an alkaloid present in coffee, tea and kola nuts, which gives them their stimulating properties

cage *noun* a housing for animals consisting of a wood or metal frame with sides made of bars or mesh, used, e.g., for keeping battery hens

cage rearing *noun* a method of rearing poultry in which birds are taken right through from day-old chicks to placement in laying cages. Chicks are started in the top tier and then spread through all the tiers of cages within a few weeks. Cages are designed so that chicks can be brooded in them and yet they may also be used as laying cages.

cage wheel *noun* a metal wheel fitted to the outside of a normal tractor wheel in order to reduce ground pressure

cake *noun* same as **compound feed**

calabrese /'kæləbriːz/ *noun* a variety of sprouting broccoli, grown as a vegetable for human consumption. It produces a large central head, and after this is cut, sprouts are produced for several months. Large quantities are grown for quick-freezing or canning.

calcareous /kæl'keəriəs/ *adjective* (*of soil or rock*) containing calcium

calcareous grassland *noun* the type of vegetation such as grasses that is typical on chalk soil. ◊ **acid grassland**

calcicole /'kælsɪkəʊl/, **calcicolous plant** /'kælsɪfaɪl/ *noun* a plant which grows well on chalky or alkaline soils. Also called **calciphile**

calcification /ˌkælsɪfɪˈkeɪʃ(ə)n/ *noun* the process of hardening by forming deposits of calcium salts

calcified /ˈkælsɪfaɪd/ *adjective* made hard ○ *Bone is calcified connective tissue.*

calcifuge /ˈkælsɪfjuːdʒ/ *noun* a plant which prefers acid soils and does not grow on chalky or alkaline soils. Also called **calciphobe**

calcimorphic soil /ˌkælsɪmɔːfɪk ˈsɔɪl/ *noun* soil which is rich in lime

calciphile /ˈkælsɪfaɪl/ *noun* same as **calcicole**

calciphobe /ˈkælsɪfəʊb/ *noun* same as **calcifuge**

calcitonin /ˌkælsɪˈtəʊnɪn/ *noun* a hormone, produced chiefly by the thyroid gland in mammals, that promotes the deposition of calcium in bones

calcium *noun* a metallic chemical element naturally present in limestone and chalk. It is essential for biological processes.

COMMENT: Calcium is essential for various bodily processes such as blood clotting and is a major component of bones and teeth. It is an important element in a balanced diet. Milk, cheese, eggs and certain vegetables are its main sources. In birds, calcium is responsible for the formation of strong eggshells. Water which passes through limestone contains a high level of calcium and is called 'hard'.

calcium borogluconate /ˌkælsiəm ˌbɔːrəʊˈgluːkəneɪt/ *noun* a chemical that is given in the form of injections to cows suffering from milk fevers as a result of calcium deficiency

calcium phosphate *noun* the main constituent of bones and bone ash fertiliser. Formula: $(Ca_3(Po_4)_2$.

calcium uptake *noun* the taking of calcium into an animal's bloodstream as it eats

calf *noun* a young of a cow, less than one year old

COMMENT: A male calf is known as a bull calf; a female calf is a heifer calf. The meat of calves fed on a milk diet is known as veal.

calf diphtheria *noun* a disease affecting the mouth and throat of a calf

calf enteric disease *noun* a disease of calves causing severe diarrhoea

calf pneumonia *noun* a disease caused by a virus, and affecting dairy-bred and suckled calves

calomel /ˈkæləmel/ *noun* same as **mercury (I) chloride**

caloric /kəˈlɒrɪk/ *adjective* referring to calories

calorie *noun* a unit of measurement of heat or energy. Symbol **cal** (NOTE: The **joule**, an SI measure, is now more usual: 1 calorie = 4.186 joules.)

COMMENT: One calorie is the amount of heat needed to raise the temperature of one gram of water by one degree Celsius. The kilocalorie (shortened to 'Calorie') is also used as a measurement of the energy content of food and to show the caloric requirement or amount of energy needed by an average person. The average adult in an office job requires about 2,500 calories per day, supplied by carbohydrates and fats to give energy and proteins to replace tissue. More strenuous physical work needs more calories. If a person eats more than the number of calories needed by his energy output or for his growth, the extra calories are stored in the body as fat.

Calorie /ˈkæləri/ *noun* same as **kilocalorie**

calorific value *noun* same as **energy value**

calve /kɑːv/ *verb* to give birth to a calf

calver /ˈkɑːvə/ *noun* a cow which has had calves

calving /ˈkɑːvɪŋ/ *noun* the act of giving birth to a calf

calving box *noun* a special pen in which a cow is put to calve

calving interval *noun* the period of time between one calving and the next

calving time *noun* the time when a cow is ready to calve

calyx /ˈkeɪlɪks/ *noun* the part of a flower made up of green sepals which cover the flower when it is in bud (NOTE: The plural is **calyces**.)

Cambridge /ˈkeɪmbrɪdʒ/ *noun* a breed of sheep

Cambridge roller *noun* a heavy roller with a ribbed surface, consisting of a number of heavy iron wheels or rings, each of which has a ridge about 4cm high. The ribbed soil surface left by the roller provides an excellent seedbed for grass and clover seeds.

Camellia /kə'miːliə/ *noun* a family of semi-tropical evergreen plants, including the tea plant

Camembert /'kæməmbeə/ *noun* a soft French cheese, produced in Normandy

Campaign to Protect Rural England *noun* full form of **CPRE**

Campden and Chorleywood Food Research Association *noun* a company which carries out research and development for food producers and suppliers. Abbr **CCFRA**

campylobacter /ˌkæmpɪlə'bæktə/ *noun* bacteria found in the gut of chickens. Dairy cattle also carry the organism. It is a cause of food poisoning in humans.

'The co-operation of the NFU and FSA on targets for the reduction of campylobacter and salmonella levels is an example of effective partnership. Greater assessment of both the success and failure of targets should be made in order to influence future policy decisions.' [*Farmers Guardian*]

can *noun* a metal container for food or drink, made of steel with a lining of tin, or made entirely of aluminium ■ *verb* to preserve food by sealing it in special metal containers

Canadian Holstein /kəˌneɪdiən 'hɒlstaɪn/ *noun* ♦ **Holstein**

canal *noun* a waterway made by people to take water to irrigate land

Canary grass *noun* a weed which produces a large cylindrical-shaped flower not unlike Timothy heads. It is troublesome in cereals.

candling /'kænd(ə)lɪŋ/ *noun* a checking process in which eggs are passed over a source of light which detects blood spots in the egg or cracks in the shell

cane *noun* a stem of large grasses such as the sugar cane and of other plants such as blackberries and raspberries

cane fruit *noun* fruit from plants belonging to the genus *Rubus* including raspberry, blackberry and loganberry. The canes need a post and wire system for support. Fruit is sold on the fresh market, as well as being used for processing, and in recent years there has been an increased interest in the PYO outlets.

cane sugar *noun* a sugar which is processed from the juice which is extracted from the stems of sugar cane in crushing mills

canker /'kæŋkə/ *noun* **1.** a disease causing lesions on a plant or on the skin of an animal ○ *a bacterial canker of fruit trees* **2.** an area of damage caused by canker ○ *cankers on the stem*

COMMENT: The chief victims of canker are fruit trees, especially apples. Cankers appear as sunken areas on the bark, or near a wound. Fungus spores infect the wound edges, laying bare the wood. The disease spreads from the infected area, causing fruit spurs to wilt.

cannibalism /'kænɪbəlɪz(ə)m/ *noun* the practice of an animal which feeds on its own species. In poultry, this may follow on from featherpecking, and may be caused by the crowded conditions in which birds are housed. Sows may eat the young of other sows in intensive breeding conditions.

canning factory /'kænɪŋ ˌfækt(ə)ri/ *noun* a factory where food is canned

cannula /'kænjʊlə/ *noun* a tube through which a trocar is inserted, used to puncture an animal's rumen to allow an escape of gas

canola /kə'nəʊlə/ *noun* oilseed rape

canopy *noun* a layer of branches and leaves of trees which shade the ground underneath. ◊ **green area index**

'Trees that grow to form the tallest part of the canopy suffer more damage than the slower growing trees forming the understorey.' [*Guardian*]

Canterbury hoe /'kæntəb(ə)ri həʊ/ *noun* a hoe which does not have a blade, but is like a three-pronged fork, with the prongs set at right angles to the handle

Canterbury lamb *noun* a lamb reared in New Zealand, mainly for export

cantle /'kænt(ə)l/ *noun* the rear bow of a saddle

CAP *abbreviation* Common Agricultural Policy

capability class *noun* the classification of the usefulness of land for agricultural purposes

capacity *noun* the amount of something which a container can hold ○ *Each cylinder has a capacity of 0.5 litres.*

caper /'keɪpə/ *noun* a Mediterranean shrub (*Capparis spinosa*) the flower buds of which are used as a flavouring

capercaillie /ˌkæpə'keɪli/ *noun* a large game bird (*Tetrao urogallus*) found in northern coniferous forests

capillarity /ˌkæpɪˈlærəti/ *noun* same as **capillary action**

capillary *noun* a tiny blood vessel between the arterioles and the venules, which carries blood and nutrients into the tissues

capillary action, capillary flow *noun* the movement of a liquid upwards inside a narrow tube or upwards through the soil

COMMENT: Capillary flow has an important effect on water in soil, as it does not drain away. Water moves through the soil by capillary action, i.e. by the surface tension between the water and the walls of the fine tubes or capillaries. It is a very slow movement, and may not be fast enough to supply plant roots in a soil which is drying out.

capital items *plural noun* items such as machinery, buildings, fences and drains used in farm production

'Mr Finnie said financial assistance for capital items, such as fencing, would also be introduced for organic converters. And up to £300/farm will be made available towards the cost of producing an organic action plan.' [*Farmers Weekly*]

capon /ˈkeɪpɒn/ *noun* a castrated edible cockerel (NOTE: A cockerel which has been treated with a sex-inhibiting hormone grows and increases in weight more rapidly than a bird which has not been treated.)

cappie /ˈkæpi/ *noun* a disease of sheep, mainly of older lambs and young sheep, associated with thinning of the skull bones. In severe cases the animal cannot eat or close its mouth.

capping /ˈkæpɪŋ/ *noun* a hard crust which sometimes forms on the surface of soil, often only about 2–3cm thick. It can be caused by heavy rain on dry soil, and also by tractors and other heavy farm machinery.

caprine /ˈkæpraɪn/ *adjective* referring to goats

caprine   arthritis-encephalitis, caprine arthritic encephalitis *noun* a disease of goats where the animal suffers loss of condition, swollen joints and pneumonia, leading eventually to death. The disease is spread by contact with saliva and milk. There is no cure. The first case in the UK was reported in April 1990. Abbr **CAE**

capsicum /ˈkæpsɪkəm/ *noun* a group of plants grown for their pod-like fruit, some of which are extremely pungent, e.g. the chilli and Cayenne peppers. Others, including the red, green or sweet peppers are less pungent and are used as vegetables. Also called **pepper**

capsid bug /ˈkæpsɪd bʌg/ *noun* a tiny insect that sucks the sap of plants

capsule *noun* **1.** a dry structure which bursts open with force releasing the seeds of flowering plants or spores of mosses **2.** a membrane round an organ

captan /ˈkæptæn/ *noun* a fungicide, used to fight apple and pear scab. It is also used in seed dressings for peas and other vegetables.

caraway /ˈkærəweɪ/ *noun* seeds of a herb *(Carum carvi)* used as a flavouring in bread and cakes

carbamate /ˈkɑːbəmeɪt/ *noun* a pesticide belonging to a large group used as insecticides, herbicides and fungicides. It is no longer approved for use in the UK.

carbohydrate *noun* an organic compound composed of carbon, hydrogen and oxygen, e.g. sugars, cellulose and starch

COMMENT: Carbohydrates are compounds of carbon, hydrogen and oxygen. They are found in particular in sugar and starch from plants, and provide the body with energy. Plants build up valuable organic substances from simple materials. The most important part of this process, which is called photosynthesis, is the production of carbohydrates such as sugars, starches and cellulose by green plants which convert carbon dioxide and water using sunlight as energy. Carbohydrates form the largest part of the food of animals.

carbon *noun* a common non-metallic element that is an essential component of living matter and organic chemical compounds

carbonate /ˈkɑːbəneɪt/ *noun* a compound formed from a base and carbonic acid

carbon dioxide /ˌkɑːbən daɪˈɒksaɪd/ *noun* a colourless odourless non-flammable atmospheric gas. It is used in photosynthesis and given off in aerobic respiration. Formula: CO_2.

COMMENT: Carbon dioxide exists naturally in air and is produced by burning or by decaying organic matter. In animals, the body's metabolism utilises carbon, which is then breathed out by the lungs as waste carbon dioxide. Carbon dioxide is removed from the atmosphere by plants when it is split by chlorophyll in photosynthesis to form carbon and oxygen. It is also dissolved from the atmosphere in water. The increasing release of carbon dioxide into the atmosphere, especially from burning fossil fuels, contributes to the greenhouse effect.

carbonic acid /kɑː'bɒnɪk 'æsɪd/ *noun* a weak acid formed in small quantities when carbon dioxide is dissolved in water. Formula: H_2CO_3.

carbon monoxide *noun* a colourless, odourless and poisonous gas found in fumes from car engines, burning gas and cigarette smoke. Formula: CO.

carbon neutral *adjective* producing and using the same amount of carbon (NOTE: Renewable plant fuels are carbon neutral. If the same numbers of plants are replanted as are harvested, the CO_2 levels in the air will remain about the same.)

'NFU deputy president Peter Kendall said: "We have reached the point where it has become imperative to break our traditional reliance on fossil fuels. Carbon neutral alternatives are needed if we are to combat global warming and if supply of those alternatives is to be guaranteed then they must be derived from domestic and renewable sources".' [*Farmers Guardian*]

carbon sequestration /,kɑːbən ,siːkwə'streɪʃ(ə)n/ *noun* the uptake and storage of carbon by trees and other plants absorbing carbon dioxide and releasing oxygen

carbon sink *noun* a part of the ecosphere such as a tropical forest which absorbs carbon

carboxyhaemoglobin /kɑː,bɒksi hiːmə[[ɑ'ɔʃ]gləʊbɪn/ *noun* a compound of carbon monoxide and the blood pigment haemoglobin formed when a person breathes in carbon monoxide from car fumes or from ordinary cigarette smoke

carcass /'kɑːkəs/, **carcase** *noun* 1. the dead body of an animal 2. especially in the meat trade, the body of an animal after removing head, limbs and offal

carcass classification scheme *noun* a system of judging the thickness of flesh on a carcass, and the fat cover over a carcass

cardoon /kɑː'duːn/ *noun* a vegetable (*Cynara cardunculus*) grown in Mediterranean areas, similar to the globe artichoke

cargo *noun* same as **brown rice**

carnallite /'kɑːnəlaɪt/ *noun* a white or pale mineral containing hydrated magnesium and potassium chlorides, used as a source of potassium and in fertilisers

carnivore *noun* an animal that eats meat. ◊ **detritivore, frugivore, herbivore, omnivore**

carnivorous *adjective* 1. referring to animals that eat meat ○ *a carnivorous animal* 2. referring to plants which trap and digest insects ○ *Sundews are carnivorous plants*.

carob /'kærəb/ *noun* a long flat dried pod of the carob tree, used in food preparations and in animal feed, where it is called 'locust bean'

carotene /'kærətiːn/ *noun* an orange or red pigment in carrots, egg yolk and some natural oils, which is converted by an animal's liver into vitamin A

carpel /'kɑːpəl/ *noun* a female part of a flower, formed of an ovary, style and stigma

carr /kɑː/ *noun* an area of wetland which supports some trees ○ *fen carr* ○ *willow carr*

carrier *noun* an organism that carries disease and infects other organisms, e.g. an insect that transmits the parasite causing malaria

carrot *noun* a vegetable root crop (*Daucus carota*) grown for human consumption. Most are grown for fresh sale, some for canning and freezing. Damaged roots which cannot be sold can be fed to cattle.

carrot fly *noun* a small fly (*Psila rosae*) with a reddish brown head and yellowish wings. It lays its eggs in the soil surface near plants and the larvae burrow into the root. The foliage wilts and dies, and in cases of severe infestation, the whole crop is lost.

carry *verb* to keep livestock such as cattle on farmland

carrying capacity *noun* the maximum number of livestock that can be supported in a given area

cartilage *noun* thick connective tissue which lines the joints and acts as a cushion and which forms part of the structure of an organ

caryopsis /,kæri'ɒpsɪs/ *noun* a seed of cereals and grasses

CAS *abbreviation* Centre for Agricultural Strategy

case *noun* **1.** an outer covering ○ *Cooling air is directed through passages in the engine case to control engine case temperature.* **2.** a single occurrence of a disease ○ *There were two hundred cases of cholera in the recent outbreak.*

case hardening *noun* the formation of a hard surface on a piece of food, by deposition of sugar or salt

casein /'keɪsiːn/ *noun* a protein found in milk

COMMENT: Casein is precipitated when milk comes into contact with an acid, and so makes milk form cheese.

caseous lymphadenitis *noun* a disease of sheep and goats caused by the bacteria *Corynebacterium pseudotuberculosis.* Symptoms include swollen lymph nodes and abscesses. Abbr **CLA**

cash crop *noun* a crop grown to be sold rather than eaten by the person who grows it, e.g. oil palm

cashmere *noun* a very fine down undercoat on a goat, less than 18 microns

cast *verb* **1.** to bear an offspring prematurely ○ *a cast calf* **2.** to place an animal on its side on the ground

cast ewes *plural noun* old breeding ewes which are sold off by hill farmers to farmers on lower ground who take one more crop of lambs from them

casting /'kɑːstɪŋ/ *noun* a method of ploughing, in which the area is ploughed in a circle, going in an anticlockwise direction

castor oil /ˌkɑːstər 'ɔɪl/ *noun* an oil derived from the seeds of the castor oil plant *(Ricinus communis)*, used as a common purgative for fowls and calves

castrate *verb* to remove the testicles of a male animal

castration /kæ'streɪʃ(ə)n/ *noun* the removal of the essential sex organs, testes and ovaries, from male or female animals. This allows bullocks and heifers to be housed together, as castrated animals are more docile.

COMMENT: Castration may be by cutting the scrotum, as in pigs, or by atrophication which follows ringing, as in sheep and cattle. In fowls, a sex-inhibiting hormone is used.

cast sheep *noun* a sheep lying on its back and unable to get up again

casual labour /ˌkæʒuəl 'leɪbə/ *noun* workers who are hired for short periods from time to time

casual labourer /ˌkæʒuəl 'leɪbərə/, **casual worker** /ˌkæʒuəl 'wɜːkə/ *noun* a worker who is hired for a short period from time to time, e.g. a student hired to pick soft fruit

CAT /kæt/ *abbreviation* Centre for Alternative Technology

catabolism /kə'tæbəlɪz(ə)m/ *noun* the breaking down of complex chemicals into simple chemicals

catch *noun* the amount of fish caught ○ *regulations to limit the herring catch* ■ *verb* to hunt and take animals, usually fish

catch crop *noun* **1.** a fast-growing crop grown in the time interval between two main crops **2.** a fast-growing crop planted between the rows of a main crop

catching pen *noun* a pen into which sheep are put while waiting to be sheared

catchment /'kætʃmənt/, **catchment area** *noun* an area of land, sometimes extremely large, that collects and drains the rainwater that falls on it, e.g. the area round a lake or the basin of a river. Also called **drainage area**, **drainage basin**

Catchment Sensitive Farming *noun* methods of farming which help reduce pollution in water catchment areas by, e.g., reducing the amount of fertiliser and pesticide used on the land

catchwater drain *noun* a type of drain designed to take rainwater from sloping ground

caterpillar *noun* a soft-bodied larva of many species of butterflies and moths. Caterpillars feed mainly on foliage, but can also attack roots, seeds and bark of crops.

caterpillar tractor *noun* a tractor with a revolving set of linked metal plates on either side in place of wheels

cation /'kætaɪən/ *noun* an ion with a positive electric charge

cation exchange capacity *noun* a measure of the fertility of soil which describes its ability to hold and supply nutrients. Abbr **CEC**

cats' faces *noun* same as **field pansy**

cattle *plural noun* domestic farm animals raised for their milk, meat and hide. Class: Bovidae. ○ *herds of cattle* ○ *dairy cattle*

COMMENT: Domesticated cattle belong to the genus Bos, of the family Bovidae, and within the genus there are a number of different species. The European domesticated cattle are *Bos taurus*; the Indian and African domesticated cattle are *Bos indicus*; the buffalo is *Bos bubalus*. The main breeds of European dairy cattle are: Ayrshire, Dairy Shorthorn, Friesian, Guernsey, Holstein and Jersey. The dairy cow is one of the most valuable domesticated animals, providing high-quality human foods. Milk is the main product, although cull cows are an important source of meat. The main breeds of beef cattle are: Aberdeen Angus, Beef Shorthorn, Charolais, Devon, Galloway, Hereford. The main dual-purpose breeds (i.e. breeds which provide both beef and dairy products) are Dexter, South Devon and Welsh Black.

cattle grid *noun* a type of grill made of parallel bars, covering a hole dug in the road. It prevents stock from crossing the grid and leaving their pasture, but allows vehicles and humans to pass.

Cattle Health Certification Standards *plural noun* an umbrella organisation for various cattle health schemes operating in the UK, which sets common standards for testing for non-notifiable diseases. Abbr **CHCS**

cattle identification document, cattle passport *noun* a document which identifies an animal and shows its movements from owner to owner. Abbr **CID**

cattleman /'kæt(ə)lmæn/ *noun* a person who looks after cattle

cattle plague *noun* a disease of cattle, eradicated from the UK in 1877, but still found in parts of Asia and Africa

cattle rustler *noun* a person who steals cattle

cattle rustling *noun* the stealing of cattle

Cattle Tracing System, Cattle Tracing Scheme *noun* a computerised system in place in the UK which registers cattle and their movements from birth to death

cauliflower *noun* a plant of the cabbage family, with a large white head made up of a mass of curds

cavings /'keɪvɪŋz/ *noun* broken pieces of straw from a threshing machine

cayenne pepper /ˌkeɪen 'pepə/ *noun* a plant (*Capsicum frutescens*) producing a pungent red pepper from ground dried pods

Cayuga /kæ'juːgə/ *noun* a breed of duck which produces dark green eggs and has shiny green-black plumage

CCFRA *abbreviation* Campden and Chorleywood Food Research Association

CCW *abbreviation* Countryside Council for Wales

Cd *symbol* cadmium

CEC *abbreviation* cation exchange capacity

cedar *noun* a large coniferous tree belonging to several genera, including *Thuya* and *Cedrus*

CEFAS *abbreviation* Centre for Environmental, Fisheries and Aquaculture Science

CEH *abbreviation* Centre for Ecology and Hydrology

celeriac /sɪ'leriæk/ *noun* a variety of celery with a thick edible root. The root is used as a vegetable in soups and salads.

celery *noun* a vegetable plant (*Apium graveolens*), with thick edible leaf stalks. The plant is grown in trenches to help growth and to blanch the stems, although some varieties are self-blanching.

cell *noun* the basic independently functioning unit of all plant and animal tissue

COMMENT: A biological cell is a unit which can reproduce itself. It is made up of a jelly-like substance (cytoplasm) which surrounds a nucleus and contains many other small organelles which are different according to the type of cell. Cells reproduce by division (mitosis) and the chemical reactions that occur with them are the basis of metabolism. The process of division and reproduction of cells is how the human body is formed.

cellular *adjective* **1.** referring to the cells of organisms **2.** made of many similar parts connected together

cellulose /'seljuləus/ *noun* **1.** a carbohydrate which makes up a large percentage of plant matter, especially cell walls **2.** a chemical substance processed from wood, used for making paper, film and artificial fibres

COMMENT: Cellulose is not digestible by humans, and is passed through the digestive system as roughage.

Celsius *noun* a scale of temperature where the freezing point of water is 0° and the boiling point is 100°. Symbol **C**. Former name **centigrade** (NOTE: It is used in many countries, but not in the USA, where the Fahrenheit system is still commonly used.)

COMMENT: To convert Celsius temperatures to Fahrenheit, multiply by 1.8 and add 32. So 20°C is equal to 68°F.

Celtic field system *noun* ♦ **field**

census *noun* a survey of a specific population to assess numbers and other features

centigrade /ˈsentɪɡreɪd/ *noun* same as **Celsius**

Central Scotland Forest Trust *noun* an alliance created with the aim of creating more and larger woodland areas across Central Scotland. Abbr **CSFT**

Central Veterinary Laboratory *noun* a veterinary service set up to assist the Government's department for agriculture and food by researching infectious diseases. Abbr **CVL**

Centre for Agricultural Strategy *noun* an organisation which carries out independent research into agricultural development issues, based at the University of Reading. Abbr **CAS**

Centre for Alternative Technology *noun* an association that advises on sustainable and environmentally sound methods in technological development. Abbr **CAT**

Centre for Ecology and Hydrology *noun* a UK organisation that does research on and monitors terrestrial and freshwater environments. Abbr **CEH**

Centre for Environmental, Fisheries and Aquaculture Science *noun* a research and advisory agency run by Defra, which investigates aquatic science and contamination

centrifugal /ˌsentrɪˈfjuːɡ(ə)l, senˈtrɪfjʊɡ(ə)l/ *adjective* going away from the centre

centrifugal sugar *noun* a type of raw sugar containing 96% to 98% sucrose, which has been isolated from sugar beet or cane by standard extraction processes

centrifugation /ˌsentrɪfjuːˈɡeɪʃ(ə)n/ *noun* the separation of the components of a liquid in a centrifuge. Also called **centrifuging**

centrifuge /ˈsentrɪfjuːdʒ/ *noun* a device which uses centrifugal force to separate or remove liquids ■ *verb* to separate liquids by using centrifugal force ○ *The rotating vanes of the breather centrifuge the oil from the mist.*

centrifuging /ˈsentrɪfjuːdʒɪŋ/ *noun* same as **centrifugation**

cep /sep/ *noun* an edible mushroom-like fungus *(Boletus edulis)*

cereal *noun* a type of grass which is cultivated for its grains. Cereals are used especially to make flour for breadmaking, for animal feed or for producing alcohol. (NOTE: The main cereals are wheat, rice, barley, maize and oats.)

COMMENT: Cereals are all members of the Graminales family. Oats, wheat, barley, maize and rye are commonly grown in colder temperate areas, and rice, sorghum and millet in warmer regions. Cereal production has considerably expanded and improved with the introduction of better methods of sowing, combine harvesters, grain driers, bulk handling and chemical aids such as herbicides, fungicides, insecticides and growth regulators.

cereal stands *plural noun* fields of standing cereal crops

cerebrocortical necrosis /ˌserəbrəʊkɔːtɪk(ə)l neˈkrəʊsɪs/ *noun* a disease of sheep caused by thiamine deficiency. The animal appears blind and fits may follow.

cerebrospinal fluid /ˌserəbrəʊspaɪn(ə)l ˈfluːɪd/ *noun* the colourless fluid that fills the spaces around and within the brain and spinal cord to cushion against injury

certificate *noun* an official paper which states something, e.g. the National Certificate in Agriculture

certificate of attestation *noun* a certificate given to an attested herd

certificate of bad husbandry *noun* a certificate issued to a tenant farmer by an Agricultural Land Tribunal if the tenant is inefficient and unable to farm to a satisfactory standard

certification /səˌtɪfɪˈkeɪʃ(ə)n/ *noun* the process of obtaining or giving approval for something such as carrying out a particular type of test, or of obtaining a certificate to prove that something is what it claims to be

certified seed *noun* seed which has been successfully tested for purity, disease and weed contamination and is granted certification for sale

'But payments for farm-saved stocks are making up an increasingly significant proportion of income, thanks to an EU decision in 1994 that allowed breeders to collect royalties on such seed, provided the rate is "sensibly lower" than the amount levied on certified seed.' [*Farmers Weekly*]

certified stock *noun* a stock of grain which has been approved for delivery

certify *verb* to give official approval to something, or to say officially that something is what it claims to be

cesspool /'sespuːl/, **cesspit** /'sespɪt/ *noun* a tank for household sewage, constructed in the ground near a house which is not connected to the main drainage system, and in which the waste is stored before being pumped out for disposal somewhere else

CFCU *abbreviation* Counter Fraud and Compliance Unit

chaff /tʃɑːf, tʃæf/ *noun* **1.** husks of corn which separate from the grain during threshing **2.** short lengths of cut straw, used as feed for ruminants and as a component of manures

chain *noun* **1.** a number of metal rings attached together to make a line **2.** a measure of length equal to 22 yards, originally measured with a chain, called Gunter's chain

chain harrow *noun* a type of harrow built in a similar way to a piece of chain-link fencing. The links may be plain or spiked, the spiked type being used to aerate grassland.

chain-link fencing *noun* material for fencing, made of an open web of thick wire links, twisted together; supplied on a roll, it is one of the easiest forms of fencing to put in place

chalaza /kə'leɪzə/ *noun* a coil of fibrous protein which holds the yolk in the centre of the egg

chalk *noun* a fine white limestone rock formed of calcium carbonate

COMMENT: Chalk is found widely in many parts of northern Europe. Formed from animal organisms it is also used as an additive (E170) in white flour. Ground chalk is used for liming soils. The sharp-edged flints of various sizes found in soils overlying some of the chalk formations in Europe, are very wearing on farm implements.

chalky *adjective* referring to soil which is contains a lot of chalk

challenge feeding *noun* the process of feeding dairy cows with concentrates to provide extra nourishment

chamomile /'kæməmaɪl/ *noun* ♦ mayweed

chandler /'tʃɑːndlə/ *noun* a person who sells or supplies goods, e.g. a seed chandler or corn chandler

channel *noun* a bed of a river or stream ■ *verb* to send water in a particular direction (NOTE: British English is **channelled – channelling**, but the US spelling is **channeled – channeling**.)

channelise /'tʃænəlaɪz/, **channelize** *verb* to straighten a stream which has many bends, in order to make the water flow faster

Channel Island breeds *plural noun* the Guernsey and Jersey breeds of dairy cattle

chaptalisation /ˌtʃæptəlaɪ'zeɪʃ(ə)n/, **chaptalization** *noun* the addition of sugar to wine in order to increase the amount of alcohol that the wine contains

charlock /'tʃɑːlɒk/ *noun* a widespread weed *(Sinapis arvensis)* mainly affecting spring cereals and other spring crops; also commonly called brasics or wild mustard

Charmoise /'ʃɑːmwɑːz/ *noun* a breed of sheep found in central France

Charolais /'ʃærəleɪ/, **Charollais** *noun* **1.** a breed of beef cattle which originated in central France, is creamy white in colour and is valued for its fast growth and lean meat **2.** a breed of sheep originating in France and having a characteristic 'red' face

chats /tʃæts/ *plural noun* small potatoes, separated from larger potatoes during the grading process

CHCS *abbreviation* Cattle Health Certification Standards

Cheddar *noun* a hard yellow cheese, originally made in the West Country

cheese *noun* food made from cow's milk curds (NOTE: Cheese is also made from goat's milk, and more rarely from ewe's milk. The curd is pressed and left to mature for a period of time (longer in the case of hard cheese). There are many varieties of both hard and soft cheese: the British Caerphilly, Gloucester, Cheddar and Cheshire are all hard cheeses; the French Brie and Camembert are soft.)

chelates /'kiːleɪts/ *plural noun* compounds of trace elements and organic substances which are water-soluble and may be safely applied as foliar sprays or to the soil

chem- /kem/ *prefix* referring to chemistry or chemicals

chemical *noun* a substance formed of chemical elements or produced by a chemical process ○ *the widespread use of chemicals in agriculture* ○ *The machine analyses the chemicals found in the collected samples.* ■ *adjective* 1. referring to chemistry or chemicals 2. made by humans from a combination of substances or chemical elements and not produced naturally (*informal*) ○ *an unpleasant chemical taste*

chemical element *noun* a substance such as iron, calcium or oxygen, which exists independently and is not formed from a combination of other substances

chemical fertiliser *noun* same as **artificial fertiliser**

chemical food poisoning *noun* poisoning by chemical substances in food, e.g. by toxic substances naturally present in some plants or insecticides in processed food

Chemical Industries Association *noun* a trade association in the UK representing the chemical and chemistry-using industries. Abbr **CIA**

Chemical of Concern *noun* ♦ List of Chemicals of Concern

chemical score *noun* comparison of the relative protein values of particular foodstuffs, tested in laboratory experiments

chemistry *noun* the study of substances, elements and compounds and their reactions with each other

chemotherapeutic /ˌkiːməʊθerə'pjuːtɪk/ *adjective* using chemicals that have been either synthesised or produced by the action of living organisms to treat diseases

chemotroph /'kiːməʊtrɒf/ *noun* an organism which converts the energy found in organic chemical compounds into more complex energy, without using sunlight. Compare **phototroph**

chemotrophic /ˌkeməʊ'trɒfɪk/ *adjective* referring to something which obtains energy from sources such as organic matter. Compare **phototrophic** (NOTE: Most animals are chemotrophic.)

Cher /ʃeə/ ♦ Berrichon du Cher

chernozem /'tʃɜːnəʊzem/ *noun* a dark fertile soil, rich in organic matter, found in the temperate grass-covered plains of Russia and North and South America. Also called **black earth**

cherry *noun* 1. a small usually sweet fruit with a single hard stone at its centre, which is usually either yellow, red, or purple in colour 2. a tree that produces cherries

cherry plum *noun* a cooking plum (*Prunus cerasifera*) which is small and usually bright red

cherry tomato *noun* a variety of tomato (*Lycopersicon esculentum*) with very small fruit

chervil /'tʃɜːvɪl/ *noun* a herb (*Anthriscus cerefolium*) used as a garnish and also in salads and soups

Cheshire /'tʃəʃə/ *noun* a crumbly hard white British cheese

Cheviot /'tʃiːviət/ *noun* a large, hardy breed of sheep which is white-faced and usually hornless. The short, thick wool is of good quality, of middle length and fairly dense.

Chewings fescue /'tʃuːɪŋz ˌfeskjuː/ *noun* a common variety of grass used for pastures in New Zealand

Chianina /ˌkiə'niːnə/ *noun* a breed of beef cattle, originating in Tuscany in Italy. One of the largest breeds of cattle, it is white in colour with black hooves, muzzle and horn tips. It is a dual-purpose breed, used both for beef production and as draught animals.

chick *noun* a young, newly hatched bird, up to the time when it is weaned from the hen or brooder

chicken *noun* 1. a young bird of a domestic fowl 2. the meat of domestic poultry

COMMENT: Chicken manure is now being used as fuel in programmes to set up renewable energy power stations.

chickpea *noun* a legume crop, grown for its large round pale-yellow seeds. It is important in India and Pakistan as a source of protein. Genus: *Cicer arietinum*. Also called **gram**

chickweed /'tʃɪkwiːd/ *noun* a widespread choking weed (*Stellaria media*) which is found in cereals and grass and germinates and flowers all the year round. Also called **white bird's-eye**

chicory /'tʃɪkəri/ *noun* a blue-flowered plant (*Cichorium intybus*) cultivated for its salad leaves and roots. Dried chicory root is used as a substitute for coffee.

Chief Veterinary Officer *noun* the managing head of a state veterinary service. Abbr **CVO**

chill *verb* to preserve by cooling to a temperature just above freezing. ◊ **cook chill**

chilli *noun* **1.** same as **chilli pepper 2. a** very pungent spice or sauce made from ground chilli peppers and other ingredients

chilli pepper *noun* a small seed pod, red or green in colour, from the plant *Capsicum frutescens* which has a very hot taste

chillshelter /'tʃɪlʃeltə/ *noun* a feeding area surrounded by a high embankment to protect the cattle against the cold

chill starvation *noun* a disease affecting very young lambs, caused by loss of body heat during severe weather conditions

Chinchilla /tʃɪn'tʃɪlə/ *noun* a small rabbit important for its soft grey fur

Chinese goose /,tʃaɪniːz 'guːs/ *noun* a breed of goose with a lighter carcass, raised for meat production

chip basket *noun* a basket woven from thin strips of wood

chisel plough *noun* a plough with a heavy-duty frame, with tines bolted to it. Rigid tines are normally used, and each tine has a point which can be replaced.

chit /tʃɪt/ *noun* a shoot or sprout ▪ *verb* to promote germination in seed before sowing, especially to set up seed potatoes to sprout before planting them ○ *About half of the potato crops in the UK are grown from chitting tubers.*

chitterlings /'tʃɪtəlɪŋz/ *plural noun* the small intestines of pigs, used for food

chitting house *noun* a storage building for trays of potatoes, where they are kept to sprout before planting

chives *noun* an onion-like herb (*Allium schoenoprasum*) of which the leaves are used as a garnish or in soups and salads

chlamydiosis /klə,mɪdi'əʊsɪs/ *noun* a bacterial infection which is transmitted by infected birds such as ducks and pigeons

COMMENT: Symptoms can be similar to those of flu, or in bad cases, pneumonia and hepatitis. The disease comes from birds and their feathers, and affects particularly poultrymen. It can also affect sheep and lambs.

chlor- /klɔːr/ *prefix* same as **chloro-**

chloride /'klɔːraɪd/ *noun* a salt of hydrochloric acid

chlorinated *adjective* treated with chlorine

chlorinated hydrocarbon *noun* a compound containing chlorine, carbon and hydrogen that remains in the environment after use and may accumulate in the food chain, e.g. an organochlorine pesticide such as lindane or DDT, an industrial chemical such as a polychlorinated biphenyl (PCB), or a chlorine waste product such as a dioxin

COMMENT: Chlorinated hydrocarbon insecticides include DDT, aldrin and lindane. These types of insecticide are very persistent, with a long half-life of up to 15 years, while organophosphorous insecticides have a much shorter life. Chlorinated hydrocarbon insecticides not only kill insects, but also enter the food chain and kill small animals and birds which feed on the insects

chlorination /,klɔːrɪ'neɪʃ(ə)n/ *noun* sterilisation by adding chlorine

COMMENT: Chlorination is used to kill bacteria in drinking water, in swimming pools and sewage farms, and has many industrial applications such as sterilisation in food processing.

chlorinator /'klɔːrɪneɪtə/ *noun* an apparatus for adding chlorine to water

chlorine *noun* a greenish chemical element, used to sterilise water and for bleaching

chloro- /klɔːrəʊ/ *prefix* **1.** chlorine **2.** green

chlorophyll /'klɒrəfɪl/ *noun* a green pigment in plants and some algae

COMMENT: Chlorophyll absorbs light energy from the Sun and supplies plants with the energy to enable them to carry out photosynthesis. It is also used as a colouring (E140) in processed food.

Chlorophyta /klɒr'ɒfɪtə/ *noun* a large group of algae that possess chlorophyll

chlorosis /klɔː'rəʊsɪs/ *noun* a reduction of chlorophyll in plants, making the leaves turn yellow

chlorpyrifos /klɔː'pɪrɪfɒs/ *noun* an organophosphate insecticide used on a wide range of crops

chocolate spot *noun* a fungal disease of winter beans, occurring in two forms, *Botrytis cinerea* or the more severe *Botrytis fabae*. The disease appears as small round brown spots on leaves and stems. In bad attacks, the symptoms move to flowers and pods.

choke *noun* **1.** stiff hairs inside the head of an artichoke. The hairs have to be removed before the heart can be eaten. **2.** a disease of grasses appearing as a white fungus that develops out of the leaves and surrounds the young flower stalk. It reduces the seed crop. ■ *verb* **1.** to kill and animal or plant by cutting off air or light ○ *The small plants were choked by weeds.* **2.** to fill something such as a channel so that nothing can move through it ○ *The drainage channels were choked with water weed.*

chop down *verb* to cut through the trunk of a tree with an axe, so that the tree falls down

chorionic gonadotrophin /kɔːri
ˌɒnɪk ˌgəʊnədəʊ'trəʊfɪn/ *noun* a hormone produced by the placenta in mammals that helps maintain a pregnancy

C horizon *noun* rock underlying the soil. ◊ **horizon**

Chorleywood bread process /ˌtʃɔːliwʊd 'bred ˌprəʊses/ *noun* a method of making bread, developed by the British Baking Industries Research Association, in which the long fermentation period is eliminated by mixing the dough vigorously by mechanical means

chromosomal /ˌkrəʊmə'səʊm(ə)l/ *adjective* referring to chromosomes

chromosome *noun* a thin structure in the nucleus of a cell, formed of DNA which carries the genes (NOTE: Different types of organism have different numbers of chromosomes.)

chronic *adjective* referring to a disease or condition which lasts for a long time ○ *The forest was suffering from chronic soil acidification.* Compare **acute**

chrysalis /'krɪsəlɪs/ *noun* **1.** a stage in the development of a butterfly or moth when the pupa is protected in a hard case **2.** the hard case in which a pupa is protected

chrysanthemum *noun* a genus of composite plants, many of which are cultivated for their flowers (NOTE: The insecticide pyrethrum is derived from *Chrysanthemum roseum*.)

churn *noun* a large metal milk container ■ *verb* to shake a liquid violently to mix it

CIA *abbreviation* Chemical Industries Association

CID *abbreviation* cattle identification document

-cide *suffix* killing

cider *noun* a fermented drink made from apple juice

cider press *noun* a device for crushing apples to extract the juice to make cider

cinnamon *noun* the aromatic inner bark from a tropical tree, *(Cinnamomum zeylanicum)*, used as a spice

circulation *noun* same as **circulation of the blood**

circulation of carbon *noun* the process by which carbon atoms from carbon dioxide are incorporated into organic compounds in plants during photosynthesis (NOTE: The carbon atoms are then oxidised into carbon dioxide again during respiration by the plants or by herbivores which eat them and by carnivores which eat the herbivores, so releasing carbon to go round the cycle again.)

circulation of the blood *noun* the movement of blood around the body from the heart through the arteries to the capillaries and back to the heart through the veins

citrus fruit *noun* the edible fruits of evergreen citrus trees, grown throughout the tropics and subtropics; the most important are oranges, lemons, grapefruit and limes. Citrus fruit have thick skin and are very acidic.

city farm *noun* a community project that uses an area of wasteland in an inner-city or urban fringe area for farming and gardening with an ecological approach to the management of land and resources

CIWF *abbreviation* Compassion in World Farming

CJD *abbreviation* Creutzfeldt-Jakob disease

Cl *symbol* chlorine

CLA *abbreviation* **1.** caseous lymphadenitis **2.** Country Land and Business Association

clamp *noun* a method of storing root crops in the open, in which the crop is heaped into a pile and covered with straw and earth ■ *verb* to store crops or silage in a clamp

COMMENT: Silage is also kept in clamps, originally trenches dug into the ground into which the crop was tipped and then covered. Silage may be clamped as a mound often covered with polythene sheeting which is weighed down with old tyres or railway sleepers.

clarts /klɑːts/ *plural noun* dung attached to fleece of sheep

classical swine fever *noun* ♦ swine fever

clay *noun* a type of heavy non-porous soil made of fine particles of silicate

clayey /'kleɪi/ *adjective* containing clay ○ *These plants do best in clayey soils.*

claying /'kleɪɪŋ/ *noun* the application of clay to sandy soils and black fen soils to improve their texture

claypan /'kleɪpæn/ *noun* a hollow on the surface of clay land where rain collects

clay soils *plural noun* soils with more than 35% clay size material (NOTE: Clay soils are sticky when wet and can hold more water than most other types of soil. They lie wet in the winter, and are liable to poaching. They are slow to warm in spring time. In long periods of dry weather, clay soils become hard and wide cracks may form.)

clean cattle *noun* cattle which have not been used for breeding

clean crop *noun* a measure of the amount of wheat or other cereal not mixed with seeds of other plants

clean land *noun* land which is free of weeds

clean pasture *noun* pasture that has been left ungrazed for four to six weeks after contamination with parasitic worm larvae, and is considered free of the parasites

cleansing /'klenzɪŋ/ *noun* same as **afterbirth**

clear *verb* to remove plants to prepare open land for cultivation ○ *They cleared hectares of jungle to make a new road to the capital.* ○ *We are clearing rainforest at a faster rate than before.*

clearance *noun* the action of clearing land for cultivation

clearcut /'klɪəkʌt/ *noun* the cutting down of all the trees in an area ■ *verb* to clear an area of forest by cutting down all the trees

clearcutting, clearfelling *noun* cutting down all the trees in an area at the same time ○ *The greatest threat to wildlife is the destruction of habitats by clearfelling the forest for paper pulp.*

COMMENT: Clearcutting is a way of managing a forest. Once the felled timber has been removed, the land is cleared of stumps and roots, and then sown with new tree seed.

clearfell /klɪə'fel/ *verb* same as **clearcut**

cleavers /'kliːvəz/ *noun* a widespread weed *(Galium aparine)* affecting winter cereals, oilseed and early-sown spring crops. It has seed pods (**burs**) with tiny hooks which catch onto the coats of passing animals. Also called **goosegrass**

cleg /kleg/ *noun* same as **horsefly**

Cleveland Bay /ˌkliːvlənd 'beɪ/ *noun* a breed of light draught horse

clevis /'klevɪs/ *noun* a U-shaped iron attachment, used to couple an implement to a tractor towbar

click beetle *noun* a brown beetle *(Agriotes* species), whose larvae are wireworms which attack cereals by eating the plants just below the soil surface

climate change *noun* a long-term alteration in global weather patterns, occurring naturally, as in a glacial or post-glacial period, or as a result of atmospheric pollution (NOTE: Sometimes climate change is used interchangeably with 'global warming', but scientists tend to use the term in the wider sense to include natural changes in the climate.)

'Keen to seize the opportunity, Mr Kendall emphasised the important role agriculture plays in sustaining rural communities, maintaining the environment, providing a safe and constant supply of food and its unique position in helping to tackle climate change.' [*Farmers Weekly*]

climate change levy *noun* a tax on the use of energy by sectors such as agriculture and industry, aimed at promoting energy efficiency and reducing emissions of greenhouse gases

climax *noun* the final stage in the development of plant colonisation of a specific site, when changes occur within a mature and relatively stable community

clingstone /'klɪŋstəʊn/ *adjective* referring to varieties of peach where the flesh is attached to the stone. Compare **freestone**

clip *verb* to cut the wool from a sheep ■ *noun* the total amount of wool obtained from a sheep or from a whole flock

clippers /'klɪpəz/ *plural noun* shears, used for clipping sheep

cloaca /kləʊ'eɪkə/ *noun* the terminal region of the gut with the intestinal, urinary and genital canals opening into it (NOTE: It is present in reptiles, amphibians, birds, many fishes, and in some invertebrates.)

cloche /klɒʃ/ *noun* a covering of either glass or plastic, which can be carried from place to place and is used to protect seedbeds

clod /klɒd/ *noun* a large lump of soil which may be difficult to break down into tilth. Clods will affect the quality of a seedbed.

clone *noun* **1.** a group of cells derived from a single cell by asexual reproduction and therefore identical to the first cell **2.** an organism produced asexually, either naturally or by means such as taking cuttings from a plant **3.** a group of organisms all of which have been derived from a single individual by asexual means ■ *verb* to reproduce an individual organism by asexual means

cloning *noun* the reproduction of an individual organism by asexual means

closed canopy *noun* a canopy which either has achieved complete cover or intercepts 95% of visible light

closewool /'kləʊswʊl/ *noun* one of the breeds of sheep with short dense wool, e.g. the Devon closewool

clostridial /klɒ'strɪdiəl/ *adjective* referring to Clostridium

clostridial disease *noun* a disease such as pasteurellosis caused by Clostridium

Clostridium /klɒ'strɪdiəm/ *noun* a type of bacterium (NOTE: Species of Clostridium cause botulism, tetanus and gas gangrene, but also increase the nitrogen content of soil.)

clot *verb* (*of blood*) to become thick and stop flowing, forming a scab if exposed to the air ■ *noun* a mass of thickened blood that forms over a wound or in a blood vessel

clotted cream *noun* a type of cream which has been heated and so becomes more solid

clove *noun* a dried flower bud of a tree (*Eugenia caryophyllata*) used for flavouring

cloven hoof /ˌkləʊv(ə)n 'huːf/ *noun* the divided hoof of animals such as cattle, sheep and pigs

clove of garlic *noun* a small bulb in a cluster of garlic

clover *noun* a large genus (*Trifolium*) of leguminous plants, with trefoil leaves and small flowers

COMMENT: Clovers are essential plants for the longer ley and permanent pasture. They are nitrogen fixing plants, and with their creeping habit of growth they knit the sward together and help keep out other weeds. The clovers of agricultural importance are the red and white clovers.

cloxacillin /ˌklɒksə'sɪlɪn/ *noun* a type of antibiotic

club root *noun* a fungal disease (*Plasmodiophora brassicae*) affecting brassicas, causing swelling and distortion of the roots and stunted growth. Also called **finger and toe**

clumping /'klʌmpɪŋ/ *noun* the gathering together of a large number of things in a mass

Clun forest /'klʌn ˌfɒrɪst/ *noun* a hardy grass hill breed of sheep which has fine dense fleece, a dark brown face and a permanent topknot which extends out over the forehead

cluster *noun* the four cup attachments of a milking machine, attached to the teats of the cow's udder

clutch *noun* a set of eggs laid by a bird

Clydesdale /'klaɪdzdeɪl/ *noun* a breed of heavy draught horse, originating in Scotland. It is brown or black in colour, with a mass of white 'feathers' at the feet.

CMPP *abbreviation* mecoprop

Co *symbol* cobalt

coarse *adjective* referring to a particle or feature which is larger than others ○ *Coarse sand fell to the bottom of the liquid as sediment, while the fine grains remained suspended.*

coarse grains *plural noun* cereal crops such as maize, barley, millet, oats, rye and sorghum which are less fine than wheat or rice

'A US Department of Agriculture report released this week cut wheat production and coarse grain stocks by more than expected, sending Chicago futures sharply higher.' [*Farmers Weekly*]

coat *noun* the hair on the body of an animal ○ *Some breeds of goat have very thick coats.*

COMMENT: Most animals have a thicker topcoat of coarse hair, which gives protection against the weather and a softer undercoat of fine wool which keeps the animal's body warm.

cob /kɒb/ *noun* **1.** a sturdy short-legged riding horse **2.** same as **corn cob 3.** a mixture of clay, gravel and straw used as a building material

cobalt *noun* a metallic element. It is used to make alloys.

cob nut *noun* a large hazel nut

coccidioidomycosis /ˌkɒkˌsɪdɪɔɪˌdəʊmaɪˈkəʊsɪs/ *noun* a lung disease caused by inhaling spores of the fungus *Coccidioides immitis*

coccidiosis /ˌkɒksɪdiˈəʊsɪs/ *noun* a parasitic disease of livestock and poultry affecting the intestines

coccus /ˈkɒkəs/ *noun* a bacterium shaped like a ball (NOTE: The plural is **cocci**.)

Cochin /ˈkɒtʃɪn/ *noun* a breed of domestic fowl originating in China

cochineal /ˌkɒtʃɪˈniːl/ *noun* a red colouring matter obtained from the dried body of an insect, the female concilla *(Coccus cacti)* found in Mexico, Central America and the West Indies

cock *noun* **1.** a male bird **2.** (*of poultry*) a male chicken over 18 months old

cockerel *noun* a young male chicken, up to 18 months old

cocksfoot /ˈkɒksfʊt/ *noun* a perennial grass (*Dactylis glomerata*). A high-yielding, deep-rooting grass, which is resistant to drought and sometimes used in pasture.

Code of Good Agricultural Practice *noun* a set of standards and guidelines for agricultural practices, which must be set up by law in each country in the EU. Abbr **COGAP**

Codex /ˈkəʊdeks/ *noun* a United Nations food standards body run by FAO and WHO to develop international food safety and quality standards

codlin /ˈkɒdlɪn/, **codling** /ˈkɒdlɪŋ/ *noun* an apple with a long tapering shape

codling moth *noun* a serious pest, the larvae of which burrow into apple fruit

codominant /kəʊˈdɒmɪnənt/ *adjective* **1.** (*of a species*) (said of two or more species) being roughly equally abundant in an area and more abundant than any other species ○ *There are three codominant tree species in this forest.* **2.** referring to alleles of a gene that are not fully dominant over other alleles in a heterozygous individual

coeliac disease /ˌsiːliæk dɪˈziːz/ *noun* a disease of the small intestine resulting from an inability to digest wheat protein (NOTE: The protein gluten causes the body's own immune system to attack and damage the gut lining.)

coffee *noun* **1.** a bush or small tree widely grown in the tropics for its seeds, which are used to make a drink. Latin name: *Coffea arabica*. **2.** the drink prepared from the seeds of coffee bushes

COMMENT: The two main varieties of coffee are Arabica and Robusta. The Arabica shrub, *(Coffea arabica)* was originally grown in the southern parts of the highlands of Ethiopia, and was later introduced into south-western Arabia. The Arabica plant only grows well on altitudes of 1,000m and above. It represents 75% of the world's total coffee production. Arabica coffee beans are generally considered to produce a higher quality drink than those obtained from the Robusta coffee plant *(Coffea canephora)* which originated in West Africa. Robusta coffee has a stronger and more bitter taste than Arabica. The most important area for growing coffee is South America, especially Bolivia, Brazil and Colombia, though it is also grown in Kenya and Indonesia.

coffee berry borer /ˈkɒfi ˌberi ˌbɔːrə/ *noun* a small beetle which lays its eggs inside green coffee berries

coffee cherry *noun* the red fruit of the coffee plant, which contains the coffee beans

coffee rust *noun* a fungus disease which attacks coffee plants

COGAP *abbreviation* Code of Good Agricultural Practice

coir /ˈkɔɪə/ *noun* a rough fibre from the outer husk of coconuts

Colbred /ˈkəʊlbred/ *noun* a breed of sheep of medium size with white face. When used in cross-breeding, they are capable of transmitting high fertility and high milking capacity.

cold frame *noun* a box construction, with a glass lid, used for raising or keeping plants out of doors but with a certain amount of protection against frost

cold shortening *noun* chilling meat too quickly after slaughter, which makes it tough

cold storage *noun* the practice of keeping perishable produce in a refrigerated room or container, before moving it to market or to a retailer. The low temperature inhibits bacterial and fungal activity.

cold treatment *noun* the use of freezing to disinfest storage containers

coleoptile /ˌkɒli'ɒptaɪl/ *noun* a sheath which protects the stem tip (**plumule**) of a germinating grass seed as it grows to the surface

colic *noun* pain in any part of the intestinal tract, especially a symptom of abdominal pain in horses

coliform /'kɒlɪfɔːm/ *adjective* referring to bacteria which are similar in shape to *Escherichia coli*

collar *noun* a leather-covered roll put round a horse's neck, to carry the weight of a plough or cart which the horse is pulling

collie /'kɒli/ *noun* Scottish breed of sheepdog

colloid /'kɒlɔɪd/ *noun* a substance with very small particles that do not settle but remain in suspension in a liquid

colloidal /kə'lɔɪd(ə)l/ *adjective* referring to a colloid

colloidally /kə'lɔɪdəli/ *adverb* □ **colloidally dispersed particles** particles which remain in suspension in a liquid

colon *noun* the large intestine, running from the caecum to the rectum

colonial animal *noun* an animal which usually lives in colonies, e.g. an ant

colonisation /ˌkɒlənaɪ'zeɪʃ(ə)n/, **colonization** *noun* the act of colonising a place ○ *Islands are particularly subject to colonisation by species of plants or animals introduced by people.*

colonise /'kɒlənaɪz/, **colonize** *verb* (*of plants and animals*) to become established in a new ecosystem ○ *Derelict city sites rapidly become colonised by plants.* ○ *Rats have colonised the sewers.*

coloniser /'kɒlənaɪzə/, **colonist** /'kɒlənɪst/ *noun* an organism that moves into and establishes itself in a new ecosystem, e.g. a plant such as a weed

colony *noun* a group of animals, plants or microorganisms living together in a place ○ *a colony of ants*

colony system *noun* a poultry rearing system in which the hens are free to move around within a large confined space

Colorado beetle /ˌkɒlərɑːdəʊ 'biːt(ə)l/ *noun* a beetle (*Leptinotarsa decemlineata*) with black and yellow stripes which eats and destroys potato plants

colostrum *noun* a yellowish fluid that is rich in antibodies and minerals produced by a mother after giving birth and prior to the production of true milk (NOTE: It

provides newborns with immunity to infections.)

colt *noun* a young male horse which is less than four years old, or in the case of thoroughbreds, less than five years old

coltsfoot /'kəʊltsfʊt/ *noun* a perennial weed (*Tussilago farfara*)

COMA /'kəʊmə/ *abbreviation* Committee on Medical Aspects of Food Policy

comb *noun* the red fleshy crest on a fowl

combine *noun* /'kɒmbaɪn/ same as **combine harvester** ■ *verb* /kəm'baɪn/ to harvest using a combine harvester ○ *The men have been combining all day during the fine weather.*

combine drill *noun* a drill which sows grain and fertiliser at the same time (NOTE: Some drills have separate tubes for the seed and the fertiliser, others have one tube for both. It is important to clean combine drills after use, as the corrosive action of the fertiliser can damage the tubes.)

combine harvester *noun* a large machine that cuts a crop, threshes it and sorts the grain or seed from the straw or chaff. Combine harvesters are used to harvest a vast range of crops such as cereals, grass, peas and oilseed rape.

COMMENT: The combine harvester cuts the crop, passes it to the threshing mechanism, then sorts the grain or seed from the straw or chaff. The straw is left in a swath behind the combine, and the chaff is blown out of the back. The grain is lifted to a hopper from which it is unloaded into trailers. Most combine harvesters are self-propelled, with a cab for the driver, power steering, and monitoring systems for the key components. Special attachments used with combines include straw spreaders, pick-up attachments for grass and clover crops, and maize pickers.

combing wool *noun* a long-stapled wool, suitable for combing and making into worsted

combining peas *plural noun* peas grown on a large scale, which are harvested with a combine harvester

comfrey /'kʌmfri/ *noun* a medicinal herb of the genus *Boraginaceae.* that is also used in salads and for composting.

Comice /'kɒmiːs/ ♦ **Doyenne du Comice**

commensal /kə'mensəl/ *noun* an organism which lives on another plant or animal but does not harm it or influence it in any way. ◊ **parasite, symbiont** ■ *adjective* referring to a commensal

commensalism /kəˈmensəlɪz(ə)m/ *noun* the state of organisms existing together as commensals

commercial *adjective* **1.** referring to business **2.** produced to be sold for profit

commercial grain farming *noun* a highly mechanised agricultural system in which large areas of mid-latitude grasslands are given over to cereal cultivation

commercial grazing *noun* same as **ranching**

commercial market *noun* a market for cattle or sheep for meat rather than for breeding. Compare **pedigree market**

commercial seed *noun* seed sold as being true to kind, but not necessarily pure

comminute /ˈkɒmɪnjuːt/ *verb* to grind meat into very small pieces

Commission of the European Union *noun* the executive body of the European Union

Committee of Professional Agricultural Organizations *noun* an organisation which represents the interests of farmers from all European member states. Abbr **COPA**

Committee on Medical Aspects of Food Policy *noun* a panel which publishes reports on medical issues relating to food preparation and packaging. Abbr **COMA**

commodity *noun* a substance sold in very large quantities, e.g. raw materials or foodstuffs such as corn, rice, butter

commodity exchange *noun* a place where commodities are bought and sold

commodity futures *plural noun* trading in commodities for delivery at a later date. The produce will often not yet have been grown or harvested.

commodity mountain *noun* a surplus of a certain agricultural product produced in the EU, e.g. the 'butter mountain.' ◊ **lake**

common *adjective* belonging to several different people or to everyone ○ *common land* ■ *noun* an area of land to which the public has access for walking

COMMENT: About 80% of common land is privately owned and, subject to the interests of any commoners, owners enjoy essentially the same rights as the owners of other land. Commoners have different types of 'rights of common', e.g. to

graze animals, or to extract sand, gravel or peat.

Common Agricultural Policy *noun* a set of regulations and mechanisms agreed between members of the European Union to control the supply, marketing and pricing of farm produce. Abbr **CAP**

COMMENT: The European Union has set up a common system of agricultural price supports and grants. The system attempts to encourage stable market conditions for agricultural produce, to ensure a fair return for farmers and reasonable market prices for the consumer, and finally to increase yields and productivity on farms in the Union. A system of common prices for the main farm products has been established with intervention buying as the main means of market support. The first major reforms in 30 years were carried out in 1992 and included arable set-aside, suckler cow quotas, ewe quotas, price reductions on oilseeds, peas, beans, cereals and beet. The second major CAP reform was in 2003 with the introduction of the Single Farm Payment Scheme (SFPS), which brought together individual subsidy schemes into a single payment calculated on the land area used.

Common Birds Census *noun* an ongoing survey of commonly occurring birds, run by the British Trust for Ornithology

common eyespot *noun* ♦ **eyespot**

common fumitory *noun* a widespread weed *(Fumaria officinalis)*. Also called **beggary**

common hemp nettle *noun* a weed *(Galeopsis tetrahit)* found in spring cereals and vegetables. Also called **day-nettle**, **glidewort**

common prices *plural noun* the prices obtained by all EU farmers for a wide range of their products, including beef, cereals, milk products and sugar. EU regulations involve control on imports and intervention buying. These prices are reviewed each year.

community *noun* a group of different organisms which live together in an area ○ *the plant community on the sand dunes*

compact *verb* to compress the ground and make it hard, e.g. by driving over it with heavy machinery or as the result of a lot of people walking on it

compaction /kəmˈpækʃ(ə)n/, **compacting** *noun* the compression of ground and making it hard, e.g. by driving over it with heavy machinery or as the result of a lot of people walking on it

companion animal *noun* an animal that is kept for company and enjoyable interaction, rather than for work or food

companion plant *noun* a plant which improves the growth of nearby plants or reduces pest infestation (NOTE: Companion plants are often used by horticulturists and gardeners because they encourage growth or reduce pest infestation in an adjacent plant.)

COMMENT: Some plants grow better when planted near others. Beans and peas help root plants such as carrots and beetroot. Most herbs (except fennel) are helpful to other plants. Marigolds help reduce aphids if they are planted near plants such as broad beans or roses which are subject to aphid infestation. The strong smell of onions is disliked by the carrot fly, so planting onions near carrots makes sense. On the other hand, most other plants (and especially peas and beans) dislike onions and will not grow well near them.

companion planting *noun* the use of plants that encourage the growth of others nearby

Compassion in World Farming *noun* a political lobby group which campaigns for the welfare of livestock and for humane practices in farming. Abbr **CIWF**

compensatory growth /ˌkɒmpənseɪt(ə)ri ˈɡrəʊθ/ *noun* growth that occurs after a period of under-feeding when the animal regains lost weight

competition *noun* the struggle for limited resources such as food, light or a mate, occurring between organisms of the same or different species

complementarity /ˌkɒmplɪmen'tærɪti/ *noun* nature conservation based on a balance between wild and domesticated species in an area

complementary feeders *plural noun* animals which feed in a way which does not compete with other animals feeding in the same area. So goats, which browse, complement sheep which graze.

complete diet *noun* same as **total mixed ration**

complex vertebral malformation *noun* a congenital condition of Holstein cattle, symptoms of which include reduced weight and a misshapen backbone. Abbr **CVM**

complications *plural noun* secondary medical problems developing as part of an existing medical condition ○ *The patient may develop complications after surgery.*

Compositae /kəmˈpɒzɪtaɪ/ *plural noun* former name for **Asteraceae**

compost *noun* **1.** rotted vegetation or organic waster, which resembles humus and is used as fertiliser or mulch **2.** a prepared soil or peat mixture in which plants are grown in horticulture ■ *verb* **1.** to allow organic material to rot and turn into compost **2.** to put compost onto soil or a particular area of ground

compost activator *noun* a chemical added to a compost heap to speed up the decomposition of decaying plant matter

compost heap *noun* a pile of organic, especially plant, waste, usually kept in a container and left to decay gradually, being turned over occasionally. It is used as a fertiliser and soil improver.

composting *noun* the controlled decomposition of organic waste, especially used for the disposal for plant waste in gardens or domestic green waste such as vegetable peelings

composting drum *noun* a cylindrical container in which organic waste is rotted down to make compost

compound *noun* a substance made up of two or more components

COMMENT: Chemical compounds are stable (i.e. the proportions of the elements in them are always the same) but they can be split into their basic elements by chemical reactions.

compounder /kɒmˈpaʊndə/ *noun* a company which produces compound feed

compound feed *noun* a type of animal feed made up of several different ingredients, including vitamins and minerals, providing a balanced diet. Compound feed is usually fed to animals in the form of compressed pellets.

compound fertiliser *noun* a fertiliser that supplies two or more nutrients. Also called **mixed fertiliser**. Compare **straight fertiliser**

compulsory *adjective* forced or ordered by an authority ○ *the compulsory slaughter of infected animals*

compulsory dipping period *noun* a period of time, usually some weeks, during which all sheep in the country must be dipped

concave *noun* part of a combine harvester, a curved dish which catches the grain after it has been threshed

concentrate *noun* **1.** the strength of a solution, or the quantity of a substance in a specific volume **2.** a strong solution which is to be diluted ○ *orange juice made from concentrate* ■ *verb* **1.** to collect in a particular place rather than spread around ○ *Most of the mass of air is concentrated at the lowest levels of the atmosphere.* **2.** to reduce the volume of a solution and increase its strength by evaporation. Opposite **dilute**

concentrates *plural noun* animal feedingstuffs with a high nutrient relative to their bulk

concentration *noun* the amount of a substance in a given volume or mass of a solution

condense *verb* to make a vapour become liquid

condition *noun* **1.** the present state of something **2.** the state of health or of cleanliness of an animal ○ *The animal was in such poor condition that the vet decided it had to be put down.* **3.** (*in breeding*) the amounts of muscle and fat present in an animal

conditioned reflex *noun* an automatic reaction by an animal to a stimulus, learned from past experience

conditioner *noun* a substance that is used to make an improvement in something else ○ *Mushroom compost can be used as a soil conditioner.*

conditioning *noun* **1.** the preparation of crops for harvesting **2.** the process of making meat more tender by keeping it for some time at a low temperature **3.** the preparation of grain for milling by adding water to it, so as to ensure that the grain has the correct moisture content

condition scoring *noun* a method of assessing the state of body condition of animals; scores range from 0–5 for cattle and 1–9 for sows. Low condition scores indicate thinness, and high scores fatness. A score of about 3 is ideal.

'Condition scoring is the ideal on-farm method of assessing cow body reserves as it requires no specialist equipment or weighing facilities and once you get a grasp of the points system it is relatively quick to do.' [*Dairy Farmer*]

conductivity /ˌkɒndʌk'tɪvɪti/ *noun* the ability of a material to conduct heat or electricity ○ *Because of the poor conductivity of air, heat is transferred from the Earth's surface upwards by convection.* ◊ **hydraulic conductivity, electrical conductivity**

cone *noun* **1.** a hard scaly structure containing seeds on such plants as conifers **2.** the fruit of the female hop plant, which is separated from leaves and other debris before being dried in an oasthouse

Confederation of European Maize Producers *noun* an organisation representing the interests of European farmers who produce maize

Conference *noun* a popular variety of dessert pear. It has a long shape and keeps very well.

conformation /ˌkɒnfɔː'meɪʃ(ə)n/ *noun* the general shape of an animal or bird ○ *Carcass conformation is very important when buying cattle at an auction.*

COMMENT: Conformation is important in the Carcass Classification System. There are five conformation classes, called EUROP: E = excellent; U (= good); R (= average); O (= below average); P (= poor).

congenital *adjective* existing at or before birth

congenital disorder *noun* a disorder which is present at birth

COMMENT: An animal may be abnormal at birth because of a genetic defect, such as misshapen heads of calves; other congenital disorders such as swayback in lambs, may be caused by deficiencies in the mother (in the case of swayback, maternal copper deficiency).

conifer *noun* a tree with long thin needle-shaped leaves and bearing seed in scaly cones. Most are evergreen.

COMMENT: Conifers are members of the order Coniferales and include pines, firs and spruce. They are natives of the cooler temperate regions, are softwoods and often grow very fast. Their tough leaves are called needles and are resistant to cold and drought. They are frequently used in timber plantations.

coniferous /kə'nɪf(ə)rəs/ *adjective* referring to conifers

connective tissue /kəˌnektɪv 'tɪʃuː/ *noun* the tissue that forms the main part of bones and cartilage, ligaments and tendons, in which a large amount of fibrous material surrounds the tissue cells

conservancy /kən'sɜːv(ə)nsi/ *noun* an official body which protects a part of the environment

conservation *noun* **1.** keeping or not wasting ○ *conservation of energy* **2.** the maintenance of environmental quality and resources by the use of ecological knowledge and principles

conservation headland *noun* an area between the edge of a crop and the first tractor tramline that is treated less intensively with pesticides so that a range of broadleaved weeds and beneficial insects survive, used as a method of encouraging biodiversity

'The achievements of The Allerton Project at Loddington in improving populations of wildlife and game are widely recognised. Until recently this has been done by improving field margins, managing set-aside for game and wildlife, installing beetle banks and by conservation headlands.' [*Arable Farming*]

conservationist *noun* a person who promotes, carries out or works for conservation

conservation of soil *noun* same as **soil conservation**

Conservation Reserve Program *noun US* a federal programme which pays farmers to let land lie fallow. Abbr **CRP**

conservation tillage *noun* a farming method which aims to plough the soil as little as possible, to prevent erosion, save energy and improve biodiversity. ◊ **minimum tillage**

conserve *verb* **1.** to keep and not waste something ○ *The sloth sleeps during the day to conserve energy.* **2.** to look after and keep something in the same state ○ *to conserve tigers' habitat*

consume *verb* **1.** to use up or burn fuel ○ *The new pump consumes only half the fuel which the other pump would use.* **2.** to eat foodstuffs ○ *The population consumes ten tonnes of foodstuffs per week.*

consumer *noun* a person or company which buys and uses goods and services ○ *Gas consumers are protesting at the increase in prices.* ○ *The factory is a heavy consumer of water.*

consumerism /kən'sjuːmərɪz(ə)m/ *noun* a movement for the protection of the rights of consumers

consumption *noun* **1.** the fact or process of using something ○ *a car with* low petrol consumption ○ *The country's consumption of wood has fallen by a quarter.* **2.** the taking of food or liquid into the body ○ *Nearly 3% of all food samples were found to be unfit for human consumption through contamination by lead.*

contact *noun* a physical connection between two or more things, especially the fact of one touching the other ○ *Don't allow the part to come into contact with water.*

contact animal *noun* an animal which has had contact with a diseased animal and which may need to be isolated

'Movement restrictions placed on the contact animals in the herd will remain in place and the animals will be subject to testing for brucellosis over a period of months.' [*Farmers Guardian*]

contact herbicide *noun* a substance which kills a plant whose leaves it touches, e.g. paraquat

contact insecticide *noun* a substance such as DDT that kills insects which touch it (NOTE: DDT is now banned in many countries because of its toxicity and ability to accumulate in the environment.)

contact weedkiller *noun* same as **contact herbicide**

contagious *adjective* referring to a disease which can be transmitted by touching an infected person, or objects which an infected person has touched. Compare **infectious**

contagious abortion *noun* Brucellosis, an infectious disease, which is usually associated with cattle where it results in reduced milk yields, infertility and abortion

contaminant /kən'tæmɪnənt/ *noun* a substance which causes contamination

contaminate *verb* to make something impure by touching it or by adding something, especially something harmful, to it ○ *Supplies of drinking water were contaminated by uncontrolled discharges from the factory.* ○ *A whole group of tourists fell ill after eating contaminated food.*

contaminated land, contaminated site *noun* an area which has been polluted as a result of human activities such as industrial processes, presenting a hazard to human health, and which needs cleaning before it can be used for other purposes

COMMENT: Contaminated land is a feature of most industrialised countries.

Careless past management of waste, lack of pollution controls and many leaks and spills have left a legacy of land contaminated by a wide variety of substances. In some cases this presents unacceptable risks to human beings, ecosystems, water resources or property and has to be dealt with by formal remedial measures.

contamination /kən,tæmɪ'neɪʃ(ə)n/ noun **1.** the action of making something impure ○ *the contamination of the water supply by runoff from the fields* **2.** the state of something such as water or food which has been contaminated and so is harmful to living organisms ○ *The level of contamination is dropping.*

continuing professional development noun the continuation of training and study throughout a person's career. Abbr **CPD**

'Experience of dairying in New Zealand and the USA indicates that both these dairying cultures place emphasis on communication, as well as the continuing professional development of their staff.' [*Farmers Weekly*]

contour /'kɒntʊə/ noun same as **contour line**

contour farming noun a method of cultivating sloping land in which the land is ploughed along a terrace rather than down the slope, so reducing soil erosion

COMMENT: In contour farming, the ridges of earth act as barriers to prevent soil being washed away and the furrows retain the rainwater.

contour line noun a line drawn on a map to show ground of the same height above sea level

contour ploughing, contour ridging noun the practice of ploughing across the side of a hill so as to create ridges along the contours of the land which will hold water and prevent erosion

contour strip cropping noun the planting of different crops in bands along the contours of sloping land so as to prevent soil erosion

contract crop noun a crop grown to order for a specific outlet

contract grower noun a grower who produces a crop to order for a specific outlet

'"Our contract growers have to grow to our specifications; we visit them every week and make a report", says Mr

Verduyn. If the crop is not to standard, and management advice has not been followed, the contract is unlikely to be renewed.' [*Farmers Weekly*]

contractor noun company or person who carries out contract work for a farmer

contract work noun work carried out by specialist firms on a contract, which involves payment for work carried out, e.g. the provision of a drainage system or combining a crop

control noun **1.** the process of restraining something or keeping something in order □ **to bring** *or* **keep something under control** to make sure that something is well regulated ○ *The authorities brought the epidemic under control.* □ **out of control** unregulated ○ *The epidemic appears to be out of control.* **2.** (in experiments) a sample used as a comparison with the one being tested ■ *verb* **1.** to direct or manage something **2.** to keep something in order ○ *The veterinary service is trying to control the epidemic.* ○ *They were unable to control the spread of the pest.*

control area noun an area where controls are operating to prevent the spread of a disease within the area, usually a larger area than the infected area

controlled atmosphere noun the conditions in which oxygen and carbon dioxide concentrations are regulated and monitored, e.g. to improve the storage of fruit and vegetables

'English plum producers are looking at the possibility of storing Victoria plums in controlled atmosphere to extend the season until at least September.' [*The Grocer*]

controlled atmosphere packaging noun the packaging of foods in airtight containers in which the air has been treated by the addition of other gases. This allows a longer shelf-life.

controlled dumping noun the disposal of waste on special sites

controlled environment noun an environment in which conditions such as temperature, atmosphere and relative humidity are regulated and monitored

controlled grazing noun a system of grazing in which the number of livestock is linked to the pasture available, with moveable fences being erected to restrict the area being grazed

controlled landfill *noun* the disposal of waste in a landfill carried out under a permit system according to the specific laws in force

controlled tipping *noun* the disposal of waste in special landfill sites. ◊ **fly-tipping**

Control of Substances Hazardous to Health (UK Regulations) *noun* full form of **COSHH**

convenience foods *plural noun* foods which have been prepared so that they are ready to be served after simply being reheated

'People are working longer hours. As a result demand for convenience foods has exploded, not just for pre-packed ready meals, but quickly cooked food like chops and steaks.' [*Farmers Guardian*]

convert *verb* to change something to a different system, set of rules or state ○ *How do you convert degrees C into degrees F?* ○ *Photochemical reactions convert oxygen to ozone.* ○ *She has converted her car to take LPG.*

converter /kən'vɜːtə/ *noun* a device which alters the form of something ○ *A backup converter converts the alternating current power into direct current.*

cook chill, cook freeze *noun* a method of preparing food for preserving, where the food is cooked to a certain temperature and then chilled or frozen

coomb /kuːm/ *noun* a measure of cereals, equalling one sack or four bushels

coop /kuːp/ *noun* a cage for poultry

cooperative *noun* a group of farmers who work together to sell their produce either for the wholesale market or in retail outlets such as farmers' markets

COPA *abbreviation* Committee of Professional Agricultural Organizations

copper *noun* a metallic trace element. It is essential to biological life and used in making alloys and in electric wiring.

copper deficiency *noun* a lack of copper in an animal's diet, sometimes caused by poisoning with molybdenum

COMMENT: Symptoms of copper deficiency vary, but can include lack of growth and change of colour, where black animals turn red or grey. In severe cases, bones can fracture, particularly the shoulder blade. Diarrhoea can also occur, as well as anaemia. Copper deficiency in ewes can cause swayback in lambs. The condition is treated with injections of copper sulphate.

coppice /'kɒpɪs/ *noun* an area of trees which have been cut down to near the ground to allow shoots to grow which are then harvested. The shoots may be used as fuel or for making products such as baskets or fencing. ■ *verb* to cut trees down to near the ground to produce strong straight shoots ○ *Coppiced wood can be dried for use in wood-burning stoves.* Compare **pollard**

COMMENT: The best trees for coppicing are those which naturally send up several tall straight stems from a bole, such as hazel and sweet chestnut. In coppice management, the normal cycle is about five to ten years of growth, after which the stems are cut back. Thick stems are dried and used as fuel, or for making charcoal. Thin stems are used for fencing. Cash aid under the set-aside scheme could be used for short-rotation energy coppicing.

coppice forest, coppice wood *noun* woodland that has regrown from shoots formed on the stumps or roots of previously cut trees, usually cut again after a few years to provide small branches for uses such as fuel

coppicing *noun* the practice of regularly cutting down trees near to the ground to produce strong straight shoots for fuel or other uses ○ *Coppicing, a traditional method of woodland management, is now of interest for producing biofuel.*

COMMENT: The best trees for coppicing are those which naturally send up several tall straight stems from a bole, such as willow, alder or poplar. In coppice management, the normal cycle is about five to ten years of growth, after which the stems are cut back.

copra /'kɒprə/ *noun* the dried pulp of a coconut, from which oil is extracted by pressing

copse /kɒps/ *noun* an area of small trees

cordon *noun* a trained fruit tree, whose growth is restricted to the main stem by pruning. Compare **espalier**

cordwood /'kɔːdwʊd/ *noun* pieces of cut tree trunks, all of the same length, ready for transporting

co-responsibility levy *noun* a levy on overproduction introduced in the EU in 1987. The levy shared the cost of disposal of surpluses between the community and the producers.

cork *noun* a protective outer layer that forms part of the bark in woody plants, taking many years to regrow once stripped (NOTE: It is used, among other things, for bottle corks, fishing net floats and flooring, but cork oaks are now attracting conservation interest.)

COMMENT: Cork is harvested by cutting large sections of bark off a cork oak tree, while still leaving enough bark on the tree to ensure that it will continue to grow.

corm /kɔːm/ *noun* a swollen underground plant stem with a terminal bud, e.g. on a crocus

COMMENT: Crocuses, gladioli and cyclamens have corms, not bulbs.

corn *noun* 1. wheat or barley (*informal*) 2. *US* maize

corn cob *noun* a seed head of maize. Also called **cob**

corn cockle *noun* a poisonous weed (*Agrostemma githago*) with a tall stem and purple flowers

corned beef /ˌkɔːnd 'biːf/ *noun* beef which has been cured in brine and is preserved in cans

cornflour /'kɔːnflaʊə/ *noun* a type of flour extracted from maize grain. It contains a high proportion of starch, and is used for thickening sauces. Also called **corn starch**

cornflower /'kɔːnflaʊə/ *noun* a common weed (*Centaurea cyanus*) with tall stems and bright blue flowers

corn marigold *noun* a common weed (*Chrysanthemum segetum*)

corn on the cob *noun* a seed head of maize when used as food

corn pansy *noun* same as **field pansy**

corn poppy *noun* a common weed (*Papaver rhoeas*) affecting cereals

corn spurrey *noun* a common weed (*Spergula arvensis*) with matted growth which makes it difficult to eradicate in row crops

corn starch *noun* same as **cornflour**

corolla /kə'rɒlə/ *noun* a set of petals in a flower

corporate social responsibility *noun* the extent to which an organisation behaves in a socially, environmentally and financially responsible way. Abbr **CSR**

'The Government agreed with the committee's criticism of the role of supermarket and that supermarkets' corporate social responsibility policies

need to address the use of labour by their suppliers.' [*Farmers Guardian*]

corpuscle /'kɔːpʌs(ə)l/ *noun* a cell in the blood

corpus luteum /ˌkɔːpəs 'luːtiəm/ *noun* a yellowish mass of tissue that forms after ovulation in the Graafian follicle of the ovary and secretes progesterone

corral /kə'rɑːl/ *noun* a pen for horses or cattle ■ *verb* to put horses or cattle in pens

Corriedale /'kɒrideɪl/ *noun* a New Zealand breed of sheep, originally from longwool rams and Merino ewes. Corriedale is now bred in Australia both for meat and its thick 27-micron wool.

Corsican pine /'kɔːsɪkən paɪn/ *noun* a fast-growing conifer (*Pinus nigra*)

cos *noun* a type of lettuce with long darker green leaves

COSHH *noun* UK regulations controlling substances with known health risks. Full form **Control of Substances Hazardous to Health (UK Regulations)**

COMMENT: Farmwork may involve exposure to many substances which can be hazardous to health. Safety in the use of pesticides extends not only to protecting the user but also the need to protect the environment.

cosset lamb /'kɒsət læm/ *noun* a lamb which has been reared by hand

cote /kəʊt/ *noun* ◆ **dove cote**

Cotentin /'kɒtəntæn/ *noun* French breed of sheep from the Cotentin peninsula of Normandy

Cotonou Agreement /'kɒtənuː əˌgriːmənt/ *noun* an agreement reached in 2000 between the European Union and the ACP states. It updates the Lomé Convention, guaranteeing free access to markets for both the EU and the ACP states.

Cotswold /'kɒtswəʊld/ *noun* a breed of sheep from the Cotswold hills, now becoming rare

cottage garden *noun* a flower garden containing old-fashioned flowers

cottage piggery *noun* a pig housing with low roofs and an open yard

cotton *noun* a white downy fibrous substance surrounding the seeds of the cotton plant, a subtropical plant (*Gossypium* sp.)

COMMENT: Cotton is widely grown in tropical and sub-tropical areas, including China, India, Pakistan, Paraguay and the southern states of the USA; it is the main crop of Egypt. It is sold packed in standard bales.

cotton gin *noun* a machine which separates the seeds from the cotton fibres

cotton grass *noun* a plant with white fluffy flower heads that grows in boggy ground. Latin name: *Eriophorum angustifolium*.

cottonseed /ˈkɒtənsiːd/ *noun* the seed of the cotton plant, one of the world's most important sources of oil

cottonseed cake, cottonseed meal *noun* a residue of cottonseed after the extraction of oil, used as a feedingstuff

cottonwood /ˈkɒtənwʊd/ *noun* a kind of poplar tree. Genus: *Populis*.

cotyledon /ˌkɒtɪˈliːd(ə)n/ *noun* the green plant structure resembling a leaf that appears as a seed germinates and before the true leaves appear, developing from the embryo of the seed

COMMENT: Cotyledons are thicker than normal leaves, and contain food for the growing plant. Plants are divided into two groups, those producing a single cotyledon (monocotyledons) and those producing two cotyledons (dicotyledons).

couch grass /ˈkuːtʃ grɑːs/ *noun* a kind of grass *(Agropyron repens)* with long creeping rhizomes, which is difficult to eradicate from cultivate crops. Also called **scutch**, **twitch**

coulter /ˈkəʊltə/ *noun* the part of the plough which goes into the soil and makes the vertical cut

COMMENT: There are several types of coulter: the disc coulter cuts the side of the furrow about to be turned; the knife coulter serves the same purpose, but is now little used; the skim coulter turns a small slice off the corner of the furrow about to be turned and throws it into the bottom of the one before; it is attached to the beam behind the disc coulters.

Council for the Protection of Rural England *noun* former name for **CPRE**

Counter Fraud and Compliance Unit *noun* an organisation which detects irregularities with CAP claims. Abbr **CFCU**

country code /ˈkʌntri kəʊd/ *noun* a voluntary code of conduct for people spending leisure time in the countryside, which indicates how to respect the natural environment and avoid causing damage to it

Country Land and Business Association *noun* an organisation representing the interests of landowners (NOTE: Formerly called the 'Country Landowners Association'.)

country planning *noun* the activity of organising how land is to be used in the countryside and the amount and type of building there will be. Also called **rural planning**

Countryside Agency *noun* a statutory body funded by Defra with the aim of making life better for people in the countryside. It is the statutory advisor on landscape issues and was formed by merging the Countryside Commission with parts of the Rural Development Commission, but is to be reorganised.

Countryside Alliance *noun* an organisation which lobbies the Government on policy and legislation affecting rural life

Countryside and Rights of Way Act *noun* legislation passed by the UK government in 2000 that gave the public greater freedom of access to privately owned areas of uncultivated land and strengthened legislation protecting wildlife. Abbr **CROW Act**

Countryside Commission *noun* a former organisation in the UK, which supervised countryside planning and recreation. It was particularly concerned with National Parks and Areas of Outstanding Natural Beauty.

Countryside Commission for Scotland *noun* an organisation in Scotland concerned with the protection of the countryside and with setting up country parks for public recreation. It is part of Scottish Natural Heritage.

Countryside Council for Wales *noun* a statutory advisory body of the UK government responsible for sustaining natural beauty, wildlife and outdoor leisure opportunities in Wales and its coastal areas. Abbr **CCW**

countryside management *noun* the study and practice of environmental conservation in association with rural enterprise, countryside access and recreational activities

Countryside Management Scheme *noun* in Northern Ireland, a system of payments designed to encourage landowners and farmers to adopt, or to continue with, environmentally sensitive farming practices

countryside recreation *noun* leisure activities that take place in the countryside. Also called **rural recreation**

'The recommendations are in line with a government conclusion two years ago that there was no case for a general ban on the recreational use of motor vehicles on byways, and argue that low-key motorised recreational use of such routes is an established form of countryside recreation.' [*Farming News*]

countryside recreation site *noun* a location visited or used by tourists in the countryside, e.g. a national park, heritage coast, cycle path or watersports facility

countryside stewardship *noun* the practice of altering farming practices to benefit wildlife and retain natural diversity. Abbr **CSS**

Countryside Stewardship Scheme *noun* formerly in England and Wales, a system of payments made to landowners and farmers who alter their farming practices to benefit the natural environment and maintain biodiversity (NOTE: The Countryside Stewardship Scheme has now been superseded by the **Environmental Stewardship** scheme.)

'The Defra-funded Countryside Stewardship and Environmentally Sensitive Areas schemes help to maintain and enhance the biodiversity and landscape value of farmed land, protect historic features and promote public access. (Delivering the evidence. Defra's Science and Innovation Strategy, 2003–06)'

country stewardship *noun* ♦ **countryside stewardship**

county, parish, holding *noun* a unique 3-part identification number for land used to keep livestock. Abbr **CPH**

coupe /kuːp/ *noun* an area of a forest in which trees have been cut down

couple *verb* to attach an implement such as a harrow to a tractor

coupling *noun* an attachment which couples an implement to a tractor

courgette *noun* marrow fruit at a very immature stage in its development, cut when between 10 and 20 cm long. It may be green or yellow in colour. Also called **zucchini**

course *noun* **1.** the development of events over a period of time ○ *the usual course of the disease* □ **in the normal course of events** usually **2.** a sequence of medical treatment given over a period of time ○ *a course of antibiotics* **3.** the length of time in a rotation, when the land is growing a particular crop ○ *The Norfolk four-course rotation has turnips, followed by spring barley, red clover and winter wheat, so that each crop will only be grown on the same land in one year out of four.*

cover *verb* to copulate with a female animal ○ *a bull covers a cow* ■ *noun* **1.** something that goes over something else completely **2.** the amount of soil surface covered with plants. ◊ **ground cover 3.** plants grown to cover the surface of the soil ○ *Grass cover will provide some protection against erosion.*

cover crop *noun* **1.** a crop sown to cover the soil and prevent it from drying out and being eroded (NOTE: When the cover crop has served its purpose, it is usually ploughed in, so leguminous plants which are able to enrich the soil are often used as cover crops.) **2.** a crop grown to give protection to another crop that is sown with it ○ *In the tropics, bananas can be used as a cover crop for cocoa.* **3.** a crop grown to give cover to game birds

'As leaving maize stubble fields bare could risk failure to meet cross-compliance, more growers are recognising cover crops could prove a worthwhile option.' [*Farmers Weekly*]

covered smut *noun* a fungal disease *(Ustilago hordei)* affecting oats and barley

cow *noun* a female bovine animal

cow beef *noun* beef from dairy cows no longer needed for milk production

cow bell *noun* a bell worn round neck of a cow to make it easier for the farmer to locate the animal

cowboy /ˈkaʊbɔɪ/ *noun* US a man who looks after cattle on a ranch

cow kennels *plural noun* a wooden building with stalls for cows

cowman /ˈkaʊmən/ *noun* a person in charge of a dairy herd (NOTE: The plural is **cowmen**.)

cowpea /ˈkaʊpiː/ *noun* a legume *(Vigna unguiculata)* grown throughout the subtropics and tropics as a pulse and green vegetable. It is grown for fodder, as a vegetable and as green manure.

cowpox /ˈkaʊpɒks/ *noun* an infectious viral disease of cattle, which can be transmitted to humans. It is used as part of the vaccine against smallpox. Also called **vaccinia**

Cox's orange pippin /ˌkɒksɪz ˌɒrɪndʒ ˈpɪpɪn/ *noun* a popular variety of dessert apple. The most important commercially grown apple in the UK.

CPA *abbreviation* Crop Protection Association

CPD *abbreviation* continuing professional development

CPH *abbreviation* county, parish, holding

CPRE *noun* a UK charity that campaigns for rural areas to be protected. Full form **Campaign to Protect Rural England**

craft food *noun* food produced according to traditional techniques or recipes

cramp *noun* a spasm of the muscles where the muscle may remain contracted for some time

cranefly /ˈkreɪnflaɪ/ *noun* a common pest *(Tipula)*. The larvae are leatherjackets, which affect cereal crops, feeding on the crops in spring, eating away the roots and stems.

cratch /krætʃ/ *noun* a rack used for feeding livestock out of doors

craving *noun* a strong desire for something, e.g. a craving for salt

crawler tractor /ˈkrɔːlə ˌtræktə/ *noun* a large powerful caterpillar tractor, used for heavy work

crazy chick disease *noun* a disease of chicks associated with a diet which is too rich in fats or deficiency of vitamin E. The symptoms include falling over and paralysis.

cream *noun* the oily part of milk, containing fats, which gathers on the top of standing milk

creamery /ˈkriːməri/ *noun* a factory where butter and other products are made from milk

creep *noun* **1.** a slow movement of soil down a slope **2.** a small entrance through which young animals can pass

creep feed *noun* feed given to small animals during creep feeding

creep feeding *noun* a process by which a young animal such as a calf is allowed access to concentrates through a small entrance, while the adult cow is unable to reach the feed

creep grazing *noun* a type of rotational grazing using creep gates, which allow the lambs access to the pasture before the ewes

creosote /ˈkriːəsəʊt/ *noun* a yellowish brown oily substance with a characteristic smell, derived from wood tar and formerly used as a wood preservative (NOTE: It is now banned in the European Union.)

cress /kres/ *noun* a plant *(Lepidium sativum)* used as a salad vegetable

crest *noun* **1.** the highest point of a hill or mountain ridge **2.** a growth on the head of a bird or other animal

crested dogstail /ˌkrestɪd ˈdɒgzteɪl/ *noun* a perennial grass *(Cynosurus cristatus)* which is not very palatable because of its wiry inflorescences, and is used in seed mixtures for lawns

Creutzfeldt-Jakob disease /ˌkrɔɪtsfelt ˈjækɒb dɪˌziːz/ *noun* a disease of the human nervous system caused by a slow-acting prion which eventually affects the brain. It may be linked to BSE in cows. Abbr **CJD**

crib *noun* a holder for fodder, with sides made of wooden bars

crimp /krɪmp/ *verb* to condition fresh cut grass, by nipping the stems and releasing the sap

crimper /ˈkrɪmpə/ *noun* a machine for crimping grass, similar to roller crushers

Criollo /krɪˈəʊləʊ/ *noun* a breed of improved Spanish longhorn cattle found in South America, used for milk production

crispbread /ˈkrɪspbred/ *noun* a dry biscuit made from rye

crisphead /ˈkrɪsphed/ *noun* a variety of lettuce with stiff leaves

croft /krɒft/ *noun* a small farm in the Highlands and Islands of Scotland

crofter /ˈkrɒftə/ *noun* a joint tenant of a divided farm in Scotland

crofting /ˈkrɒftɪŋ/ *noun* a system of farming in Scotland, where the arable land of small farms, which was previously held in common, was divided among the joint tenants into separate crofts, while the pasture remains in common

crook *noun* a long-handled staff with a hooked end, used by shepherds to catch sheep

crop *noun* **1.** a plant grown for food **2.** a yield of produce from plants ○ *The tree has produced a heavy crop of apples.* ○ *The first crop was a failure.* ○ *The rice crop has failed.* **3.** the bag-shaped part of a bird's throat where food is stored before digestion ■ *verb* (*of plants*) to produce fruit ○ *a new strain of rice which crops heavily*

crop breeder *noun* a person who specialises in developing new varieties of crops ○ *Crop breeders depend on wild plants to develop new and stronger strains.*

crop breeding *noun* the development of new varieties of crops

crop circles *plural noun* usually circular patterns occurring in cereal stands, where crops have been flattened

crop dusting *noun* the practice of applying insecticide, herbicide or fungicide to crops in the form of a fine dust or spray. Also called **crop spraying**

crop growth rate *noun* the rate of increase in dry weight per unit area of all or part of a sward

crop husbandry *noun* the practice of growing and harvesting crops

cropland /'krɒplænd/ *noun* agricultural land which is used for growing crops

Crop Protection Association *noun* an association which promotes best practice in food safety with regard to the use of pesticides on crops. Abbr **CPA**

crop relative *noun* a wild plant that is genetically related to a crop plant

crop rotation *noun* a system of cultivation where crops such as cereals and oilseed rape that need different nutrients and/or management are grown one after the other

COMMENT: The advantages of rotating crops are firstly that pests particular to one crop are discouraged from spreading, and secondly that some crops actually benefit the soil. Legumes (peas and beans) increase the nitrogen content of the soil if their roots are left in the soil after harvesting. If the rotation includes a ley, the system is known as alternative husbandry or mixed farming. One of the best-known rotations was the Norfolk four-course system. A rotation should increase and maintain soil fertility, control weeds and pests, decrease the risk of crop failure and employ labour throughout the year.

crop sprayer *noun* a machine or aircraft which sprays insecticide, herbicide or fungicide onto crops, or a company that performs this service

crop spraying *noun* same as **crop dusting**

crop standing *noun* herbage growing in a field before harvest

crop year *noun* a period of twelve months calculated as the time from the sowing and harvesting of one crop until the next sowing season

cross *verb* to produce a new form of plant or animal from two different breeds, varieties or species ○ *They crossed two strains of rice to produce a new strain which is highly resistant to disease.* ■ *noun* **1.** an act of crossing two plants or animals ○ *made a cross between two strains of cattle* **2.** a new form of plant or animal bred from two different breeds, varieties or species

crossbar *noun* a bar which goes across something to make it more solid, as the crossbar of a rake

crossbred /'krɒsbred/ *adjective* having been bred from two parents with different characteristics ○ *a herd of crossbred sheep*

crossbreed /'krɒsbriːd/ *noun* an animal bred from two different pure breeds ■ *verb* to produce new breeds of animals by mating animals of different pure breeds

crossbreeding /'krɒsˌbriːdɪŋ/ *noun* mating or artificial insemination of animals of different breeds in order to combine the best characteristics of the two breeds

cross-compliance *noun* the setting of environmental conditions that must be met when developing agricultural support policies, especially in the European Union. Also called **environmental conditionality**

'While there is some evidence to suggest that farmers in Nitrate Vulnerable Zones are taking greater account of the value of manures it is becoming increasingly apparent that cross-compliance, and therefore the Single Payment Scheme, also requires farmers to make allowances for the nutrients in any organic matter applied.' [*Farmers Guardian*]

cross-fertilisation *noun* the fertilising of one individual plant by another of the same species

cross-infection *noun* an infection of other animals in a herd or flock from an infected animal

crossing *noun* the breeding of plants or animals from two different breeds or varieties

cross-pollination *noun* the pollination of a flower with pollen from another plant of the same species. Compare **self-pollination** (NOTE: The pollen goes from the anther of one plant to the stigma of another.)

COMMENT: Cross-pollination, like cross-fertilisation and cross-breeding, avoids inbreeding, which may weaken the species. Some plants are self-fertile (i.e. they are able to fertilise themselves) and do not need pollinators, but most benefit from cross-fertilisation and cross-pollination.

croup /kruːp/ noun part of the back of a horse, near the tail

CROW Act abbreviation Countryside and Rights of Way Act

crow garlic noun same as **wild onion**

crown noun 1. the top part of a plant where the main growing point is ○ protecting the crowns from frost ○ The disease first affects the lower branches, leaving the crowns still growing. 2. the perennial rootstock of some plants

crown graft noun a type of graft where a branch of a tree is cut across at right angles, slits are made in the bark around the edge of the stump, and shoots are inserted into the slits

crown rust noun a fungal disease affecting oats, causing the grain to shrivel

CRP abbreviation Conservation Reserve Program

crucifer noun a plant such as cabbage whose flowers have four petals. Family: Cruciferae.

Cruciferae /kruːˈsɪfəriː/ noun former name for **Brassicaceae**

crucifer crop noun crops such as broccoli, cabbage, turnips and spinach belonging to the Cruciferaceae family. ○ **brassica**

crude fibre noun a term used in analysing foodstuffs, as a measure of digestibility. Fibre is necessary for good digestion, and lack of it can lead to diseases in the intestines.

crude protein noun an approximate measure of the protein content of foods

cruels /kruːlz/ noun same as **actino-bacillosis**

crumb noun 1. the soft inside part of baked bread, surrounded by the harder crust 2. arrangement of soil particles in a group. ○ **ped**

crumbly /ˈkrʌmbli/ adjective referring to something which falls apart into particles ○ a crumbly soil

crupper /ˈkrʌpə/ noun a strap fixed to the back of a saddle and looped under the horse's tail

crush noun a steel or wood appliance like a strong stall, used to hold livestock when administering injections or when the animal is being inspected by a veterinary surgeon ■ verb to press something with a heavy weight, as when crushing seeds to extract oil

crusher /ˈkrʌʃə/ noun a company or factory which specialises in crushing seed to extract oil

crushing mill noun a machine used to flatten grain before feeding it to livestock

crushing subsidy noun payment made in the EU to oil producers to compensate for the difference between vegetable oil prices in the EU and those outside

crush margin noun the difference in price between the unprocessed seed and the product extracted after crushing

crust noun a hard layer which forms on the surface of something, e.g. the crust of salts formed on soil after evaporation

crutch noun a forked pole, used when dipping sheep, to push the animal's shoulders down into the liquid

cryophilous /kraɪˈɒfɪləs/ adjective referring to a plant that needs a period of cold weather to grow properly

COMMENT: Cryophilous crops need a period of cold weather in order to produce flowers later in the growing period. If such crops do not undergo this cold period, their growth remains vegetative, or they only form abortive flowers with no seeds. Wheat, barley, oats, peas, sugar beet and potatoes are all cryophilous.

cryophyte /ˈkraɪəfaɪt/ noun a plant which lives in cold conditions such as in snow

cryptosporidiosis /ˌkrɪptəʊspəˌrɪdiˈəʊsɪs/ noun a disease of humans, caused by bacteria found in animals and in contaminated water

Cs symbol caesium

CSF abbreviation 1. Catchment Sensitive Farming 2. classical swine fever

CSFT abbreviation Central Scotland Forest Trust

CSR abbreviation corporate social responsibility

CSS abbreviation Countryside Stewardship Scheme

CTS abbreviation Cattle Tracing Scheme

Cu symbol copper

cube *noun* a small square block or pellet ■ *verb* to press animal feed into small square pellets

cubed concentrates *plural noun* concentrates for livestock in the form of small cubes

cuber /'kjuːbə/ *noun* a machine used for making cubes or pellets from meal. Meal mixed with molasses is forced through small holes and cut into various lengths.

cubicle *noun* a compartment, similar to a stall, for housing a single cow or bull, the floor of each cubicle being covered with straw or sawdust for bedding. Cubicles are usually arranged in rows backed by a dunging passage.

cucumber *noun* a creeping plant (*Cucumis sativus*) with long green fleshy fruit, used as a salad vegetable

COMMENT: Cucumbers are native to India, and are used as a cooked vegetable in oriental cooking; they are a major glasshouse crop.

Cucurbitaceae /ˌkʊkɜːbɪ'teɪsɪiː/ *noun* the Latin name for vine crops, the family of plants including melons, marrows and gourds. Also called **the cucurbits**

cud /kʌd/ *noun* food that ruminating animals bring back from the first stomach into the mouth to be chewed again

cull *noun* **1.** killing a certain number of living animals to keep the population under control or to remove excess animals from a herd or flock □ **deer cull**, **dairy cow cull** the act of killing a certain number of deer or dairy cows **2.** an animal that has been separated from the herd or flock and killed, usually because it is old or of poor quality ■ *verb* to reduce the numbers of wild animals by killing them in a controlled way ○ *Deer may have to be culled each year to control the numbers on the hills.*

COMMENT: In the management of large wild animals without predators, such as herds of deer in Europe, it is usual to kill some mature animals each year to prevent a large population forming and overgrazing the pasture. Without culling, the population would seriously damage their environment and in the end die back from starvation. In management of dairy cattle, animals are culled from herds to eradicate disease.

cull cows, cull ewes, cull sows *plural noun* cows, ewes or sows which are removed from the herd or flock and sold for slaughter

culm /kʌlm/ *noun* the stem of a grass which bears flowers

cultivable acreage /ˌkʌltɪvəb(ə)l 'eɪkərɪdʒ/ *noun* the number of acres on which crops can be grown

cultivar /'kʌltɪvɑː/ *noun* a variety of a plant that has been developed under cultivation and that does not occur naturally in the wild

cultivate *verb* **1.** to grow crops ○ *Potatoes are cultivated as the main crop.* **2.** to dig and manure the soil ready for growing crops ○ *The fields are cultivated in the autumn, ready for sowing wheat.*

cultivated land *noun* land that has been dug or prepared for growing crops

cultivation /ˌkʌltɪ'veɪʃ(ə)n/ *noun* the action of cultivating land or plants

cultivator /'kʌltɪveɪtə/ *noun* **1.** a person who cultivates land **2.** an instrument or small machine for cultivating small areas of land

COMMENT: A cultivator has a frame with a number of tines which break up and stir the soil as the implement is pulled across the surface. There are several types of tine, both rigid and spring-loaded. Cultivators can also be used for cleaning stubble and general weed control; the tines can be grouped together so that they pass easily between the rows of growing plants.

cultural *adjective* referring to agricultural techniques

cultural control *noun* the control of pests using various agricultural techniques such as crop rotation

culture *noun* a microorganism or tissues grown in a culture medium ■ *verb* to grow a microorganism or tissue in a culture medium. ◊ **subculture**

cultured milk products *plural noun* products such as yoghurt made from milk which has been exposed to harmless bacteria

culvert /'kʌlvət/ *noun* a covered drain for water

cumin /'kjuːmɪn/ *noun* a Mediterranean aromatic herb (*Cuminum cyminum*) used for flavouring

curative fungicide /ˌkjuərətɪv 'fʌŋgɪsaɪd/ *noun* a fungicide that is applied to plants once they have been infected with a fungus rather than as a preventative measure

curb *noun* a strap passing under the lower jaw of a horse, used as a check

curd *noun* a coagulated substance formed by action of acid on milk, used to make into cheese

curds *plural noun* tight flower heads formed on brassicas such as cauliflowers and broccoli

cure *verb* to preserve meat by salting or smoking

COMMENT: Meat is cured by keeping in brine for some time; both salting and smoking have a dehydrating effect on the meat, preventing the reproduction and growth of microorganisms harmful to man.

currant *noun* **1.** a small dried seedless black grape **2.** a small round juicy fruit from a bush, e.g. a blackcurrant

curry *verb* to rub down or dress a horse

Curry Report *noun* a UK government report published in 2002 after a major outbreak of foot and mouth disease. It recommended radical changes to the agriculture and food industries, looking forward to a profitable sustainable future for farming in providing good food for consumers who place increasing emphasis on a healthy diet as well as caring for the environment.

cut *noun* **1.** the act of cutting hay or other plants ○ *It is necessary to get enough silage from three cuts to see the herd through the winter.* **2.** the act of cutting down trees ■ *verb* to fell trees with a saw or an axe

cuticle /'kjuːtɪk(ə)l/ *noun* a thin continuous waxy layer that covers the aerial parts of a plant to prevent excessive water loss

cutter *noun* a pig finished for both the fresh meat and the processing markets at weights similar to bacon pigs, i.e. 80–90kg live weight

cutter bar *noun* a device on a mower or combine harvester, formed of a number of metal fingers which support the knife

cutter bar mower *noun* a machine used to cut grass and other upright crops (NOTE: The knife cutter bar mower has mostly been replaced by the rotary mower.)

cutting *noun* a small piece of a plant from which a new plant will grow

COMMENT: Taking cuttings is a frequently used method of propagation which ensures that the new plant is an exact clone of the one from which the cutting was taken.

cutworm /'kʌtwɜːm/ *noun* a caterpillar of the turnip moth and the garden dart moth, which attacks plants such as turnips, swedes and potatoes by eating their roots and stems

CVL *abbreviation* Central Veterinary Laboratory

CVM *abbreviation* complex vertebral malformation

CVO *abbreviation* Chief Veterinary Officer

cwt *abbreviation* hundredweight

cyanocobalamin /ˌsaɪənəʊkəʊ'bæləmɪn/ *noun* vitamin B_{12}

cycle *noun* a series of actions which end at the same point as they begin ○ *With the piston engine, the cycle is intermittent, whereas in the gas turbine, each process is continuous.*

cyclical /'sɪklɪkl/ *adjective* occurring in cycles ○ *Off-shore and on-shore wind patterns are cyclical.*

cypress /'saɪprəs/ *noun* a tree (*Cupressus sempervirens*) of the Mediterranean region, the wood of which is used for furniture

cyst *noun* an unusual growth in the body or on a plant, shaped like a pouch and containing liquid or semi-liquid substances

cyst nematodes *plural noun* dark brown lemon-shaped cysts, which live and breed in the roots of cereals, mainly oats. The crops will show patches of stunted yellowish-green plants.

cytoplasm /'saɪtəʊplæz(ə)m/ *noun* a jelly-like substance inside the cell membrane which surrounds the nucleus of a cell

D

DA *abbreviation* disadvantaged area

daddy-long-legs *noun* a popular name for the cranefly

dag /dæg/ *noun* a tuft of dirty wool round the tail of a sheep ■ *verb* to remove dirty wool from the hindquarters of a sheep

dairy *noun* **1.** a building used for cooling milk at the farm, before it is taken to a commercial factory **2.** a company which receives milk from farms and bottles it and distributes it to the consumer **3.** a company which produces cream, butter, cheese and other milk products

dairy cows *plural noun* cows and heifers kept for milk production and for rearing calves to replace older cows in a dairy herd

dairy farm *noun* a farm which is principally engaged in milk production

COMMENT: The UK is Europe's 3rd largest milk producer, and is limited to an annual production quota of 14.2 billion litres. Although it is largely self-sufficient in milk, related products such as cheese, milk powder, cream and butter are heavily imported and exported between the UK and other EU countries.

dairy farming *noun* keeping cows for milk production

dairy followers *plural noun* young dairy cattle, intended to replace older cows in due course

dairy herd *noun* a herd of dairy cows

dairying /ˈdeərɪɪŋ/ *noun* an agricultural system which involves the production of milk and other dairy products from cows kept on special farms

dairyman /ˈdeərɪmən/ *noun* **1.** a person who works with dairy cattle **2.** a person employed in a commercial dairy

dairy products *plural noun* foods prepared from milk, e.g. butter, cream, cheese or yoghurt

Dairy Shorthorn *noun* a dual-purpose breed of cattle; the colour may be red, white or red and white

Dalesbred /ˈdeɪlzbred/ *noun* a local sheep of the Swaledale type. It has a white spot on either side of a black face, with a grey muzzle, and provides a long coarse fleece.

dam *noun* **1.** a construction built to block a river in order to channel the flow of water into a hydroelectric power station or to regulate the water supply to an irrigation scheme **2.** the female parent of an animal, usually a domestic animal

COMMENT: Dams are constructed either to channel the flow of water into hydroelectric power stations or to regulate the water supply to irrigation schemes. Dams can have serious environmental effects. The large lake behind the dam may alter the whole climate of a region. The large heavy mass of water in the lake may trigger earth movements if the rock beneath is unstable. In tropical areas, dams encourage the spread of bacteria, insects and parasites, leading to an increase in diseases such as bilharziasis. Dams may increase salinity in watercourses and retain silt which otherwise would be carried down the river and be deposited as fertile soil in the plain below. They may also deprive downstream communities or countries of water, leading to regional tensions.

damp off *verb* to die from a fungus infection which spreads in warm damp conditions and attacks the roots and lower stems of seedlings

COMMENT: Damping off is a common cause of loss of seedlings in greenhouses.

damson /ˈdæmzən/ *noun* a small dark purple plum *(Prunus damascena)*

dandelion *noun* a yellow weed *(Taraxacum officinale)* found in grassland and also sometimes eaten as salad

Danish red /ˌdeɪnɪʃ 'red/ *noun* a dual-purpose breed of cattle, originating in Jutland, Denmark

danthonia /dæn'θəʊniə/ *noun* a tufted pasture grass found in Australia and New Zealand

DAPP *abbreviation* Deadweight Average Pig Price

dapple /'dæp(ə)l/ *noun* rounded patches of colour, especially on a horse

DARD *abbreviation* Department of Agriculture and Rural Development

DARDNI *abbreviation* Department of Agriculture and Rural Development, Northern Ireland

darnel /'dɑːnəl/ *noun* a common weed *(Lolium temulentum)* which affects cereals and is poisonous to animals

Dartmoor /'dɑːtmɔː/ *noun* a breed of large moorland sheep, white-faced with black spots, and a long curly fleece

date *noun* the fruit of a date palm

COMMENT: The biggest producers of dates are Saudi Arabia, Iraq and Algeria, though most Middle Eastern countries produce small quantities.

Daucus /'daʊkəs/ *noun* the Latin name for the family of plants which includes the carrot

day-nettle *noun* same as **common hemp nettle**

day-old chick *noun* a chick up to 24 hours old, sent from a breeder or hatchery to a buyer

DCS *abbreviation* Deer Commission for Scotland

DDT /ˌdiː diː 'tiː/ *noun* an insecticide that was formerly used especially against malaria-carrying mosquitoes. It is now banned in many countries because of its toxicity and ability to accumulate in the environment. Formula: $C_{14}H_9Cl_5$. Full form **dichlorodiphenyltrichloroethane**

dead heading *noun* the process of cutting the dead flower heads from a plant, so as to prevent the formation of seeds

dead-in-shell *adjective* referring to chicks which die in the egg, because they cannot break out, or can only break part of the way out of the shell

deadly nightshade /ˌdedli 'naɪtʃeɪd/ *noun* a poisonous plant *(Atropa belladonna)* sometimes eaten by animals

deadnettle /'dednet(ə)l/ *noun* ♦ **red deadnettle**

dead stock *noun* a comprehensive term for all implements, tools, appliances and machines used on a farm. It can also be used to include seed, fertiliser and feedingstuffs.

deadweight /'dedweɪt/ *noun* the weight of a dressed carcass

Deadweight Average Pig Price *noun* the average price for pigs, calculated each week from reports by abattoirs on the price they paid for the pigs they have slaughtered. It replaced the Adjusted Eurospec Average price report in 2004. Abbr **DAPP**

decay *noun* a process by which tissues become rotten and decompose, caused by the action of microorganisms and oxygen ■ *verb (of organic matter)* to rot or decompose ○ *The soft leaves will gradually decay on the compost heap.*

deciduous *adjective* referring to trees that shed all their leaves in one season ○ *beech, oak and other deciduous trees* ○ *deciduous woodlands*

decompose *verb (of organic material)* to break down into simple chemical compounds by the action of sunlight, water or bacteria and fungi

decomposer /ˌdiːkəm'pəʊzə/ *noun* an organism which feeds on dead organic matter and breaks it down into simple chemicals, e.g. a fungus or bacterium

decomposition /ˌdiːkɒmpə'zɪʃ(ə)n/ *noun* the process of breaking down into simple chemical compounds

decortication /diːˌkɔːtɪ'keɪʃ(ə)n/ *noun* the process of removing husks from seeds

decoupling /diː'kʌplɪŋ/ *noun* the breaking of the link between the amount of money paid to farmers as a subsidy and the amount they produce. ◊ **Single Payment Scheme**

decumbent /dɪ'kʌmbənt/ *adjective* referring to plant stems which lie on the surface of the soil for part of their length, but turn upwards at the end

deep-freezing *noun* long-term storage at temperatures below freezing point (NOTE: Many crops such as peas and beans are grown specifically for commercial deep-freezing.)

deep-litter *noun* a system of using straw, wood shavings, sawdust or peat moss for bedding poultry or cattle

COMMENT: For poultry an inch of well-composted horse manure is laid down first, on which wood shavings, peat moss or cut straw are placed. The litter is changed after each crop of birds. Deep litter also has value as a manure. For cattle, straw, shavings and sawdust form a deep litter. Warmth is given off as faeces in the litter ferment, and additions of fresh litter can be made on top of the old.

deep ploughing *noun* ploughing very deep into the soil, used when reclaiming previously virgin land for agricultural purposes

deep-rooted /,diːp 'ruːtɪd/, **deep-rooting** *adjective* referring to a plant with long roots which go deep into the soil. Compare **surface-rooting**

deer *noun* a ruminant animal, the males of which have distinctive antlers (NOTE: The meat of deer is venison.)

COMMENT: There are three wild species in the UK: the fallow deer *(Dama dama)*, the roe deer *(Capreolus capreolus)* and the red deer *(Cervus elaphus)*, which is also raised commercially. Deer are hardy animals, and are well adapted to severe winters. They can suffer from tuberculosis, and the British government has introduced a compulsory slaughter scheme for animals suffering from the disease. According to the 2000 agricultural survey, the total number of farmed deer holdings in England, Wales, Scotland and Northern Ireland is 300, farming approximately 36,000 deer. They represent less than 0.6% of the farmed animals in the UK (excluding poultry).

Deer Commission for Scotland *noun* an association in Scotland which advises on best practices in wild deer management. Abbr **DCS**

deer farming *noun* the commercial farming of deer to be sold as venison

deer forest *noun* an extensive tract of upland, usually treeless, but managed by keepers to provide deer-stalking

deer-stalking *noun* the hunting of deer in the wild

deficiency payment *noun* payment made to a producer, where the price for a commodity at the market does not reach a preset guaranteed price

deficient *adjective* lacking something essential ○ *The soil is deficient in important nutrients.* ○ *Scrub plants are well adapted to this moisture-deficient habitat.* ○ *She has a calcium-deficient diet.*

definite inflorescence *noun* a type of inflorescence in which the main stem ends in a flower and stops growing when the flower is produced. Compare **indefinite inflorescence**

definitive host *noun* a host on which a parasite settles permanently

deflector plate /dɪ'flektə pleɪt/ *noun* an attachment in a slurry spreader which spreads the slurry over a wide area

deflocculation /diː,flɒkjʊ'leɪʃ(ə)n/ *noun* a state in which clay particles repel each other instead of sticking together ○ *Deflocculation may occur, when clays are worked in a wet condition or if the soil becomes saline.*

defoliant /diː'fəʊliənt/ *noun* a type of herbicide which makes the leaves fall off plants

defoliate /diː'fəʊlieɪt/ *verb* to make the leaves fall off a plant, especially by using a herbicide or as the result of disease or other stress

defoliation /diː,fəʊli'eɪʃ(ə)n/ *noun* the loss of leaves from a plant, especially as the result of using a herbicide or because of disease or other stress

deforest /diː'fɒrɪst/ *verb* to cut down forest trees from an area for commercial purposes or to make arable land ○ *Timber companies have helped to deforest the tropical regions.* ○ *About 40000 square miles are deforested each year.*

deforestation *noun* the cutting down of forest trees for commercial purposes or to make arable or pasture land

Defra /'defrə/, **DEFRA** *abbreviation* Department for Environment, Food and Rural Affairs

degradable /dɪ'greɪdəb(ə)l/ *adjective* referring to a substance which can be broken down into its separate elements. ◊ **biodegradable**

degradation /,degrə'deɪʃ(ə)n/ *noun* the decomposition of a chemical compound into its elements

degrade *verb* **1.** to reduce the quality of something ○ *The land has been degraded through overgrazing.* ○ *Ozone may worsen nutrient leaching by degrading the water-resistant coating on pine needles.* **2.** to make a chemical compound decompose into its elements

degressivity /,diːgre'sɪvɪti/ *noun* a proposed reduction in the amount of subsidies paid under the CAP. This proposal was rejected as a method of keeping the CAP budget under control.

dehair /diːˈheə/ *verb* to remove hard hairs from fine goat fibres such as angora

dehisce /dɪˈhɪs/ *verb* (*of a ripe seed pod, fruit or capsule*) to burst open to allow seeds or spores to scatter

dehiscence /dɪˈhɪs(ə)ns/ *noun* the sudden bursting of a seed pod, fruit or capsule when it is ripe, allowing the seeds or spores to scatter

dehiscent /dɪˈhɪs(ə)nt/ *adjective* referring to seed pods, fruit or capsules which burst open to allow the seeds or spores to scatter. Compare **indehiscent**

dehorn /diːˈhɔːn/ *verb* to remove the horns of an animal, done by disbudding when the animal is young

dehusk /diːˈhʌsk/ *verb* to remove the husk from seeds such as corn

dehydrate *verb* to remove water from something in order to preserve it

COMMENT: Food can be dehydrated by drying in the sun (as in the case of dried fruit), or by passing through various industrial processes, such as freeze-drying.

dehydrated milk *noun* milk which has been dried and reduced to a powder

dehydration /ˌdiːhaɪˈdreɪʃ(ə)n/ *noun* the process of removing water from something in order to preserve it

deintensified **farming** /ˌdiːɪntensɪfaɪd ˈfɑːmɪŋ/ *noun* farming which was formerly intensive, using chemical fertilisers to increase production, but has now become extensive. ◊ **extensification**

demonstration farm *noun* a farm used as a means of spreading best practice to other farmers

denature /diːˈneɪtʃə/ *verb* **1.** to add a poisonous substance to alcohol to make it unsuitable for humans to drink **2.** to change the natural structure of a protein or nucleic acid by high temperature, chemicals or extremes of pH **3.** to make something change its nature **4.** to convert a protein into an amino acid

denatured wheat *noun* wheat which has been stained to make it unusable for human consumption

denaturing *noun* the process of staining wheat grain with a dye, so as to make it unusable for human consumption. Denatured grain may be used as animal feed.

dendrochronology /ˌdendrəʊkrɒˈnɒlədʒi/ *noun* a scientific method of finding the age of wood by the study of tree rings

denitrification /diːˌnaɪtrɪfɪˈkeɪʃ(ə)n/ *noun* the releasing of nitrogen from nitrates in the soil by the action of bacteria

dental *adjective* referring to teeth

dentition /denˈtɪʃ(ə)n/ *noun* the arrangement of teeth in an animal's mouth (NOTE: An examination of an animal's teeth may help in estimating its age.)

denudation /ˌdɪnjuːˈdeɪʃ(ə)n/ *noun* the process of making land or rock bare by cutting down trees or by erosion

denude /dɪˈnjuːd/ *verb* to make land or rock bare by cutting down trees and other plants or by erosion ○ *The timber companies have denuded the mountains.*

Department for Environment, Food and Rural Affairs *noun* the UK government department responsible for farming, the environment, animal welfare and rural development in England and Wales. Abbr **Defra**

'Defra was created to focus and lead the Government's wider approach to sustainable development and specifically to address this aim for the environment, the food industry and rural economies and communities. (Delivering the evidence. Defra's Science and Innovation Strategy, 2003–06)'

Department of Agriculture and Rural Development *noun* the government department responsible for farming, the environment, animal welfare and rural development in Scotland. Abbr **DARD**

Department of Agriculture and Rural Development, Northern Ireland *noun* the department of regional government which deals with farming, the environment, animal welfare and rural development in Northern Ireland. Abbr **DARDNI**

depress *verb* to make a price lower ○ *Overproduction of some items in the EU may depress the price level in the open market.*

depression *noun* an area of low atmospheric pressure. Also called **low**

Derbyshire Gritstone /ˌdɜːbiʃə ˈɡrɪtstəʊn/ *noun* a blackfaced, hornless hardy breed of sheep, which produces a soft fleece of high quality (NOTE: The name comes from a type of rock, millstone grit, found in the Peak District of Derby-shire)

derelict *adjective* **1.** referring to land which has been damaged and made ugly by mining or other industrial processes, or which has been neglected and is not used for anything ○ *a plan to reclaim derelict inner city sites* **2.** referring to a building which is neglected and in ruins ○ *derelict barns*

derris /'derɪs/ *noun* a powdered insecticide extracted from the root of a tropical plant, used against fleas, lice and aphids. ◊ **rotenone**

desalinate /diː'sælɪneɪt/ *verb* to remove salt from a substance such as sea water or soil

desalination /ˌdiːsælɪ'neɪʃ(ə)n/ *noun* the removal of salt from a substance such as sea water or soil

descending aorta /dɪˌsendɪŋ eɪ'ɔːtə/ *noun* the second section of the aorta as it turns downwards

desert *noun* an area of land with very little rainfall, arid soil and little or no vege-tation

COMMENT: A desert will be formed in areas where rainfall is less than 25 cm per annum whether the region is hot or cold. About 30% of all the land surface of the Earth is desert or in the process of becoming desert. The spread of desert conditions in arid and semi-arid regions is caused not only by climatic conditions, but also by human pressures. So over-grazing of pasture and the clearing of forest for fuel and for cultivation both lead to the loss of organic material, a reduction in rainfall by evaporation and soil erosion.

desertification /dɪˌzɜːtɪfɪ'keɪʃ(ə)n/ *noun* the process by which an area of land becomes a desert because of a change of climate or because of the action of humans, e.g. through intensive farming ○ *Changes in the amount of sunlight reflected by different vegetation may contribute to desertification.* ○ *Increased tilling of the soil, together with long periods of drought, have brought about the desertification of the area.*

'Desertification, broadly defined, is one of the principal barriers to sustainable food security and sustainable livelihoods in our world today' [*Environmental Conservation*]

desertify /dɪ'zɜːtɪfaɪ/ *verb* to make land into a desert ○ *It is predicted that half the country will be desertified by the end of the century.*

desiccant /'desɪkənt/ *noun* **1.** a substance which dries something **2.** a type of herbicide which makes leaves wither and die

desiccate /'desɪkeɪt/ *verb* **1.** to preserve food by removing moisture from it **2.** to dry out

desiccation /ˌdesɪ'keɪʃ(ə)n/ *noun* **1.** the act or process of removing water **2.** the act of drying out the soil ○ *The greenhouse effect may lead to climatic changes such as the desiccation of large areas.*

dessert fruit *noun* fruit which are sweet and can be eaten raw, as opposed to being cooked

determination *noun* the process of finding something out by calculation or experiment ○ *determination of the maximum safe dose*

detritivore /dɪ'traɪtɪvɔː/ *noun* an organism which feeds on dead organic matter and breaks it down into simple chemicals, e.g. a fungus or bacterium. Also called **detrivore**, **scavenger**

Devon /'devən/ *noun* a breed of fine-boned dual-purpose cattle. North and South Devons are dark red, and belong to a type of red cattle bred for centuries in England. They thrive on pasture which would not be sufficient for larger breeds, and provide both meat and milk. (NOTE: They are commonly known as **Red Rubies**.)

Devon and Cornwall Longwool /ˌdevən ən ˌkɔːnwəl 'lɒŋwʊl/ *noun* a breed of sheep with long curly, high-quality fleece; the lambs have a fine soft white wool

Devon Closewool *noun* a breed of medium-sized sheep, the product of crosses between the Devon Longwool and the Exmoor Horn

dew *noun* drops of condensed moisture left on surfaces overnight in cool places

dewatering /diː'wɔːtərɪŋ/ *noun* the extraction of water from a crop by pressing, reducing the cost of artificial drying

dew claw *noun* a rudimentary fifth digit found on the heels of dogs, pigs and cattle

dewlap /'djuːlæp/ *noun* a fold of loose skin hanging from the throat of cattle

dew pond *noun* a small pond of rainwater which forms on high ground in chalky soil

COMMENT: Dew ponds are found in areas of chalk or limestone country. To make a dew pond, a hollow is scooped out and lined with clay. The pond is kept full by rainwater.

Dexter /ˈdekstə/ *noun* a rare breed of cattle, originating from the west of Ireland. The animals are small in size, coloured black or red.

dextrose /ˈdekstrəʊz/ *noun* a simple sugar found in fruit and also extracted from corn starch

Diamonds disease *noun* ♦ **erysipelas**

diarrhoea *noun* a condition where an animal frequently passes liquid faeces. Also called **scouring**

dibber /ˈdɪbə/ *noun* a hand tool for making holes in soil to plant small plants

dichlorodiphenyltrichloroethane /daɪˌklɒrəʊdaɪˌfiːnɪltraɪ[[ðɪʃç],klɒːrəʊ ˈiːθeɪn/ *noun* full form of **DDT**

dichotomous branching /daɪ ˌkɒtəməs ˈbrɑːntʃɪŋ/ *noun* a pattern of plant growth that develops when a growing point forks into two points that later divide into two

dicotyledon /ˌdaɪkɒtɪˈliːdən/ *noun* a plant with seeds that have a cotyledon with two parts ○ *Dicotyledons form the largest group of plants.* Compare **monocotyledon**. ◊ **cotyledon**

die back *verb* (*of plants*) to be affected by the death of a branch or shoot ○ *Roses may die back after pruning in frosty weather.*

dieback /ˈdaɪbæk/ *noun* **1.** a fungal disease of some plants which kills shoots or branches **2.** a gradual dying of trees starting at the ends of branches ○ *Half the trees in the forest are showing signs of dieback.*

COMMENT: There are many theories explaining the environmental cause of dieback. Sulphur dioxide, nitrogen oxides and ozone have all been suggested as causes, as well as acidification of the soil or acid rain on leaves.

die down *verb* (*of plants*) to stop growing before the winter and keep only the parts below ground until spring ○ *Herbaceous plants die down in autumn.*

dieldrin /ˈdiːldrɪn/ *noun* an organochlorine insecticide which kills on contact (NOTE: It is very persistent and can kill fish, birds and small mammals when it enters

the food chain. It is banned in the European Union.)

diet *noun* the amount and type of food eaten (NOTE: Animal welfare codes lay down rules about the quality of diet that should be provided for animals or birds to ensure their good health and welfare.)

dietary *adjective* referring to diet

dietary fibre *noun* same as **roughage**

COMMENT: Dietary fibre is found in cereals, nuts, fruit and some green vegetables. It is believed to be necessary to help digestion and to avoid developing constipation, obesity and appendicitis.

dietary reference values *plural noun* the nutrients that are essential for health, published as a list by the UK government

dietetic /ˌdaɪəˈtetɪk/ *adjective* referring to diet

dietetics /ˌdaɪəˈtetɪks/ *noun* the study of food, nutrition and health, especially when applied to food intake

diet formulation *noun* the combining of different types of feedstuffs or nutrients so as to form a healthy and balanced diet for an animal

'Probably the most likely area for reform was diet formulation, and with feed accounting for 70 per cent of production costs, it was essential to examine management practices such as phase feeding to more precisely tailor feed inputs to requirements.' [*Farming News*]

diffuse water pollution *noun* water pollution which is caused by several small sources such as runoff from farms

dig *verb* to turn over ground with a fork or spade

digest *verb* **1.** to break down food and convert it into elements which can be absorbed by the body **2.** to use bacteria to process waste, especially organic waste such as manure, in order to produce biogas ○ *55% of UK sewage sludge is digested.* ○ *Wastes from food processing plants can be anaerobically digested.*

digester /daɪˈdʒestə/ *noun* a device that produces gas such as methane from refuse

digestibility /daɪˌdʒestɪˈbɪlɪti/ *noun* the proportion of food which is digested and is therefore of value to the animal which eats it

digestibility coefficient *noun* the proportion of food digested and not excreted, shown as a percentage of the total food eaten

digestibility trial *noun* a test to measure the digestibility of a known food by recording the weight of food eaten, and then excreted

digestibility value *noun* the amount of digestible organic matter in the dry matter of plants. Abbr **D value**

digestible /daɪˈdʒestɪb(ə)l/ *adjective* able to be digested ○ *Glucose is an easily digestible form of sugar.*

digestible organic matter *noun* an organic substance which can be processed to produce biogas, e.g. manure. Abbr **DOM**

digestion *noun* **1.** the process by which food is broken down and converted into elements which can be absorbed by the body **2.** the conversion of organic matter into simpler chemical compounds, as in the production of biogas from manure. ◊ **bacterial digestion**

digestive *adjective* referring to digestion

digestive enzymes *plural noun* enzymes which speed up the process of digestion

digestive juices *plural noun* juices in an animal's digestive tract which convert food into a form which is absorbed into the body

digestive system *noun* the set of organs in the body associated with the digestion of food

digger /ˈdɪɡə/ *noun* a type of plough body with a short, sharply curved mouldboard. Diggers are used for deep ploughing, especially to prepare for root crops or for land reclamation.

digging stick *noun* one of the earliest agricultural implements, still used in areas where shifting cultivation is practised. The stick has a sharpened end, sometimes with a metal tip, and is used to dig holes to plant crops.

dill /dɪl/ *noun* a common aromatic herb *(Anethum graveolens)* used in cooking and in medicine

dioecious /ˌdaɪəʊˈiːʃəs/ *adjective* referring to a plant species in which male and female flowers occur on different individuals. ◊ **monoecious**

dip *noun* a chemical which is dissolved in water, used for dipping animals, mainly sheep, to remove lice and ticks ■ *verb* to plunge an animal into a dip, for about thirty seconds

diphtheria *noun* a serious infectious disease where a membrane forms in the throat passages of an animal such as in calf diphtheria

diploid /ˈdɪplɔɪd/ *adjective* referring to an organism that has two matched sets of chromosomes in a cell nucleus, one set from each parent (NOTE: Each species has a characteristic diploid number of chromosomes.)

dipper /ˈdɪpə/ *noun* a deep trench into which sheep are guided to be dipped

dipping /ˈdɪpɪŋ/ *noun* the process of plunging an animal in a chemical solution to remove ticks, etc.

COMMENT: Sheep are dipped to eradicate parasites such as lice and ticks, and to prevent sheep scab. Dipping varies from region to region according to custom, breed and climate. Dipping may be ordered by Defra to control outbreaks of disease, and in certain cases it has to be witnessed by a local authority inspector.

dipping bath *noun* same as **dipper**

dipterous /ˈdɪptərəs/ *adjective* referring to an insect such as a fly with two wings

direct drilling *noun* a form of minimal cultivation, where the seed is sown directly into the field without previous cultivation. Several types of drill are used, with heavy discs for cutting narrow drills, or strong cultivator tines.

directive *noun* an order from the European Union, referring to a particular problem

direct proportional application *noun* a system of making sure that the output from a sprayer is proportional to the speed at which it moves forward. Abbr **DPA**

direct reseeding *noun* the process of sowing grass seed without a cover crop

direct sowing *noun* the process of sowing grass seed on a prepared seed bed

dirt tare *noun* the percentage of dirt and waste material lifted with a crop such as sugar beet when it is harvested

disadvantaged area *noun* a name for land in mountainous and hilly areas, which is capable of improvement and use as breeding and rearing land for sheep and cattle. These areas are divided into Disadvantaged or Severely Disadvantaged Areas. The EU recognises such areas and gives financial help to farmers in them.

disbud /dɪsˈbʌd/ *verb* **1.** to remove the horn buds from calves, soon after birth **2.** to remove small flower buds from a plant, to allow the main flower to develop more strongly, e.g. when growing chrysanthemums or paeonies

disc *noun* one of the heavy round metal plates, used in harrow and ploughs to cultivate the soil

disc coulter *noun* the part of a plough which cuts the side of the furrow about to be turned

disc harrow *noun* a type of harrow, with two or more sets of saucer-shaped discs fixed to a frame. The disc angle can be changed and the working depth varied.

discoloration /dɪskʌlə'reɪʃ(ə)n/, **discolouration** *noun* a change of colour, especially one caused by deterioration

discolour /dɪs'kʌlə/ *verb* to change the colour of something, especially through deterioration, usually making it paler (NOTE: The US spelling is **discolor**.)

disc plough *noun* a type of plough with large rotating discs in place of the mouldboard. Disc ploughs are used for deep cultivation, but not common in Great Britain.

disease *noun* an illness of people, animals or plants ○ *He is a specialist in plant diseases.*

disease control *noun* the systems put in place by a farm or a government to prevent diseases from spreading within the area under their supervision

'Dr Reynolds said DEFRA had considered the impact on the poultry sector and believed close observation, biosecurity, movement restrictions and swift culling of infected flocks was the most appropriate form of disease control.' [*Farmers Weekly*]

diseased /dɪ'ziːzd/ *adjective* affected by a disease and so not functioning as usual or not whole ○ *a diseased kidney* ○ *To treat dieback, diseased branches should be cut back to healthy wood.*

disease dynamics *noun* the study of the change, growth or activity of a disease

Diseases of Animals Act (1950) *noun* an Act of Parliament covering the diseases that are listed as notifiable

disease status *noun* an assessment of how many animals are diseased and which diseases are present in a herd or flock

Dishley Leicester /ˌdɪʃli 'lestə/ *noun* a breed of improved Leicester sheep, used by Sir Robert Bakewell in the 18th century

disinfect *verb* to make something or somewhere free from microorganisms

such as bacteria ○ *All utensils must be thoroughly disinfected.* (NOTE: **Disinfect**, **disinfection** and **disinfectant** are used for substances which destroy germs on instruments, objects or the skin.)

disinfectant *noun* a substance used to kill microorganisms such as bacteria

disinfection /ˌdɪsɪn'fekʃən/ *noun* the process of making something or somewhere free from microorganisms such as bacteria

COMMENT: Disinfection is a necessary process affecting buildings such as stables, and implements, after infection has been present. It may involve removing litter and dung, and cleaning floors and partitions. Implements and tools should also be treated. Methods of disinfection include the use of approved chemical solutions, steam cleaning and fumigation with powerful antiseptics.

disorder *noun* **1.** a disruption of a system or balanced state **2.** an illness ○ *a stomach disorder*

dispatcher /dɪ'spætʃə/ *noun* an implement used to kill chickens

dispersal *noun* the moving of individual plants or animals into or from an area ○ *seed dispersal by wind* ○ *Aphids breed in large numbers and spread by dispersal in wind currents.*

disperse *verb* **1.** (*of organisms*) to separate and move away over a wide area **2.** to send something out over a wide area ○ *Some seeds are dispersed by birds.* ○ *Power stations have tall chimneys to disperse the emissions of pollutants.*

dispersing agent *noun* a chemical added to a fungicide/bactericide formulation to allow particles of the active agent to be distributed effectively

dispersion /dɪ'spɜːʃ(ə)n/ *noun* the pattern in which animals or plants are found over a wide area

distil *verb* to produce a pure liquid by heating a liquid and condensing the vapour, as in the production of alcohol or essential oils

distillation /ˌdɪstɪ'leɪʃ(ə)n/ *noun* the process of producing a pure liquid by heating a liquid and condensing the vapour, as in the production of alcohol or essential oils

distillers' grains /dɪs'tɪləz greɪnz/ *plural noun* by-product of whisky production, which consists of the remains of malted barley, used as a valuable cattle food which may be fed wet or dry

distort *verb* to change the shape of something, so that it does not look normal, as when brassica roots are distorted by club root disease

distribution channel *noun* **1.** the route by which a product reaches a customer after it leaves the producer or supplier **2.** an area where controlled amounts of feed are made available to livestock

ditch *noun* a channel to take away rainwater ■ *verb* to dig channels for land drainage

COMMENT: Ditches may be enough to drain an area by themselves, but usually they serve as outlets for underground drains. Ditches can deal with large quantities of water in very wet periods. They should be kept cleaned to their original depth, a process carried out usually once a year. Many types of machine are now available for making new ditches or cleaning neglected ones.

ditcher /'dɪtʃə/ *noun* a mechanical excavator used in ditching

dithiocarbamates /ˌdɪθiəʊ'kɑːbəmeɪts/ *plural noun* fungicides formerly used on fruit, vegetables and arable crops but no longer approved for use in the UK

diversification /daɪˌvɜːsɪfɪ'keɪʃ(ə)n/ *noun* the expansion of a farm or other enterprise into new areas of business, e.g. allowing land to be used for leisure activities or introducing new crops or livestock

COMMENT: The main alternative enterprises undertaken by farmers are: farm holidays and bed-and-breakfast; farm shops, selling produce from the farm; camping and caravan sites; country sports, such as horse riding, pony-trekking and fishing.

diversify *verb* **1.** to develop something in different ways ○ *Farmers are encouraged to diversify land use by, for example, planting woodlands or creating recreational facilities.* **2.** to start doing several different things ○ *Farmers are being encouraged to diversify into other areas of business, such as rural tourism.*

Divisional Veterinary Manager *noun* the manager of an Animal Health Divisional Office. Abbr **DVM**

Divisional Veterinary Officer *noun* a trained technician working for an Animal Health Divisional Office. Abbr **DVO**

DM *abbreviation* dry matter

DMI *abbreviation* dry matter intake

docile *adjective* quiet and easy to handle

dock *verb* to cut off the tail of an animal (NOTE: Lowland breeds of sheep are often docked to prevent dirt and faeces accumulating on the tail.) ■ *noun* a broadleaved or curled weed *Rumex* with a long tap root, making it difficult to remove

dockage /'dɒkɪdʒ/ *noun* waste material which is removed from grain as it is being processed before milling

docking disorder *noun* a disorder of sugar beet, caused by eelworms, found on sandy soils in East Anglia, causing irregularly stunted plants with split root growth

doe *noun* the female of deer, goat, rabbit or hare

dogdaisy /'dɒgdeɪsi/ *noun* same as **mayweed**

Dogs Trust *noun* a UK charity that campaigns for the welfare of dogs

DOM /ˌdiː əʊ 'em/ *abbreviation* **1.** digestible organic matter **2.** dry organic matter

domestic *adjective* **1.** referring to the home ○ *domestic waste* **2.** kept as a farm animal or pet

domestic animal *noun* **1.** an animal such as a dog or cat which lives with human beings as a pet **2.** an animal such as a pig or goat which is kept by human beings for food or other uses

domesticate /də'mestɪkeɪt/ *verb* **1.** to breed wild animals so that they become tame and can fill human needs **2.** to breed wild plants, selecting the best strains so that they become useful for food or decoration

domesticated *adjective* **1.** referring to a wild animal which has been trained to live near a house and not be frightened of human beings **2.** referring to a species which was formerly wild but has been selectively bred to fill human needs

domestication /dəˌmestɪ'keɪʃ(ə)n/ *noun* the action of domesticating wild animals or plants

domestic livestock *noun* pigs, goats, sheep, cows and other animals which are kept by human beings

dominance *noun* **1.** a state where one species in a community is more abundant than others **2.** the priority for food and reproductive mates that one animal has over another in a group **3.** the characteristic of a gene form (**allele**) that leads to the trait which it controls being shown in any individual carrying it. Compare **recessiveness**

dominance hierarchy *noun* the system of priority given to specific individuals in terms of access to food and reproductive mates ○ *In many species a male is at the top of the dominance hierarchy.*

dominant *adjective* **1.** important or powerful **2.** (*of an allele*) having the characteristic that leads to the trait which it controls being shown in any individual carrying it. Compare **recessive 3.** (*of a species*) being more abundant than others in a community. ◊ **codominant, subdominant** ■ *noun* a plant or species which has most influence on the composition and distribution of other species

COMMENT: For physical characteristics controlled by two alleles, if one allele is dominant and the other recessive, the resulting trait will be that of the dominant allele. Traits governed by recessive alleles appear only if alleles from both parents are recessive.

Dorking /'dɔːkɪŋ/ *noun* **1.** a breed of fowl, with dark and silver-grey plumage **2.** a silver-grey breed of bantam

dormancy /'dɔːmənsi/ *noun* an inactive period ○ *see dormancy*

dormant *adjective* not actively growing

Dorset Down /ˌdɔːsət 'daʊn/ *noun* a medium-sized down breed of sheep with a brown face and wool growing over the forehead. It provides a good-quality fine stringy fleece.

Dorset Horn *noun* a breed of sheep in the south-west of England, both rams and ewes of which have long curly horns. It produces a fine white clear wool, and is unique among British breeds in that it can lamb at any time of the year.

Dorset wedge silage *noun* a method of storing silage in wedge-shaped layers, usually covered with polythene sheeting. The first loads are tipped against the end wall and further loads are built up with a buckrake to form a wedge.

dose *noun* the amount of medicine given to an animal to cure it of a disorder ■ *verb* to give an animal medicine

dosing gun *noun* a device used to give an animal medicine in the form of pellets. The pellet is forced into the back of animal's throat.

double chop harvester *noun* a type of forage harvester, which chops the crop into short lengths rather than just lacerating it.

The chopping unit is a vertical rotating disc, usually with three knives and three fan blades. ◊ **precision chop forage harvester**

double cropping *noun* a type of multicropping, taking more than one crop off a piece of land in one year

double digging *noun* a cultivation technique, where a spit is dug out, the soil placed on one side, and a second spit dug. This loosens the soil at a deeper level than normal digging.

double flower *noun* a flower with two series of petals as opposed to a single flower

Double Gloucester *noun* a rich orange-coloured British cheese

double lows *plural noun* varieties of oilseed rape with low erucic acid and glucosinolate contents

doubles *plural noun* twins of animals, especially lambs

double suckling *noun* a method of raising beef calves, where a second calf is placed with the cow's own calf and allowed to suckle

Douglas fir *noun* a North American softwood tree widely planted throughout the world, and producing strong timber. Latin name: *Pseudotsuga menziesii.*

dove *noun* a white domesticated pigeon

dove cote /'dʌv kəʊt/ *noun* a small shelter for doves

down *noun* **1.** the small soft feathers of a young bird, or soft feathers below the outer feathers in some adult birds **2.** an undercoat of very soft hair on a goat

Down breeds *plural noun* breeds of short-wooled sheep, giving wool of a creamy colour. They have dark faces and legs, and are hornless. They are found in hilly areas, and include the Southdown, Hampshire Down, Dorset Down and Suffolk.

down-calver *noun* a cow or heifer about to calve

downer animal /'daʊnə ˌænɪməl/, **downer** *noun* a farm animal that is unable to stand or walk because of injury or disease

downland /'daʊnlænd/ *noun* an area of grassy treeless hills

downy *adjective* referring to something such as plumage which is very soft

downy mildew *noun* a disease (*Perenospora brassica*) which causes white bloom on the under surface of leaves, most damaging to Brassica seedlings

Doyenne du Comice /dɔɪ,en dʊ kɒ'miːs/ *noun* a variety of dessert pear, originating in France. The fruit are very round and mature slowly.

DPA *abbreviation* direct proportional application

draft ewe *noun* an ewe sold from a breeding flock of sheep while still young enough to produce lambs

draft off *verb* to remove certain animals from a herd or flock

drag harrow *noun* a heavy type of harrow, used in the preparation of seedbeds

drain *noun* 1. an underground pipe which takes waste water from buildings or from farmland 2. an open channel for taking away waste water 3. a device to allow fluid to escape from its container ■ *verb* 1. to remove liquid from somewhere 2. (*of liquid*) to flow into something ○ *The stream drains into the main river.* 3. to remove water from farmland (NOTE: On most types of farmland, except free-draining soils, some sort of artificial drainage is necessary to carry away surplus water and so keep the water table at a reasonable level.)

drainage *noun* the removal of water by laying drains in or under fields

COMMENT: The main methods of drainage are open channels (ditches) and underground pipe drains and mole drains. Signs of bad drainage include machinery getting bogged down in mud, poaching by stock in grazing pastures, water lying in pools after heavy rain, weeds such as rushes, sedges and horsetail appearing in grassland, young plants being pale green or yellow and subsoil being various shades of blue or grey.

drainage area, drainage basin *noun* same as **catchment**

drainage channel *noun* a small ditch made to remove rainwater from the soil surface

drainage ditch *noun* a channel to take away rainwater

draining pen *noun* a pen for sheep to go in after dipping, where surplus liquid can drain off the wet fleece and go back into the sheep dip

drake /dreɪk/ *noun* a male duck

draught *noun* the effort needed to pull an implement through the soil

draught animal *noun* an animal used to pull vehicles or carry heavy loads

COMMENT: Considerable use is made of draught animals in many areas of the world. Oxen, buffaloes, yaks, camels, elephants, donkeys, horses are all used as draught animals. The advantages of using animals are many. They produce young so do not always have to be bought. They are cheaper to buy than machines. They do not use expensive fuel, even though they eat large quantities of food. They may be slower than machines but they can work in difficult terrains. Their most important advantage is that they are appropriate to the local conditions.

draught control system *noun* a system of preventing damage to an implement such as a harrow, as it is being pulled through the soil. When the draught reaches a set level, the implement is automatically raised out of the soil.

drawbar /'drɔːbɑː/ *noun* a metal bar at the back of a tractor, used to pull trailed implements. Some tractors have a drawbar which can be attached to the hydraulic linkage.

drawbar power *noun* the power available to pull an implement, as opposed to the brake horsepower of a tractor. Under field conditions, not all brake horsepower will be available to pull implements, because some of it is needed to make the tractor itself move forwards and overcome the resistance of the bearings and the soil on the wheels.

draw hoe *noun* a hoe whose blade is at right angles to the handle and is pulled backwards towards the worker

dray /dreɪ/ *noun* a flat cart without sides

dredge corn *noun* a mixture of cereals grown together and used for livestock feeding. The commonest type is a mixture of barley and oats, and sometimes cereals and pulses are mixed.

drench *noun* a method of applying a liquid medicine, by passing it into the stomach through a tube ■ *verb* to soak with a liquid, as when spraying with a coarse nozzle

dress *verb* to clean or prepare the carcass of something such as a chicken so that it is ready for cooking and eating

dressing *noun* the process of treating seeds before sowing, to control disease. ◊ **top dressing**

dribble bar *noun* an attachment which applies a liquid top dressing to a crop through trailing pipes from a boom

dried /draɪd/ *adjective* referring to foodstuffs which are preserved by dehydration

dried blood *noun* an organic fertiliser with a nitrogen content of 10% – 13%. It is a soluble quick-acting fertiliser, used mainly by horticulturists.

dried fruit *noun* fruit that has been dehydrated to preserve it for later use

dried grass *noun* grass which has been artificially dried and is used as an animal feed of high nutritional value

dried milk *noun* milk powder produced by removing water from liquid milk. The techniques involved include roller-drying and spray-drying.

drier *noun* a machine used to dry a crop, usually grain

drift *verb* to float in the air onto areas which are not to be sprayed

drill *noun* 1. an implement used to sow seed. A drill consists of a hopper carried on wheels, with a feed mechanism which feeds the seed into seed tubes. 2. a little furrow for sowing seed ■ *verb* to sow seed in drills

COMMENT: Some drills have a hopper divided into two parts: one contains the seed, the other fertiliser. This is the combine drill which drills grain and fertiliser at the same time.

drill coulter *noun* a coulter which makes a furrow for sowing seed

drip irrigation *noun* an irrigation system where water is supplied by ground-level pipes and released slowly at the base of each plant. Also called **trickle irrigation**

drone *noun* a male bee

drop *noun* 1. a small amount of liquid that falls ○ *a drop of water* ○ *a few drops of rain* 2. a fall of immature fruit ■ *verb* to give birth to a lamb

droppings *plural noun* excreta from animals ○ *The grass was covered with rabbit and sheep droppings.*

droppings board *noun* a bench, under the perches in smaller poultry houses, on which bird droppings collect

drought *noun* a long period without rain at a time when rain usually falls

drought order *noun* legislation which permits water companies to place restrictions on the use of water for a specific period when there is a drought

drought stress *noun* a lack of growth caused by drought

drove *noun* a number of cattle or sheep being driven from one place to another

drover /ˈdrəʊvə/ *noun* a person in charge of a flock or herd which is being moved from one place to another

drove road *noun* a track along which sheep or cattle are regularly driven

drum *noun* the cylinder of a combine harvester, which rotates and has rasp-like beater bars which thresh the grain

drupe /druːp/ *noun* a fruit with a single seed and a fleshy body (NOTE: Stone fruits such as cherries or plums are drupes.)

dry cow *noun* a cow which is between lactations and is therefore not giving milk

dry curing *noun* the process of curing meat in salt, as opposed to brine

dry farming *noun* a system of extensive agriculture, producing crops in areas of limited rainfall, without using irrigation

dry feeding *noun* the feeding of meal to animals without the addition of water. This may cause problems with pigs and poultry.

drying *noun* a method of preserving food by removing moisture, either by leaving it in the sun, as for dried fruit, or by passing it through an industrial process

drying off *noun* a gradual reduction in the quantity of milk taken from a cow, so as to make it stop lactating

dry matter *noun* the matter remaining in a biological sample or in animal feed after the water content has been removed. Abbr **DM**

dry matter intake *noun* the amount of feed that an animal consumes or requires, discounting its water content. Abbr **DMI**

'Maintaining dry matter intake (DMI) at turnout and maximising it throughout the grazing season is key to feeding the high yielding dairy cow at grass during the summer.' [*Dairy Farmer*]

dry organic matter *noun* organic matter such as sewage sludge or manure which has been dried out and may be used as a fertiliser. Abbr **DOM**

dry period *noun* in cattle, a period of six to eight weeks between lactations, when a cow is rested from giving milk

dry pluck *noun* the process of removing the feathers when the bird is dry, so avoiding harming the skin

dry rot *noun* a fungal disease causing rot in wood, potatoes or fruit

dry roughage *noun* dry bulky food-stuffs, e.g. hay or straw

Drysdale /ˈdraɪzdeɪl/ *noun* a breed of New Zealand sheep, a crossbreed from Romney and Cheviot

dry-stone wall *noun* a wall made of stones carefully placed one on top of the other without using any mortar

dual-purpose breed *noun* a breed of animal valuable for more than one product

dubbin /ˈdʌbɪn/ *noun* prepared grease used for waterproofing and softening leather

duck *noun* a bird reared for both egg and meat production. The male is a 'drake'.

duckling *noun* a young duck

dug *noun* a teat or udder of an animal, especially of a cow

dump *noun* a place where waste, especially solid waste, is thrown away ○ *The mine is surrounded by dumps of excavated waste.* ■ *verb* **1.** to throw away waste, especially without being subject to environmental controls **2.** to get rid of large quantities of excess farm products cheaply in an overseas market

dump box *noun* a large hopper on wheels with a floor conveyor which receives silage from trailers and from which the crop is discharged into the silo

dumping /ˈdʌmpɪŋ/ *noun* **1.** the disposal of waste ○ *illegal dumping* **2.** the sale of agricultural products at a price below the true cost, to get rid of excess produce cheaply, usually in an overseas market

dunes *plural noun* an area of sand blown by the wind into small hills and ridges which may have plants growing on them ○ *The village was threatened by encroaching dunes.* ○ *The dunes were colonised by marram grass.*

dung *noun* solid waste excreta from animals, especially cattle, often used as fertiliser

COMMENT: In some areas of the world dried dung is used as a cooking fuel, which has the effect of preventing the dung from being returned to the soil and leads to depletion of soil nutrients.

dunging passage *noun* a passage at the back of a cow shed, into which dung can be washed with water

dungleweed /ˈdʌŋgəlwiːd/ *noun* same as **orache**

dung weed *noun* same as **fat hen**

duramen /djuːˈrɑːmən/ *noun* same as **heartwood**

Durham /ˈdʌrəm/ *noun* a breed of dairy shorthorn cattle, developed in the Tees valley of County Durham

Duroc /djuːˈrɒk/ *noun* a breed of pig, originating in the eastern USA, imported into the UK for cross-breeding. The pigs are red in colour.

durum /ˈdjuərəm/ *noun* a type of wheat grown in southern Europe and the USA and used in making semolina for processing into pasta. Latin name: *Triticum durum*.

dust *noun* a fine powder made of particles, e.g. dry dirt or sand

dusting /ˈdʌstɪŋ/ *noun* the act of using dry powdered fungicide or insecticide on crops

Dutch barn /ˈdʌtʃ bɑːn/ *noun* a type of farmyard building used for storage of hay, loose or baled, corn crops and agricultural implements (NOTE: The older types of Dutch barn were built of iron with no enclosing side walls. Modern designs incorporate precast concrete, asbestos-cement sheeting with curved roofs. The sides may be partly or completely covered.)

Dutch elm disease *noun* a fungal disease that kills elm trees, caused by *Ceratocystis ulmi* and spread by a bark beetle

Dutch harrow *noun* an implement with metal or wooden frame, with heavy tine bars almost at right angles to the direction of travel. The tines loosen the soil and the heavy bars level the surface. Also called **float**

Dutch hoe *noun* an implement with a long handle and a more or less straight D-shaped blade, used with a push-pull action

duty of care *noun* a duty which every citizen and organisation has not to act negligently, especially the system for the safe handling of waste, introduced by the UK Environmental Protection Act 1990

'Growers have a "duty of care" to ensure that their waste is disposed of in a legal manner and must retain proof that they have fulfiled this requirement.' [*Farmers Guardian*]

D value *abbreviation* digestibility value

DVM *abbreviation* Divisional Veterinary Manager

DVO *abbreviation* Divisional Veterinary Officer

dwarf bean *noun* a term used for French or kidney beans, which make a bushy plant, as opposed to runner beans which climb

dwarfing rootstock /ˌdwɔːfɪŋ ˈruːtstɒk/ *noun* a plant which is normally low-growing so causing the plant grafted on to it to grow smaller than it would otherwise

dyke *noun* **1.** a long wall of earth built to keep water out **2.** a ditch for drainage ■ *verb* to build walls of earth to help prevent water from flooding land

COMMENT: Dyke pond farming is a system of organic agriculture combining crop growing on the dykes which surround ponds in which fish are bred. It is common in China.

dysentery *noun* an infection and inflammation of the colon causing bleeding and diarrhoea

dystocia /dɪsˈtəʊsiə/ *noun* difficulty in the process of giving birth

dystrophic /dɪsˈtrɒfɪk/ *adjective* referring to a pond or lake that contains very acidic brown water, lacks oxygen, and is unable to support much plant or animal life because of excessive humus content. ◊ **eutrophic, mesotrophic, oligotrophic**

E

E *symbol* Excellent (*in the EUROP carcass conformation classification system*)

EA *abbreviation* Environment Agency

EAGGF *abbreviation* European Agricultural Guidance and Guarantee Fund

ear *noun* the flower head of a cereal plant such as wheat or maize where the grains develop

ear emergence *noun* the main stage used in determining the heading date of a crop. In the case of a sward, this is the date at which 50% of the inflorescences have appeared.

earlies /'ɜːliz/ *plural noun* potatoes grown for harvesting early in the season. Compare **maincrop potatoes**

early bite *noun* grazing in the spring, provided by new growths of grass which sprout when the weather gets warmer

early weaning *noun* the practice of removing young from the dam earlier than is usual

earmarking /'ɪəmɑːkɪŋ/ *noun* the process of identifying an animal by attaching a tag to its ear

earth *noun* **1.** soil **2.** the ground or land surface

earth up *verb* to move soil to make a ridge, in which a crop such as potatoes or celery can grow

COMMENT: Plants are earthed up to protect the tender stems from frost, or to make them white. Potatoes are earthed up to prevent the tubers from turning green and tasting bitter.

earthworm /'ɜːθwɜːm/ *noun* an invertebrate animal with a long thin body divided into many segments, living in large numbers in the soil

COMMENT: Earthworms aerate the soil as they tunnel. They also eat organic matter and help increase the soil's fertility. They help stabilise the soil structure by compressing material and mixing it with organic matter and calcium. It is believed that they also secrete a hormone which encourages rooting by plants.

easement /'iːzmənt/ *noun* the right of someone who does not own a piece of land to use it, especially for access to another place

East Friesland *noun* a breed of sheep introduced into the UK from Holland. It is a large long slim-bodied breed and is much valued for its high milk yield.

easy feed *noun* a means of feeding livestock which allows easy access to feed by means of hoppers or feeding passages

EBLEX /'ebleks/ *abbreviation* English Beef and Lamb Executive

EBV *abbreviation* estimated breeding value

ECN *abbreviation* Environmental Change Network

ecoagriculture /'iːkəʊˌægrɪkʌltʃə/ *noun* the practice of productive agriculture using methods designed to maintain natural resources, biodiversity and the landscape

E. coli *abbreviation* Escherichia coli

ecological *adjective* referring to ecology

ecological corridor *noun* a strip of vegetation allowing the movement of wildlife or other organisms between two areas

ecological diversity *noun* a variety of biological communities that interact with one another and with their physical and chemical environments

ecological efficiency *noun* a measurement of how much energy is used at different stages in the food chain or at different trophic levels

ecological engineering *noun* a design process that aims to integrate human activities with the natural environment for the benefit of both, taking ecological impact into account in the construction of roads or harbours, the introduction of new plants or animals, or other actions

ecological factors *plural noun* factors which influence the distribution of a plant species in a habitat

ecologically sustainable development *noun* development which limits the size of the human population and the use of resources, so as to protect the existing natural resources for future generations

ecological recovery *noun* the return of an ecosystem to its former favourable condition

ecological restoration *noun* the process of renewing and maintaining the health of an ecosystem

ecologist /ɪ'kɒlədʒɪst/ *noun* **1.** a scientist who studies ecology **2.** a person who is in favour of maintaining a balance between living things and the environment in which they live in order to improve the life of all organisms

ecology *noun* the study of the relationships among organisms as well as the relationships between them and their physical environment

ecoparasite /'iːkəʊˌpærəsaɪt/ *noun* a parasite which is adapted to a specific host. Compare **ectoparasite**, **endoparasite**

ecosphere *noun* the part of the Earth and its atmosphere where living organisms exist, including parts of the lithosphere, the hydrosphere and the atmosphere. Also called **biosphere**

ecosystem *noun* a complex of plant, animal and microorganism communities and their interactions with the environment in which they live ○ *European wetlands are classic examples of ecosystems that have been shaped by humans.*

'Agricultural pesticides in the water can cause two problems; they affect aquatic ecosystems, and impact on drinking water abstraction.' [*Farmers Guardian*]

COMMENT: An ecosystem can be any size, from a pinhead to the whole ecosphere. The term was first used in the 1930s to describe the interdependence of organisms among themselves and their relationships with the living and non-living environment.

ecotax /'iːkəʊtæks/ *noun* a tax that is used to encourage people to change from an activity that damages the environment or to encourage activities with beneficial

environmental effects. Also called **environmental tax**

ecotone /'iːkəʊtəʊn/ *noun* an area between two different types of vegetation which may share the characteristics of both, e.g. the border between forest and moorland

ecotourism /'iːkəʊˌtʊərɪz(ə)m/ *noun* a form of tourism that increases people's understanding of natural areas, without adversely affecting the environment, and gives local people financial benefits from conserving natural resources

ecotoxicity /ˌiːkəʊtɒk'sɪsɪti/ *noun* the degree to which a chemical released into an environment by human activities affects the organisms that live or grow there

ECS *abbreviation* Energy Crops Scheme

ecto- /ektəʊ/ *prefix* outside. Compare **endo-**

ectoparasite /ˌektəʊ'pærəsaɪt/ *noun* a parasite which lives on the skin or outer surface of its host but feeds by piercing the skin. Compare **endoparasite**

ectoparasite disease *noun* a disease caused by lice and other insects, usually characetrised by intense irritation

edaphic /ɪ'dæfɪk/ *adjective* referring to soil

edge effect *noun* an increase in growth and yield seen in crop plants growing at the edge of a plot or field

edible snails *plural noun* snails reared for human consumption

eelworm /'iːlwɜːm/ *noun* a minute worm-like animal *(Nematode)* which attacks a great variety of food crops. ◊ **stem eelworm**

EFA *abbreviation* essential fatty acid

EFB *abbreviation* European foul brood

effective field capacity *noun* the actual average rate of work achieved by a machine, usually expressed in acres or hectares per hour

effluent *noun* liquid, semisolid or gas waste from industrial processes or material such as slurry or silage effluent from a farm

EFSA *abbreviation* European Food Safety Authority

egg *noun* **1.** a reproductive cell produced in a female mammal by the ovary which, if fertilised by male sperm, becomes an embryo **2.** a fertilised ovum of an animal such as a bird, fish, reptile, amphibian or insect, protected by a membrane layer in which the embryo continues developing outside the mother's body until it hatches **3.** a round object laid by female birds, with a hard calcareous shell forming a case containing albumen and yolk. The young bird grows inside the egg until it hatches.

COMMENT: The average hen's egg weighs about 60g, of which about 20g is yolk, 35g white and the rest shell and membranes. Eggs contain protein, fat, iron and vitamins A, B, D and E. In 2004, total egg consumption in the UK was 8.961 million, and the annual egg consumption per capita was 174. The percentage of egg production from intensive systems was 66% (a significant drop from the 87% of 1993), 27% were from free range (11% in 1993) and 7% came from other sources. The value of retail sales of eggs in 2004 was £568 million, with each bird producing an average yield of 292 eggs per year.

egg binding noun an unsuccessful attempt by a hen to lay an egg

eggbound /'egbaʊnd/ adjective referring to a hen that produces an egg but is unable to lay it

egg classes plural noun the grading of eggs under EU regulations, into Class A (fresh eggs), Class B (preserved eggs) and Class C (eggs for use in food processing)

egg eating noun a form of behaviour by intensively housed poultry in which birds eat their own eggs. It may be due to eggs being broken because of thin shells.

eggplant noun a plant with purple fruit (Solanum melongena), used as a vegetable. It is a native of tropical Asia. Also called **aubergine**

EH abbreviation English Heritage

EHF abbreviation experimental husbandry farm

EHS abbreviation experimental horticulture station

EIA abbreviation environmental impact assessment

EID abbreviation electronic identification

eject verb to throw something out

ejector /ɪ'dʒektə/ noun a mechanism at the back of a farmyard manure spreader, which throws out the manure over a wide area

elder noun a small tree (Sambucus nigra) with little black berries used to make wine

electrical conductivity noun a measurement of salt concentration in soils

electric dog noun an electric wire at the side of the fence at the entrance to a milking parlour, which encourages the cows to go into the parlour

electric fence noun thin wires supported by posts, the wires being able to carry an electric current. This type of fence is easily moved around the farm, and makes strip grazing on limited areas possible.

electronic identification noun a way of marking animals with tags containing a readable chip, which identifies them. Abbr **EID**

element noun a chemical substance that cannot be broken down to a simpler substance (NOTE: There are 110 named elements.)

elephant grass noun same as **miscanthus**

elevator noun **1.** a machine for carrying grain or silage to the top of a storage unit **2.** a very large storage unit for grain, in the US and Canadian prairies

elevator digger noun a machine for harvesting crops such as potatoes, which can be adapted to harvest carrots, onions or flower bulbs

elm noun a large hardwood tree that grows in temperate areas. Genus: Ulmus. ◊ **Dutch elm disease**

ELS abbreviation Entry Level Stewardship

elt /elt/ noun a young sow (NOTE: This is not a common word.)

emaciation noun becoming extremely thin ○ scab causes emaciation in sheep

'Johne's Disease affects ruminants, its symptoms being diarrhoea, emaciation and loss of body condition.' [Farming News]

Embden /'emdən/ noun a heavy white breed of goose, with blue eyes

embryo noun an organism that develops from a fertilised egg or seed, e.g. an animal in the first weeks of gestation or a seedling plant with cotyledons and a root (NOTE: After eight weeks an unborn baby is called a **fetus**.)

embryonic adjective **1.** referring to an embryo **2.** in the first stages of development

embryo transfer noun the transplanting of an embryo from one animal into the womb of another, used as a method of improving breeding quality. Abbr **ET**

emerge *verb* to come out of the soil

emergence *noun* **1.** the germination of a seed **2.** a stage in the growth of a plant, when the new shoot or stalk appears through the surface of the soil

emission *noun* a substance discharged into the air by an internal combustion engine or other device ○ *Exhaust emissions contain pollutants.* ○ *Gas emissions can cause acid rain.*

emission charge *noun* a fee paid by a company to be allowed to discharge waste into the environment

emission standard *noun* the amount of an effluent or pollutant that can legally be released into the environment, e.g. the amount of sewage which can be discharged into a river or the sea, or the amount of carbon monoxide that can legally be released into the atmosphere by petrol and diesel engines

Emmer /'emə/ *noun* a species of wheat (*Triticum dicoccoides*) which is a natural hybrid of wild wheat and a goat grass. A further crossing between Emmer and a goat grass produced wheat (*Triticum aestivum*).

emulsifier /ɪ'mʌlsɪfaɪə/ *noun* a substance added to mixtures of food such as water and oil to hold them together. ◊ **stabiliser** (NOTE: Emulsifiers are used in sauces and added to meat to increase the water content so that the meat is heavier. In the European Union, emulsifiers and stabilisers have E numbers E322 to E495.)

emulsify *verb* to mix two liquids so thoroughly that they will not separate

emulsifying agent *noun* same as **emulsifier**

EN *abbreviation* **1.** endangered species **2.** English Nature

encephalopathy /en,kefə'lɒpəθi/ *noun* ♦ **BSE**

enclosure *noun* **1.** an area surrounded by a fence, often to contain animals. Compare **exclosure 2.** the action of enclosing open land. ◊ **field** (NOTE: The term **enclosure** is used in England to refer especially to the enclosure of common land in the 16th and 18th centuries, when rights to common land were removed and major landowners used ditches, fences, hedgerows and walls to mark the boundaries of land which they owned freehold.)

encroach on *verb* to come close to and gradually cover something ○ *The town is spreading beyond the by-pass, encroaching on farming land.* ○ *Trees are spreading down the mountain and encroaching on the lower more fertile land in the valleys.*

endangered species *noun* a species that is facing a risk of extinction in the wild, usually taken to be when fewer than 250 mature individuals exist. Abbr **EN** (NOTE: The plural is **endangered species.**)

endemic *adjective* **1.** referring to an organism that exists or originated from a specific area ○ *The isolation of the islands has led to the evolution of endemic forms.* **2.** referring to a disease that occurs within a specific area ○ *This disease is endemic to Mediterranean countries.* ◊ **epidemic, pandemic** ■ *noun* an endemic disease

endive /'endɪv/ *noun* a salad plant (*Cichorium endiva*)

endo- /endəʊ/ *prefix* inside or within. Compare **ecto-**

endocarp /'endəʊkɑːp/ *noun* the innermost of the layers of the wall (**pericarp**) of a fruit (NOTE: Sometimes it is toughened or hardened, as in a cherry stone or peach stone.)

endocrine gland /'endəʊkraɪn glænd/ *noun* a gland such as the pituitary gland which produces hormones introduced directly into the bloodstream

endoparasite /,endəʊ'pærəsaɪt/ *noun* a parasite that lives inside its host. Compare **ectoparasite**

'The first step in putting together an effective endoparasite control strategy is to look at pasture management, to see if there are any practical ways to reduce the risk of infection. In the case of liver fluke, fencing off wet land and not grazing wetter pasture during the key infection period between September and November can have a significant impact.' [*Farmers Guardian*]

endosperm /'endəʊspɜːm/ *noun* a storage tissue in plant seeds that provides nourishment for the developing embryo

endotoxin /,endəʊ'tɒksɪn/ *noun* a poison from bacteria which passes into the body when contaminated food is eaten

energy *noun* **1.** the force or strength to carry out activities ○ *You need to eat carbohydrates to give you energy.* **2.** electricity or other fuel ○ *We have to review our energy requirements regularly.*

energy balance *noun* a series of measurements showing the movement of energy between organisms and their environment (NOTE: In farming a common use of the energy balance is to assess the ratio between the amount of energy used to grow a crop and the amount of energy that crop produces.)

energy crop *noun* a crop which is grown to be used to provide energy, e.g. a fast-growing tree

'Biomass itself can come from a variety of sources. Those include energy crops such as miscanthus and short-rotation willow, forestry trimmings, off-cuts and roots, tree surgeons' chips, reclaimed timber and municipal solid waste, some industrial waste, sewage sludge, liquid animal waste and the enormous amount of food waste.' [*Farmers Weekly*]

Energy Crops Scheme *noun* a system set up by Defra under which farmers can apply for grants to establish energy crops on their land

energy value *noun* the heat value of a substance measured in joules. Also called **calorific value**

England Rural Development Programme *noun* a set of schemes run by Defra in the UK to develop sustainable farming methods in rural areas. Abbr **ERDP**

English Beef and Lamb Executive *noun* an association providing market information for beef and lamb producers and suppliers. Abbr **EBLEX**

English Heritage *noun* an organisation partly funded by government that is responsible for maintaining buildings and monuments of historical interest in England. Abbr **EH**

English Leicester *noun* a breed of sheep derived from Robert Bakewell's flock, used for breeding of many other longwool breeds. It produces a heavy fleece and is now a rare breed.

English Nature *noun* the UK government agency that is responsible for nature conservation in England. Abbr **EN** (NOTE: It was formerly part of the Nature Conservancy Council and is about to undergo another reorganisation.)

enhancer /ɪnˈhɑːnsə/ *noun* an artificial substance that increases the flavour of food or of an artificial flavouring that has been added to food (NOTE: In the European

Union, flavour enhancers added to food have the E numbers E620 to E637.)

enrich *verb* **1.** to make something richer or stronger, e.g. soil can be enriched by adding humus **2.** to improve the nutritional quality of food ○ *enrich with vitamins* **3.** to improve the living conditions of farm animals, e.g. by providing them with larger living areas

enriched cage *noun* a type of cage in which battery hens are kept, where the bird's living conditions have been improved by an increase in the size of the cage and the inclusion of perches, nests and litter so that the bird can peck and scratch

'Enriched cages will replace existing battery cages in 2012, when the EU Directive on the welfare of laying hens comes into force. However, a review of the Directive is underway, with a report expected next year.' [*The Grocer*]

enrichment /ɪnˈrɪtʃmənt/ *noun* the increase in nitrogen, phosphorous and carbon compounds or other nutrients in water, especially as a result of a sewage flow or agricultural run-off, which encourages the growth of algae and other water plants. ◊ **environmental enrichment**

ensilage /ɪnˈsaɪlɪdʒ/, **ensiling** /ɪnˈsaɪlɪŋ/ *noun* the process of making silage for cattle by cutting grass and other green plants and storing it in silos

ensile /ɪnˈsaɪl/ *verb* to make silage from something ○ *We ensile lush young material.*

enter- /ˈentə/ *prefix* same as **entero-**

enteric /enˈterɪk/ *adjective* referring to the intestine

enteritis /ˌentəˈraɪtɪs/ *noun* an inflammation of the mucous membrane of the intestine

entero- /ˈentərəʊ/ *prefix* referring to the intestine

Enterobacteria /ˌentərəʊbækˈtɪəriə/ *plural noun* a family of bacteria, including *Salmonella* and *Escherichia*

Enterobius /ˌentəˈrəʊbiəs/ *noun* a threadworm or nematode which infests the intestine

enterotoxin /ˌentərəʊˈtɒksɪn/ *noun* a bacterial exotoxin which particularly affects the intestine

enterovirus /ˌentərəʊˈvaɪrəs/ *noun* a virus which prefers to live in the intestine

COMMENT: The enteroviruses are an important group of viruses, and one causes Teschen disease in pigs.

entire *adjective* referring to an animal which has not been castrated

entomological /ˌentəməˈlɒdʒɪk(ə)l/ *adjective* referring to insects

entomologist /ˌentəˈmɒlədʒɪst/ *noun* a scientist who specialises in the study of insects

entomology /ˌentəˈmɒlədʒi/ *noun* the study of insects

Entry Level Stewardship *noun* one of the categories under the Environmental Stewardship scheme, where farmers with any size holding can apply for funding in return for implementing certain environmental management schemes on their land. Abbr **ELS**

E number *noun* a classification of additives to food approved by the European Union

COMMENT: Additives are classified as follows: colouring substances E100 – E180; preservatives E200 – E297; antioxidants E300 – E321; emulsifiers and stabilisers E322 – E495; acids and bases E500 – E529; anti-caking additives E530 – E578; flavour enhancers and sweeteners E620 – E637.

environment *noun* the surroundings of any organism, including the physical world and other organisms. ◊ **built environment, natural environment**

COMMENT: The environment is anything outside an organism and in which the organism lives. It can be a geographical region, a climatic condition, a pollutant or the noises which surround an organism. The human environment includes the country or region or town or house or room in which a person lives. A parasite's environment includes the body of the host. A plant's environment includes the type of soil at a specific altitude.

Environment Agency *noun* in England and Wales, the government agency responsible for protection of the environment, including flood and sea defences. Abbr **EA**

environmental *adjective* referring to the environment

environmental assessment *noun* the identification of the expected environmental effects of a proposed action

Environmental Change Network *noun* an association which monitors trends

in climate change and produces statistics. Abbr **ECN**

environmental conditionality /ɪn ˌvaɪrənment(ə)l kənˌdɪʃ(ə)ˈnælɪti/ *noun* same as **cross-compliance**

environmental degradation *noun* a reduction in the quality of the environment

environmental directive *noun* an EU policy statement on the appropriate ways of dealing with a specific environmental issue

environmental enrichment *noun* the practice of improving the living conditions and welfare of animals, such as by increasing the amount of space they have to live in

environmental ethics *noun* the examination and discussion of people's obligations towards the environment

environmental impact *noun* the effect upon the environment of actions or events such as large construction programmes or the draining of marshes

'In the meantime, the world needs more, higher quality food grown in a climate that is likely to be wetter in some areas, warmer with more drought in others, and more volatile. Water protection and environmental impact will continue to be important considerations.' [*Farmers Weekly*]

environmental impact assessment *noun* an evaluation of the effect upon the environment of an action such as a large construction programme. Abbr **EIA**

environmentalist *noun* a person who is concerned with protecting the environment

environmentally friendly *adjective* intended to minimise harm to the environment, e.g. by using biodegradable ingredients. Also called **environment-friendly**

Environmentally Sensitive Area *noun* in the UK, a rural area designated by Defra as needing special protection from modern farming practices. Abbr **ESA** (NOTE: This programme has now been superseded by the **Environmental Stewardship** scheme.)

'The Defra-funded Countryside Stewardship and Environmentally Sensitive Areas schemes help to maintain and enhance the biodiversity and landscape value of farmed land, protect historic features and promote public access. (Delivering the evidence. Defra's

Science and Innovation Strategy, 2003–06)'

environmental management *noun*
1. the idea of humans interacting with the environment in a responsible and ethically sound way, without sacrificing productivity **2.** guidelines or practices which support this aim

'Donald Curry, the government adviser who first suggested introducing a "broad and shallow" style agri-environment scheme, said the launch represented a fundamental step in farmers committing themselves to sound environmental management. "It is essential that as many farmers as possible participate in the schemes", he added.' [*Farmers Weekly*]

environmental pollution *noun* the pollution of the environment by human activities

environmental protection *noun* the activity of protecting the environment by regulating the discharge of waste, the emission of pollutants and other human activities. Also called **environment protection**

Environmental Protection Act 1990 *noun* a UK regulation to allow the introduction of integrated pollution control, regulations for the disposal of waste and other provisions. Abbr **EPA**

Environmental Protection Agency *noun* an administrative body in the USA which deals with pollution. Abbr **EPA**

environmental quality standard *noun* a limit for the concentration of an effluent or pollutant which is accepted in a specific environment, e.g. the concentration of trace elements in drinking water or of additives in food

environmental set-aside *noun* a scheme of suspending cultivation of food crops for a period with clearly defined environmental aims and designed appropriately for local conditions

Environmental Stewardship *noun* a system under which farmers and land managers receive funding for implementing schemes to protect the environment, such as preventing soil erosion or protecting wildlife

environmental tax *noun* same as **ecotax**

environment-friendly *adjective* same as **environmentally friendly**

environment protection *noun* same as **environmental protection**

Envirowise /enˈvaɪrəʊwaɪz/ *noun* a government programme providing advice to businesses in industry and commerce on improving efficiency in the use of resources and reducing waste

enzootic abortion /ˌenzəʊɒtɪk əˈbɔːʃ(ə)n/ *noun* a virus infection of sheep causing abortion about two weeks before lambing

enzootic bovine leucosis /ˌenzəʊɒtɪk ˌbəʊvaɪn luˈkəʊsɪs/ *noun* a blood cancer disease of cattle. It is a notifiable disease.

enzootic disease *noun* an outbreak of disease among certain species of animals in a certain area. Compare **epizootic disease**

enzootic pneumonia *noun* a disease of pigs, previously thought to be caused by a virus. Symptoms include coughing and stunted growth. Abbr **EP**

enzyme *noun* a protein substance produced by living cells which promotes a biochemical reaction in living organisms (NOTE: The names of enzymes mostly end with the suffix **-ase**.)

COMMENT: Many different enzymes exist in organisms, working in the digestive system, metabolic processes and the synthesis of certain compounds. Some pesticides and herbicides work by interfering with enzyme systems or by destroying them altogether.

EP *abbreviation* enzootic pneumonia

EPA *abbreviation* **1.** Environmental Protection Act 1990 **2.** Environmental Protection Agency

ephemeral *noun* a plant or insect that has a short life cycle and may complete several life cycles within a year ○ *Many weeds are ephemerals.*

epicarp /ˈepɪkɑːp/ *noun* the outer skin of a fruit. Also called **exocarp**

epidemic *noun* **1.** an infectious disease that spreads quickly through a large part of the population ○ *The health authorities are taking steps to prevent an epidemic of cholera* or *a cholera epidemic.* **2.** a rapidly spreading infection or disease. ◊ **endemic, pandemic**

epidermis /ˌepɪˈdɜːmɪs/ *noun* an outer layer of cells of a plant or animal

epiphyte /ˈepɪfaɪt/ *noun* a plant that lives on another plant for physical support, but is not a parasite of it (NOTE: Many orchids are epiphytes.)

epiphytic /ˌepɪˈfɪtɪk/ *adjective* attached to another plant for support, but not parasitic

episodic /ˌepɪˈsɒdɪk/ *adjective* happening sometimes but not regularly

epizoon /ˌepɪˈzəʊɒn/ *noun* an animal which lives on another animal

epizootic disease /ˌepɪzəʊɒtɪk dɪ ˈziːz/ *noun* a disease which spreads to large numbers of animals over a large area. Compare **enzootic disease**

equine /ˈekwaɪn/ *adjective* relating to horses

eradicate *verb* to remove something completely ○ *international action to eradicate glaucoma*

eradication /ɪˌrædɪˈkeɪʃ(ə)n/ *noun* **1.** the complete removal of something **2.** the total extinction of a species

eradication area *noun* an area from which a particular animal disease is eradicated, usually involving the slaughter of infected animals

ERDP *abbreviation* England Rural Development Programme

erect habit *noun* the habit of a plant which grows upright, and does not lie on the ground

ergot /ˈɜːgət/ *noun* a fungus that grows on cereals, especially rye, producing a mycotoxin which causes hallucinations and sometimes death if eaten. Genus: *Claviceps.*

ergotamine /ɜːˈgɒtəmiːn/ *noun* the toxin that causes ergotism

ergotism /ˈɜːgətɪz(ə)m/ *noun* poisoning by eating cereals or bread contaminated by ergot

erode *verb* to wear away gradually, or to wear something away ○ *The hills have been eroded by wind and rain.*

erosion /ɪˈrəʊʒ(ə)n/ *noun* the wearing away of soil or rock by rain, wind, sea or rivers or by the action of toxic substances ○ *Grass cover provides some protection against soil erosion.*

COMMENT: Accelerated erosion is caused by human activity in addition to the natural rate of erosion. Cleared land in drought-stricken areas can produce dry soil which may blow away. Felling trees removes the roots which bind the soil particles together and so exposes the soil to erosion by rainwater. Ploughing up and down slopes as opposed to contour ploughing, can lead to the formation of rills and serious soil erosion.

erucic acid /ɪˌruːsɪk ˈæsɪd/ *noun* a fatty acid found in rape oil, which is linked to heart disease. Varieties of oilseed rape with low erucic acid content are considered the best.

erysipelas /ˌerɪˈsɪpələs/ *noun* an infectious disease mainly affecting pigs and also turkeys. In pigs, the symptoms are reddish inflammations on the skin and a high fever. It may cause infertility or abortion and manifests itself in three forms: acute, subacute and chronic. Also called **Diamonds disease**

erythromycin /ɪˌrɪθrəˈmaɪsɪn/ *noun* an antibiotic used to combat bacterial infections

ES *abbreviation* Environmental Stewardship

ESA *abbreviation* Environmentally Sensitive Area

Escherichia coli /ˌeʃərɪkiə ˈkəʊlaɪ/ *noun* a Gram-negative bacterium commonly found in faeces and associated with acute gastroenteritis if it enters the digestive systems of humans or animals

espalier /ɪˈspæliei/ *noun* **1.** a method of training a fruit tree, in which its branches are made to grow flat against a wall or other support. Compare **cordon 2.** a tree, especially apple or pear, trained in this way

COMMENT: From a vertical trunk pairs of branches are usually trained horizontally about 50cm apart.

esparto /ɪsˈpɑːtəʊ/ *noun* a species of grass which yields fibres used mainly in making paper. It originally came from North Africa and Southern Spain.

essence *noun* a concentrated oil extracted from a plant, used in food, cosmetics, analgesics and antiseptics ○ *vanilla essence*

essential amino acid *noun* an amino acid necessary for growth but which cannot be synthesised by monogastric animals and has to be obtained from the food supply (NOTE: The essential amino acids are: isoleucine, leucine, lysine, methionine, phenylalanine, threonine, tryptophan, valine, arginine and histidine.)

essential fatty acid *noun* an unsaturated fatty acid essential for growth but which cannot be synthesised by the body and has to be obtained from the food supply. Abbr **EFA** (NOTE: The two essential fatty acids are linoleic acid and linolenic acid.)

'Camelina is a fast growing drought tolerant break crop which is spring (or occasionally winter sown). A few hundred acres are grown yearly for both food and industrial use, and the oil contains a range of essential fatty acids.' [*Farmers Guardian*]

Essex Saddleback /ˌesɪks ˈsæd(ə)lbæk/ *noun* a breed of pig which has been bred with the Wessex Saddleback to form the British Saddleback

establish *verb* **1.** to work out or calculate something □ **to establish a position** to find out where something is **2.** to start or set up something ○ *We established routine procedures very quickly.* □ **to establish communication** to make contact □ **to establish control** to get control **3.** to settle or grow permanently ○ *The starling has become established in all parts of the USA.* ○ *Even established trees have been attacked by the disease.*

established *adjective* living or growing successfully

establishment *noun* **1.** the germination and emergence of seedlings ○ *There was a good crop establishment.* **2.** a period when a newly seeded sward is becoming established

establishment grant *noun* an amount of money given to farmers under the Energy Crops Scheme which covers some of the costs of planting energy crops

estate *noun* **1.** a rural property consisting of a large area of land and a big house **2.** a plantation

estate village *noun* a planned village built within an estate

estimated breeding value *noun* the value of an animal, calculated using an estimate of how many offspring it will have and what they will be worth. Abbr **EBV**

'From its foundation the Tyddewi herd has participated in the MLC beef recording scheme and has an estimated breeding value of 31 for sires and 20 for females – the highest figures in the scheme's directory.' [*Farmers Guardian*]

estimated transmitting ability *noun* the value of an animal, calculated using an estimate of how many offspring it will have and how much genetic material it will transfer to each one. It is equal to half the animal's estimated breeding value. Abbr **ETA**

ET *abbreviation* embryo transfer

ETA *abbreviation* estimated transmitting ability

ethene /ˈiːθiːn/ *noun* same as **ethylene**

ethical trading *noun* business practices which are socially responsible and protect the environment and the rights of workers

ethnobotany /ˈeθnəʊˌbɒtəni/ *noun* the study of the way plants are used by humans

ethology /iːˈθɒlədʒi/ *noun* the study of the behaviour of living organisms

ethylene /ˈeθəliːn/ *noun* a hydrocarbon occurring in natural gas and ripening fruits. It is used in the production of polythene and as an anaesthetic. Also called **ethene**

etiolation /ˌiːtiəˈleɪʃ(ə)n/ *noun* the process by which a green plant grown in insufficient light becomes yellow and grows long shoots

EU *abbreviation* European Union

eucalyptus /ˌjuːkəˈlɪptəs/ *noun* an Australian hardwood tree (*Eucalyptus* spp.) with strong-smelling resin. The trees are quick-growing and often used for afforestation, but are susceptible to fire.

euro *noun* a unit of currency adopted as legal tender in several European countries from January 1st, 1999

EUROP /ˈjʊərəp/ *noun* letters which make up the conformation classes in the Carcass Classification System

European Agricultural Guidance and Guarantee Fund *noun* a fund set up to cover the costs of administering the CAP, financed by the European Union budget. Abbr **EAGGF**

European Food Safety Authority *noun* a consultative body, funded by the European Community, which advises policymakers on health and food safety issues. Abbr **EFSA**

European foul brood *noun* a disease affecting bees that is caused by a bacterial parasite of the Streptococcacaea family that infests the larvae. Abbr **EFB**

European Union *noun* an alliance of 25 European countries, originally established with six members in 1957 by the Treaty of Rome. Among its powers are those for environmental and agricultural policy in its member states. Abbr **EU** (NOTE: Formerly called the **European Community** or **European Economic Community**.)

euthanasia *noun* the act of killing a sick animal in a humane way

eutherian /juːˈθɪəriən/ *noun* a mammal whose young develop within the womb attached to maternal tissues by a placenta. Subclass: Eutheria. Also called **placental mammal**

eutrophic /juːˈtrɒfɪk/ *adjective* referring to water which is high in dissolved mineral nutrients. ◊ **dystrophic**, **mesotrophic**, **oligotrophic**

eutrophication /ˌjuːtrɒfɪˈkeɪʃ(ə)n/, **eutrophy** /ˈjuːtrəfi/ *noun* the process by which water becomes full of phosphates and other mineral nutrients which encourage the growth of algae and kill other organisms

evaporate *verb* to change from being a liquid to being a vapour, or to change a liquid into a vapour ○ *In the heat of the day, water evaporates from the surface of the earth.* ○ *The sun evaporated all the water in the puddle.* Opposite **condense**

evaporated milk *noun* milk which has been made thick and rich by evaporating some of its water content

evaporation /ɪˌvæpəˈreɪʃ(ə)n/ *noun* the process of changing from a liquid into a vapour

evapotranspiration /ɪˌvæpəʊtrænspɪˈreɪʃ(ə)n/ *noun* the movement of water from soil through a plant until it is released into the atmosphere from leaf surfaces

evapotranspire /ɪˌvæpəʊˈtrænspaɪə/ *verb* to lose water into the atmosphere by evaporation and transpiration

evening primrose *noun* a biennial plant with hairy leaves and seeds that produce an oil which is used by the pharmaceutical industry

evergreen *adjective* referring to a plant which has leaves all year round ■ *noun* a tree or shrub which has leaves all year round (NOTE: Yew trees and holly are evergreens.) ▶ compare (all senses) **deciduous**

eviscerate /ɪˈvɪsəreɪt/ *verb* to remove the intestines and offal from a carcass

ewe *noun* an adult female sheep

ewe lamb *noun* a female lamb less than six months old

excavator /ˈekskəveɪtə/ *noun* a large machine for digging holes, as for laying drainage pipes

exceed *verb* to be more than expected, needed or allowed ○ *The concentration of radioactive material in the waste exceeded the government limits.* □ **it is dangerous to**

exceed the stated application rate do not apply more than the recommended amount

excess *noun* an amount or quantity greater than what is expected, needed or allowed ■ *adjective* more than is expected, needed or allowed

excessive *adjective* more than expected, needed or allowed ○ *Excessive ultraviolet radiation can cause skin cancer.*

exclosure /eksˈkləʊʒə/ *noun* an area fenced to prevent animals from entering. Compare **enclosure**

excrement *noun* faeces

excreta /ɪkˈskriːtə/ *plural noun* the waste material excreted from the body of an animal, e.g. faeces, urine, droppings or sweat

excrete *verb* to pass waste matter out of the body ○ *The urinary system separates waste liquids from the blood and excretes them as urine.* Compare **secrete**

excretion /ɪkˈskriːʃ(ə)n/ *noun* the passing of the waste products of metabolism such as faeces, urine, sweat or carbon dioxide out of the body. Compare **secretion**

excretion rate *noun* the rate at which a substance such as nitrogen is excreted by an animal

ex-farm *adverb* referring to a price for a product which does not include transport from the farm to the buyer's warehouse

exhausted fallow *noun* fallow land which is no longer fertile

exhaustive *adjective* complete and thorough ○ *an exhaustive reply to the safety concerns* ○ *an exhaustive search for the information*

existing chemicals *plural noun* the chemicals listed in the European Inventory of Existing Commercial Chemical Substances between January 1971 and September 1981, a total of over 100,000. Compare **new chemicals**

Exmoor Horn /ˈeksmɔː hɔːn/ *noun* a stock fat sheep, with a broad head, curled horns and dense fleece. Mainly found on Exmoor, the breed has been crossed with the Devon Longwool to create the Devon Closewool.

exocarp /ˈeksəʊkɑːp/ *noun* same as **epicarp**

exotic *adjective* referring to an organism or species that is not native and has been introduced from another place or region ■ *noun* an organism or species that is not native to its current environment ▶ also called (all senses) **alien**

exotoxin /ˌeksəʊˈtɒksɪn/ *noun* a poison produced by bacteria which affects parts of the body away from the place of infection

expander /ɪkˈspændə/ *noun* a device behind a mole which widens a drain

experimental farm *noun* a farm which is used to experiment with new farming techniques, rather than being run as a commercial enterprise

experimental horticulture station *noun* an experimental farm which specialises in plants, rather than livestock. Abbr **EHS**

experimental husbandry farm *noun* an experimental farm which specialises in livestock, rather than plants. Abbr **EHF**

exploit *verb* 1. to take advantage of something ○ *Ladybirds have exploited the sudden increase in the numbers of insects.* 2. to use a natural resource ○ *exploiting the natural wealth of the forest* 3. to treat something or someone unfairly for personal benefit

exploitation /ˌeksplɔɪˈteɪʃ(ə)n/ *noun* 1. the action of taking advantage of something 2. the utilisation of natural resources 3. the unfair use of something or treatment of someone for personal benefit

export /ˈekspɔːt/ *noun* produce or a crop which is sold to a foreign country ■ *verb* to send and sell crops or produce to foreign countries

export quotas *plural noun* limits set to the amount of a type of produce which can be exported

export refunds *plural noun* refunds made by the EU to farmers to compensate for a lower export price for produce

exposed *adjective* 1. referring to something or someone not covered or hidden 2. not protected from environmental effects ○ *left in an exposed position on the hillside*

exposure *noun* the harmful effect of having no protection from the weather ○ *suffering from exposure after spending a night in the snow*

extender /ɪkˈstendə/ *noun* a food additive which makes the food bigger or heavier without adding to its food value

extensification /ɪkˌstensɪfɪˈkeɪʃ(ə)n/ *noun* 1. the use of less intensive farming methods. Compare **intensification** 2. a payment made to farmers to encourage them to farm less intensively

COMMENT: Less intensive use of farming involves using fewer chemical fertilisers, leaving uncultivated areas at the edges of fields, reducing sizes of herds of cattle, etc. This allows lower yields from the same area of farmland, which is necessary if production levels are too high (as they are in the EU).

Extensification Payments Scheme *noun* until 2005, a system of payments made to farmers who received payments under the Beef Special Premium Scheme or Suckler Cow Premium Scheme and met specific stocking densities (NOTE: Now superseded by the Single Payment Scheme.)

extensification schemes *plural noun* pilot schemes for beef cattle and sheep which were begun in 1990 to offer compensation to farmers who reduced their beef output or the number of sheep by at least 20% and maintained this reduction over a 5 year period. The schemes were aimed at a less intensive use of land and reduction in use of pesticides and fertilisers.

extensive agriculture, **extensive farming** *noun* a way of farming which is characterised by a low level of inputs per unit of land. Compare **intensive agriculture**

extensive system *noun* a farming system which uses a large amount of land per unit of stock or output ○ *an extensive system of pig farming*

'The basic regime will persist when the new herd is established, as Mr Dugdale is convinced that low input, extensive systems are the best way to maximise returns.' [*Farmers Weekly*]

extract *verb* 1. to take something out of somewhere ○ *Vanilla essence is extracted from an orchid.* 2. to produce a substance from another ○ *Coconut oil is extracted from copra.*

extraction *noun* the action of producing a substance out of another ○ *the extraction of sugar from cane*

extraction rate *noun* the percentage of flour produced as a result of milling grain

extreme *adjective* referring to breeds of cattle which are traditionally kept for dairy or for meat, and not for a combination of the two

eye *noun* 1. a growth bud which has not developed, e.g. the bud of a potato tuber which develops shoots when the tuber is planted **2.** the instinctive action of a sheepdog when working sheep

eye-bright *noun* same as **ivy-leaved speedwell**

eyespot /ˈaɪspɒt/ *noun* a disease of cereals (*Cercosporella herpotrichoides*), which causes lesions to form on the stem surface and grey mould inside the stem. Compare **sharp eyespot**

F

F *symbol* Fahrenheit

F₁ *noun* (*in breeding experiments*) the first generation of offspring from a cross between two plants or animals

F₁ hybrid *noun* an animal or plant that is the result of a cross between two different plants or animals (NOTE: F₁ hybrids can be crossbred to produce F₂ hybrids and the process can be continued for many generations.)

FAC *abbreviation* Food Advisory Committee

factory farm *noun* a farm that uses intensive methods of rearing animals

factory farming *noun* a highly intensive method of rearing animals characterised by keeping large numbers of animals indoors in confined spaces and feeding them processed foods, with the use of drugs to control diseases

faecal /ˈfiːk(ə)l/ *adjective* referring to faeces (NOTE: The US spelling is **fecal**.)

faeces *plural noun* solid waste matter passed from the bowels of a human or other animal after food has been eaten and digested (NOTE: The US spelling is **feces**.)

Fagopyrum /ˌfægəʊˈpaɪrəm/ *noun* the Latin name for buckwheat

Fagus /ˈfeɪgəs/ *noun* the Latin name for beech

Fahrenheit *noun* a scale of temperatures on which the freezing and boiling points of water are 32° and 212°, respectively. Compare **Celsius**. Symbol **F** (NOTE: The Fahrenheit scale is still commonly used in the USA.)

COMMENT: To convert Fahrenheit temperatures to Celsius, subtract 32, multiply by 5 and divide by 9. So 68°F is equal to 20°C. As a quick rough estimate, subtract 30 and divide by two.

fair *noun* a regular meeting for the sale of goods or animals, often with sideshows and other entertainments. Fairs can be specialised, e.g. horse fairs or cheese fairs, or can cover all types of farm animals.

fair average quality *noun* the average quality of agricultural produce based on samples taken from bulk. Abbr **FAQ**

fair trade *noun* an international system where food companies agree to pay producers in developing countries a fair price for their products

'The scheme is the first initiative to apply the principles of fair trade to the UK market for food grown by British farmers. It offers an open trading structure where the prices paid to producers must meet their costs of production, provide them with a profit, and be guaranteed by a written contract.' [*Farmers Weekly*]

fairy ring *noun* a circle of darker coloured grass in a pasture, which is caused by fungi

fallen stock *noun* dead animals

falling time *noun* the time taken for wheat grain to fall to the bottom of a container of water, measured by the Hagberg test

fallopian tube /fəˈləʊpiən ˌtjuːb/ *noun* in mammals, a tube that conveys eggs from an ovary to the womb

fallow /ˈfæləʊ/ *noun* a period when land is not being used for growing crops for a period so that nutrients can build up again in the soil or to control weeds ○ *Shifting cultivation is characterised by short cropping periods and long fallows.* ■ *adjective* referring to land that is not being used for growing crops for a period □ **to let land lie fallow** to allow land to remain without being cultivated for a period

fallow crop *noun* a crop grown in widely spaced rows, so that it is possible to hoe and cultivate between the rows

fallow cultivation *noun* a type of cultivation in which the period under crops is increased and the length of the fallow is reduced

fallowing /'fæləʊɪŋ/ *noun* the process of allowing land to lie fallow for a period

fallow length *noun* the period of time between cultivation periods. As population density increases and land becomes scarce, food supply can be increased by making the period under crops longer and the length of fallow shorter.

false seedbed *noun* a seedbed prepared to allow weed seeds to germinate. These are then killed by cultivation before sowing root crops.

false staggers *noun* a disease of sheep caused by maggots which cause inflammation of nostrils and head, making the sheep appear dazed

family *noun* a group of genera which have some characteristics in common ○ *the plant family Orchidaceae* ○ *Tigers and leopards are members of the cat family.* (NOTE: Scientific names of families of animals end in **-idae** and those of families of plants end in **-ae**.)

family farm *noun* a farm unit which is owned and operated by one family

famine *noun* a period of severe shortage of food ○ *When the monsoon failed for the second year, the threat of famine became more likely.*

fancy breed *noun* a breed reared for decoration or show, rather than for produce

FAO *abbreviation* Food and Agriculture Organization

FAQ *abbreviation* frequently asked questions

farm *noun* **1.** an area of land used for growing crops and keeping animals to provide food and the buildings associated with it **2.** an area of land or water where particular animals or crops are raised commercially ○ *a fish farm* ○ *a butterfly farm* ■ *verb* **1.** to run a farm **2.** to grow crops and keep animals on a particular piece or size of land ○ *He farms 100 acres in Devon.* **3.** to raise a particular animal or crop commercially ○ *He intends to farm salmon and sea bass.*

FARMA *abbreviation* National Farmer's Retail and Markets Association

Farm and Wildlife Advisory Group *noun* an organisation in the UK which advises farmers on environmental and conservation issues. Abbr **FWAG**

Farm Animal Welfare Council *noun* an agency established by the British government in 1979 to review the welfare of farm animals on agricultural land, at markets, in transit and at the place of slaughter. Abbr **FAWC**

farm assurance *noun* a scheme whereby specific criteria are applied in order to guarantee quality control for farm produce

farm assured *adjective* produced in accordance with the quality standards of a farm assurance scheme

Farm Assured *noun* a UK government scheme to inform the public that farm produce is of good quality according to a set of standards. It is symbolised by a little red tractor on the packaging.

Farm Business Survey *noun* a survey carried out by Defra on the financial performance of different types of farms. Abbr **FBS**

farm consolidation /,fɑːm kənˌsɒlɪ'deɪʃ(ə)n/ *noun* the process of joining small plots of land together to form larger farms or bringing scattered units together to form large fields

farmed *adjective* **1.** grown or produced commercially and not in the wild ○ *farmed salmon* **2.** referring to meat from fish grown in a fish farm or from animals kept on a farm ○ *farmed venison*

farmed deer *plural noun* deer which are raised on deer farms on land enclosed by deer-proof barriers, for meat or skins or other by-products, or as breeding stock

Farm Environment Plan *noun* a survey carried out of all of a farm's environmental features, such as its natural resources, landscape, resident wildlife and points of access. Abbr **FEP**

'Environmental Stewardship replaces existing conservation schemes next year and to join the highest tier, which attracts additional payments, applicants must complete a Farm Environment Plan (FEP) prior to submitting a claim. An FEP is a whole-farm assessment of all the environmental features on the farm including biodiversity, historic, woodland and landscape features.' [*Farmers Guardian*]

Farm Environment Record *noun* a basic plan of a farm describing its landscape and main features. It is more simple than a Farm Environment Plan. Abbr **FER**

farmer-controlled business *noun* a farm which is owned and controlled by the farmer who also manages the land

'With traditional grain marketing businesses under increasing financial pressure from declining tonnages and margins, the future belongs to well managed farmer controlled businesses which adopt a proactive, long-term approach to marketing and return value to members through innovative contracts.' [*Farmers Guardian*]

farmers' list *noun* a list of veterinary medicines which can be obtained from agricultural merchants with a prescription from a veterinary surgeon

farmer's lung *noun* a type of asthma caused by an allergy to rotting hay

farm fragmentation /ˌfɑːm ˌfrægmən ˈteɪʃ(ə)n/ *noun* a situation where the fields of a farm are scattered over an area, so that the holding is not made up of a single unit of land. It can be the result of inheritance practices where the land of a person who has died is split between all the children, of land reclamation schemes where the land is reclaimed piece by piece, or because old open-field systems have been kept.

farm fresh eggs *plural noun* term used in the EU to describe Class A eggs

farm gate prices *plural noun* prices which a farmer receives for his or her produce

farmhand /ˈfɑːmhænd/ *noun* a person who works on a farm

farm health planning *noun* an official set of guidelines from Defra for preventing, managing and treating diseases in farm animals

'Prevention was better than cure, making farm health planning crucial, said DEFRA's chief vet Debby Reynolds at the strategy's launch last week. "The UK's previous policy of fire-fighting disease outbreaks is no longer an option", she warned.' [*Farmers Weekly*]

farming *noun* running a farm, including activities such as keeping animals for sale or for their products and growing crops

farming community *noun* a group of families living near to each other and having farming as their main source of income

farming systems *plural noun* different types of farming and methods of cultivation, e.g. shifting cultivation systems, ley systems, systems with permanent upland cultivation, fallow systems, grazing systems and systems with perennial crops

farmland *noun* land on a farm which is used for growing crops or rearing animals for food

farmland bird *noun* a bird that nests in an agricultural environment. Many are declining in numbers because of changes in agricultural practices.

farmland bird indicator *noun* a standard way of measuring the frequency of birds found in agricultural areas

'The farmland bird indicator is calculated on the breeding populations of 20 species (including skylark, grey partridge and goldfinch), which have collectively declined by 40% since the mid-1970s. (Delivering the evidence. Defra's Science and Innovation Strategy 2003–06)'

farm manager *noun* a person who runs a farm on behalf of the owner

farm produce *noun* food such as fruit, vegetables, meat, milk and butter, which is produced on a farm

farm rent *noun* rent paid by a tenant farmer to a landlord on a regular basis for the use of the farm holding

farm-saved seed *noun* seed kept from the previous year's harvest and replanted on the same farm. Also called **home-saved seed**

farm-scale *adjective* relating to trials or evaluations carried out on farms using regular farming practices rather than on small experimental plots

farmscape /ˈfɑːmskeɪp/ *noun* a landscape dominated by agriculture. Farmland is the main element in farmscape, though non-agricultural uses may be included.

Farm Service Agency *noun* a government agency which provides information and support for farmers in the United States, part of the USDA. Abbr **FSA**

farmstead /ˈfɑːmsted/ *noun* a farmhouse and the farm buildings around it

Farm Support Scheme *noun* Article 39 of the Treaty of Rome provides the framework of the Common Agricultural Policy. Each member state contributes to the European Agricultural Guarantee and Guidance Fund. Payments are made for structural changes under the guidance fund and much larger payments under the guarantee section.

farm to fork *noun* the chain of food supply, from the farm where it is produced to the consumer

farm trail *noun* a walking trail around a farm which is open to the public

Farm Watch Scheme *noun* a scheme which organises networks of farmers to be in touch with one another and the local police to report suspicious activity

farmworker /'fɑːmwɜːkə/ *noun* a person who works on a farm

farmyard *noun* the area around farm buildings

farmyard manure *noun* manure formed of cattle excreta mixed with straw, used as a fertiliser. Abbr **FYM**

farmyard manure spreader *noun* a machine for spreading manure, basically a trailer with a moving floor conveyor and a combined shredding and spreading mechanism which distributes the material: there are two types, the rear ejector and the side delivery spreader

Farrand test /'færənd test/ *noun* a method for determining the alpha amylase content of milling wheat. The amount of alpha amylase enzyme present in wheat is important for making bread. Excessive alpha amylase in flour results in poorer loaves.

farrier /'færiə/ *noun* a person who makes and fits shoes for horses. ◊ **blacksmith**

farrowing /'færəʊɪŋ/ *noun* the act of giving birth to piglets

farrowing crate *noun* a steel frame which holds the sow during farrowing and helps to prevent the overlying of the piglets

farrowing fever *noun* a disease of pigs caused by inflammation of the womb. Pigs suffer high temperatures and loss of appetite. Also called **MMA**

farrowing rails *plural noun* rails which prevent the sow from overlying the piglets

fasciation /ˌfeɪʃiˈeɪʃ(ə)n/ *noun* **1.** an abnormal plant growth in which several stems become fused together **2.** the production of several shoots from the crown of a plant such as a pineapple

fascioliasis /fəˌsiəˈlaɪəsɪs/ *noun* a disease caused by an infestation of parasitic liver flukes

fat *noun* **1.** a white oily substance in the body of mammals, which stores energy and protects the body against cold (NOTE:

Fat has no plural when it means the substance in the body of mammals; the plural **fats** is used to mean different types of fat.) **2.** a type of food which supplies protein and Vitamins A and D, especially that part of meat which is white and solid substances (like lard or butter) produced from animals and used for cooking or liquid substances like oil ■ *adjective* referring to an animal which has been reared for meat production and which has reached the correct standard for sale in a market

fat class *noun* the amount of external fat present on a beef or sheep carcass, classified from lean (class 1) to very fat (class 5)

'Although lambs by the high-index sire had lighter carcasses, a higher proportion were within the 21kg weight limit. Many also earned a bonus for fat class and conformation, with most of them grading U for conformation.' [*Farmers Weekly*]

fat hen *noun* a weed (*Chenopodium album*) which affects spring cereals, peas and row crops. Fat hen is found especially in rich soils and muck heaps. Also called **dung weed**, **muck weed**

fatness /'fætnəs/ *noun* the amount of fat on an animal

fatstock /'fætstɒk/ *noun* livestock which has been fattened for meat production

fatten *verb* to give animals food so as to prepare them for slaughter ○ *He buys lambs for fattening and then sells them for meat.*

fatty *adjective* containing a lot of fat ○ *Fatty foods contribute to the risk of obesity.*

fatty liver, fatty liver syndrome *noun* a condition in older cows, where the animal absorbs calcium too slowly and the liver is affected. Goats are also affected.

faults *plural noun* the proportion of rotten or diseased items in a quantity of produce sold

fauna *noun* the wild animals and birds which live naturally in a specific area. Compare **flora**

Faverolle /'fævərɒl/ *noun* a breed of poultry for table consumption

FAWC *abbreviation* Farm Animal Welfare Council

fawn *noun* a young deer

FBS *abbreviation* Farm Business Survey

FCE *abbreviation* feed conversion efficiency

FCR *abbreviation* feed conversion rate

Fe *symbol* iron

feather *noun* an outgrowth of the epidermis on a bird's body and wings providing insulation. The feathers on the wings and those forming the tail are important in flight. ◊ **feathers**

feathereating /ˈfeðərˌiːtɪŋ/, **feather-pecking** /ˈfeðəˌpekɪŋ/, **featherpulling** /ˈfeðəˌpʊlɪŋ/ *noun* the pulling of the feathers of a bird by another bird

COMMENT: Feathereating occurs in poultry where the birds are penned up and lack exercise and facilities for scratching. It may also be caused by wrong nutrition. Once the birds start pulling feathers, an outbreak of cannibalism may occur.

feathers *plural noun* the fringes of hair that some dogs and horses have on their legs or tails

febrile /ˈfiːbraɪl/ *adjective* referring to a fever

febrile disease *noun* a disease such as Newcastle disease which is accompanied by a fever

feces /ˈfiːsiːz/ *plural noun* US spelling of **faeces**

fecundity /fɪˈkʌndəti/ *noun* **1.** the fertility of a plant or animal **2.** a measurement of the number of offspring born and reared by a dam

fee *noun* money paid to a professional for a service

feed *verb* **1.** to take food ○ *The herd feeds here at dusk.* **2.** to give food to a person or an animal **3.** to provide fertiliser for plants or soil **4.** to supply or add to something ○ *Several small streams feed into the river* ■ *noun* **1.** food given to animals and birds ○ *Traces of pesticide were found in the cattle feed.* **2.** fertiliser for plants or soil ○ *Tomato plants need liquid feed twice a week at this time of year.*

feed additive *noun* a supplement added to the feed of farm livestock, particularly pigs and poultry, to promote growth, e.g. an antibiotic or hormone

'Weight for weight chlorides are four times more efficient than sulphates at acidifying cows urine. Recently, feed additives based on chloride (hydrochloric acid mixed with soybean meal) have been available with much improved palatability.' [*Dairy Farmer*]

feed block *noun* a block of foodstuff left out in the pasture, especially on hill farms, used by sheep to prevent loss of condition

feed compounds *plural noun* a number of different ingredients including major minerals, trace elements and vitamins, mixed and blended to provide properly balanced diets for stock

feed concentrate *noun* an animal feed which has a high food value relative to volume

feed conversion efficiency, feed conversion rate *noun* the number of kilograms of feed required to produce a kilogram of weight gain in an animal such as a pig. Abbr **FCE, FCR**

feeder /ˈfiːdə/ *noun* **1.** something which supplies or adds to something else of the same type **2.** a container from which livestock are fed ■ *adjective* referring to livestock which are being fed to be slaughtered

feed grain *noun* a cereal which is fed to animals and birds, e.g. wheat or maize

feeding *noun* the action of giving animals food to eat (NOTE: Animal welfare codes lay down rules on when animals and birds should be fed, especially in relation to when they are transported or slaughtered.)

feeding face *noun* the area allowed to each animal to feed from under controlled conditions. Each cow needs 150mm of feeding face, or less if continuous access is provided. This refers to the self-feeding method of silage.

feedingstuff /ˈfiːdɪŋstʌf/ *noun* same as **feedstuff**

feeding value *noun* the nutritional value of feedingstuffs

feed intake *noun* the amount of food eaten by an animal

feedlot /ˈfiːdlɒt/ *noun* an area of land where livestock are kept at a high density, with small pens in which the animals are fattened. All feed is brought into the feedlot from outside sources. ◊ **chillshelter**

COMMENT: A new type of feedlot is an area of land surrounded by an earth embankment, which protects the cattle from cold winds while they are being fed intensively.

feedmill /ˈfiːdmɪl/ *noun* a mill for preparing animal feed

feed off *noun* the practice of allowing stock to feed on a crop while it is still in the ground

feed passage *noun* **1.** the rate at which feed pass through an animal's digestive system **2.** the area in a livestock shed where the feed is placed so that the animals can access it

feed preparation *noun* the milling and crushing of grain, mixing of the ingredients and making into cubes or pellets

feed ratio *noun* the ratio showing the price of an animal sold on the market against the cost of feeding it

feed refusal *noun* the amount of allotted feed that a farm animal does not eat. This should be monitored as it can give an early indication that the animal is unwell.

feed ring *noun* a circular container for forage, from which livestock can feed

feed stance *noun* an open stall in a building where animals can feed

feedstuff /'fiːdstʌf/ *noun* food for farm animals. Also called **feedingstuff**

feed wheat *noun* wheat which is used as an animal feed, and not for human consumption

'Cereal variety choice for whole crop has more often than not been dictated by price and availability as there has been an assumption that one feed wheat or barley variety is pretty much like any other.'
[*Dairy Farmer*]

Feeke's large scale /ˌfiːks 'lɑːdʒ ˌskeɪl/ *noun* a method of determining the growth stage of a crop relying on comparing plant size and leaf arrangement when the plant is young. It is not a very reliable method and has been replaced by Zadoks scale.

fell *noun* a high area of open moorland in the north of England ■ *verb* to cut down a tree

felling licence *noun* permission from the Forestry Commission to fell trees. There are some exceptions to this: e.g. trees in gardens and fruit trees in orchards do not need a licence to be felled.

female *adjective* **1.** referring to an animal that produces ova and bears young **2.** referring to a flower which has carpels but not stamens, or a plant that produces such flowers

fen /fen/ *noun* an area of flat marshy land, with plants such as reeds and mosses growing in alkaline water

fenbendazole /fen'bendəzəʊl/ *noun* a medicinal substance used to worm cattle

fence *noun* a barrier put round a field, either to mark the boundary or to prevent animals entering or leaving ■ *verb* to put a fence round an area of land

COMMENT: Various methods are used to fence field boundaries, most commonly woven wire and wooden fence posts. Movable electric fences are an efficient way of limiting areas of a field for grazing purposes.

fenland /'fenlænd/ *noun* a large area of flat marshy land with alkaline water

fenland rotation *noun* a system of crop rotation developed on the Fens of East Anglia, using potatoes, sugar beet and wheat in rotation

fennel *noun* an aromatic herb (*Foeniculum vulgare*) of Mediterranean origin, used to flavour fish dishes and soups

FEP *abbreviation* Farm Environment Plan

FEPA /'fepə/ *abbreviation* Food and Environmental Protection Act

FER *abbreviation* Farm Environment Record

ferment *verb* to produce alcohol and heat under the effect of yeast

fermentation /ˌfɜːmen'teɪʃ(ə)n/ *noun* the process whereby carbohydrates are broken down by enzymes from yeast and produce heat and alcohol. In making silage, fermentation is essentially the breaking down of carbohydrates and proteins by aerobic and then anaerobic bacteria.

fertile *adjective* **1.** referring to an animal or plant that is able to produce offspring by sexual reproduction. Opposite **sterile 2.** referring to soil with a high concentration of nutrients that is able to produce good crops

fertilisation /ˌfɜːtɪlaɪ'zeɪʃ(ə)n/, **fertilization** *noun* the joining of an ovum and a sperm to form a zygote and so start the development of an embryo

fertilise /'fɜːtəlaɪz/, **fertilize** *verb* **1.** (*of a sperm*) to join with an ovum **2.** (*of a male*) to make a female pregnant **3.** to put fertiliser on crops or soil

fertiliser /'fɜːtəlaɪzə/, **fertilizer** *noun* a chemical or natural substance spread and mixed with soil to stimulate plant growth

COMMENT: Organic materials used as fertilisers include manure, slurry, rotted vegetable waste, bonemeal, fishmeal and seaweed. Inorganic fertilisers such as powdered lime or sulphur are also used. In commercial agriculture, artificially prepared fertilisers (manufactured compounds containing nitrogen, potassium and other chemicals) are most often used, but if excessive use of them is made, all the chemicals are not taken up by plants and the excess is leached out of the soil into ground water or rivers where it may cause algal bloom.

fertiliser distributor *noun* a machine used to spread fertiliser

fertility *noun* **1.** the state of being fertile **2.** the proportion of eggs which develop into young **3.** a measure of the ability of a female to conceive and produce young or of the male to fertilise the female (NOTE: The opposite is **sterility.**)

fescue /ˈfeskjuː/ *noun* a common name for about 100 species of grasses, including many valuable pasture and fodder species

fetlock /ˈfetlɒk/ *noun* the thicker back part of a horse's leg near the hoof (NOTE: The thin part between the fetlock and the hoof is called the **pastern.**)

fetus /ˈfiːtəs/ *noun* another spelling of **foetus**

FFB *abbreviation* Food From Britain

fibre *noun* **1.** a long narrow plant cell with thickened walls that forms part of a plant's supporting tissue, especially in the outer region of a stem **2.** a hair, e.g. of a rabbit or goat

fibrous /ˈfaɪbrəs/ *adjective* made of a mass of fibres

fibrous rooted plants *plural noun* plants with roots which are masses of tiny threads, with no major roots like tap roots

field *noun* **1.** an area of land, usually surrounded by a fence or hedge, used for growing crops or for pasture **2.** an area of interest or activity ○ *He specialises in the field of environmental health.*

COMMENT: Some older types of field can still be seen in the UK. **Celtic fields** are small rectangular fields, mainly on chalk uplands, surrounded by banks and walls, dating back to as early as the 6th century B. C. **Open fields** with large furrows were used by the Saxons who had heavier ploughs and divided the fields into strips, separated by banks of turf or walls of stone. **Enclosure fields** are rectangular fields with regular hedgerows, formed after the enclosures of the 16th to 18th centuries. In recent years the removal of many field boundaries in the interest of farm consolidation has led to an increase in the size of the average British field. Hedges have been removed to allow large farm machinery to be used more economically, and the loss of hedgerows has had a marked effect on the wildlife in the countryside.

field beans *plural noun* used for stock feeding, or for producing broad beans, which are the immature seeds used for human consumption. Field beans are usually grown as a break crop. ◊ **broad bean**

field bindweed *noun* a deep-rooted perennial weed (*Convolvulus arvensis*) which causes great problems because of its mass of clinging growths

field book *noun* an annual record of field utilisation and other operations, kept on a farm

field capacity *noun* the maximum possible amount of water remaining in the soil after excess water has drained away

field crop *noun* a crop grown over a wide area, e.g. most agricultural crops and some market-garden crops

field drainage *noun* building drains in or under fields to remove surplus water

field-grown *noun* referring to a crop which is grown in a field as opposed to in a greenhouse

field observation *noun* an examination made in the open air, looking at organisms in their natural habitat, as opposed to in a laboratory

field pansy *noun* a widespread flower (*Viola arvensis*) increasingly found in winter crops, especially cereals. Also called **corn pansy**, **love-in-idleness**, **cats' faces**

field trial *noun* a trial that tests the ability of a crop variety to perform under normal cultivation conditions

fig *noun* a tree (*Ficus* spp) with soft sweet fruit with many small seeds, grown mainly in Mediterranean countries

filament /ˈfɪləmənt/ *noun* the stalk of a stamen, carrying an anther

filbert /ˈfɪlbət/ *noun* the Kentish cob, a commercially grown hazel-like nut (*Corylus maxima*)

fill-belly *noun* feed which fills the animal's stomach, without providing any useful nutrients

filly /ˈfɪli/ *noun* a young female horse (NOTE: A **filly** is so called for a year or so, after which she becomes a 'mare'.)

fine grains *plural noun* high quality grains such those of a wheat and rice. Compare **coarse grains**

fineness count /ˈfaɪnnəs ˌkaʊnt/ *noun* a scale used to assess the fineness of wool fibres

fine wool *noun* wool of very good quality

finger and toe *noun* same as **club root**

fingers *plural noun* pieces of metal which project from and are attached to the cutter bar of a binder, mower or harvester. The knives pass across the fingers, cutting the grass or cereal crop as they do so.

finger wheel swath turner *noun* a machine used for raking

finish *verb* to feed cattle or sheep at a rate of growth which increases the ratio of muscle to bone, and increases the proportion of fatty tissue in the carcass to a level at which the animal is considered to be fit for slaughter ■ *noun* the furrow left at the edge of a ploughed field

finishing *noun* the action of feeding cattle or sheep until they are ready for slaughter

finishing ration *noun* feed given to animals to prepare them for slaughter ○ *A finishing ration includes silage, beet pulp and by-products such as outsize carrots.*

Finncattle /fɪnˈkæt(ə)l/ *noun* a breed of dairy cattle derived from three Finnish breeds. The animals are medium-sized and brown.

Finnish Ayrshire /ˌfɪnɪʃ ˈeəʃə/ *noun* a breed of cattle found in northern Finland, which is similar to the Ayrshire and mainly reared for milk

fir *noun* a common evergreen coniferous tree. Genus: *Abies*.

fir cone *noun* a hard oval or round structure on a fir tree containing the seeds (NOTE: The term is sometimes applied to the cones of other trees such as pines.)

fireblight /ˈfaɪəblaɪt/ *noun* a disease of apples and pears, which is characterised by dead flowers and branches. It is caused by the bacterium *Erwinia amylovora.*

firebreak /ˈfaɪəbreɪk/ *noun* an area kept free of vegetation, so that a fire cannot pass across and spread to other parts of the forest or heath

firm *adjective* solid ○ *The soil is firm and not too crumbly.* ■ *verb* to make cultivated soil solid, before sowing ○ *A roller is used to firm the soil.*

first calf heifer *noun* a heifer which has borne its first calf

first calver *noun* a cow which has produced its first calf

fish *noun* a cold-blooded vertebrate that lives in water (NOTE: Some species are eaten for food. Fish are high in protein, phosphorus, iodine and vitamins A and D. White fish have very little oil.)

fishery *noun* **1.** the commercial activity of catching fish **2.** an area of sea where fish are caught ○ *The boats go each year to the rich fisheries off the north coast.*

fish farm *noun* a place where edible fish are bred or reared in special pools for sale as food

fish farming *noun* the commercial activity of keeping fish in ponds or fenced areas of the sea for sale as food. Also called **aquaculture, aquafarming, aquiculture**

fishmeal /ˈfɪʃmiːl/ *noun* a powder of dried fish, used as an animal feed or as a fertiliser

FISS *abbreviation* Food Industry Sustainability Strategy

five freedoms, the *plural noun* a set of guidelines which should be considered when looking after the welfare of farm animals (NOTE: The five freedoms are: freedom from hunger or thirst; freedom from discomfort; freedom from pain, injury or disease; freedom to express normal behaviour; and freedom from fear and distress.)

'Over the past 18 months, Dairy Crest has introduced its own farm assured milk scheme, encompassing the five freedoms of animal welfare in a code of practice.' [*Farmers Guardian*]

fixation *noun* the act of fixing something. ◊ **nitrogen fixation**

fixed costs *plural noun* costs such as rent which do not increase with the quantity of a product produced

flail *noun* **1.** a wooden hand tool used for beating grain to separate the seeds from the waste parts **2.** a type of hedgecutter, with rapidly turning cutting arms ■ *verb* to beat grain with a flail

flail forage harvester *noun* a type of forage harvester which uses a high-speed flail rotor. The cut crop passes through a vertical chute and is discharged into a trailer.

flail mower *noun* a mowing machine with a high-speed rotor fitted with swinging flails which cut the grass and leave it bruised in a fluffy swath

flaked maize /ˌfleɪkd ˈmeɪz/ *noun* a type of animal feedingstuff made from maize which has been treated with steam, rolled and dried. It is highly digestible, rich in starch and often given to pigs.

flat deck piggery *noun* a piggery used for rearing weaned piglets from between two and eight weeks of age. It has a mesh floor, self-feed hoppers and controlled heating and ventilation.

flat rate feeding *noun* a system of feeding concentrates to dairy cows, involving few changes to the level of concentrate input. It lasts from calving to turnout.

flatten, flatten out *verb* **1.** to become flat or make something flat **2.** to make plants lie flat on the ground ○ *The stems are flattened by a roller-crusher.* ○ *Harvesting is difficult after the entire crop has been flattened by rain.*

flatworm /ˈflætwɜːm/ *noun* a worm with a flat body, a single gut opening and no circulatory system. Phylum: Platyhelminthes. (NOTE: Flatworms include both free-living species and parasites such as flukes and tapeworms.)

flax /flæks/ *noun* the linseed plant

flaxseed /ˈflækssiːd/ *noun* seed from the flax plant, crushed to produce linseed oil

flea *noun* a small jumping insect which lives as a parasite on animals, sucking their blood and possibly spreading disease. Order: Siphonaptera. (NOTE: Historically, bubonic plague was spread by fleas.)

flea beetle *noun* a small dark beetle which causes damage to Brassica seedlings, especially during hot dry weather between April and mid-May

fleece *noun* a coat of wool covering a sheep or goat ■ *verb* to shear a sheep, to cut the fleece off an animal

COMMENT: The fleece is both the wool growing on the animal and the wool which has been cut off, usually in one piece, when shearing.

fleeced *adjective* covered with a coat of wool

flex-fuel *adjective* referring to a vehicle which is designed to run on petrol, an alcohol-based fuel such as ethanol or any combination of the two

flies plural of **fly**

flight feathers *plural noun* the main feathers on a bird's wing, properly called the 'primaries'

flightless bird *noun* a bird which has small wings and cannot fly, e.g. an ostrich or a penguin

flights *plural noun* same as **flight feathers**

flint *noun* a hard stone of nearly pure silica, found in lumps in chalk soils

COMMENT: Sharp-edged flints of various sizes found in some calcareous soils are very wearing on farm implements and tractor tyres; they can also cause damage to machinery if they are picked up during harvesting. Flints are very durable and are used to form a weather-resistant facing to farm buildings and walls.

flixweed /ˈflɪkswiːd/ *noun* a common annual weed *(Descurainia sophia)*

float *noun* same as **Dutch harrow**

flocculation /ˌflɒkjʊˈleɪʃ(ə)n/ *noun* the grouping of small particles of soil together to form larger ones ○ *The flocculation of particles is very important in making clay soils easy to work.*

flock *noun* a large group of birds or some farm animals such as sheep and goats ○ *a flock of geese* ○ *a flock of sheep* (NOTE: The word used for a group of cattle, deer or pigs is **herd**.)

flock book *noun* a record of the pedigree of a particular breed of sheep or goat, kept by the breed society

flockmaster /ˈflɒkmɑːstə/ *noun* a farm worker in charge of a flock of sheep or goats

flock mating *noun* a mating system which uses several males to mate with the females of a flock

flood *noun* a large amount of water covering land that is usually dry, caused by phenomena such as melting snow, heavy rain, high tides or storms ■ *verb* to cover dry land with a large amount of water ○ *The river bursts its banks and floods the whole valley twice a year in the rainy season.*

flood irrigation *noun* a method of irrigation using water brought down by a river in flood. The floodwaters are led off into specially prepared basins. ◊ **basin irrigation**

flood plain *noun* a wide flat part of the bottom of a valley which is usually covered with water when the river floods

floodwater /ˈflʌdwɔːtə/ *noun* water that spreads uncontrolled onto land that is usually dry ○ *After the floodwater receded the centre of the town was left buried in mud.*

flora *noun* the wild plants that grow naturally in a specific area. Compare **fauna**

floral *adjective* referring to plants or flowers

floret /'flɒrət/ *noun* a little flower that forms part of a larger flower head

flour *noun* a soft fine powder made from ground cereal grains, and used for making bread

COMMENT: Flour is made by grinding grain, and removing impurities. White flour is flour which has had all bran and germ removed: the best quality of white flour is patent flour, which is very fine. Self-raising flour is white flour with baking powder added. Brown flour is not refined, and still contains the wheat germ and parts of the bran: if it contains all the grain, it is called 'wholemeal' and if it contains most of the grain it is 'wheatmeal'. Some expensive flours are 'stoneground', that is they are made in the traditional way with millstones.

flour corn *noun* a variety of maize with large soft grains and friable endosperm, making it easy to grind to flour

flourish *verb* to live or grow well and increase in numbers ○ *The colony of rabbits flourished in the absence of any predators.*

floury /'flaʊri/ *adjective* soft and powdery, like flour

floury potatoes *plural noun* varieties of potato which, when cooked, turn easily into flour

flower *noun* the reproductive part of a seed-bearing plant (NOTE: Some flowers are brightly coloured to attract pollinating insects and birds and usually consist of protective sepals and bright petals surrounding the stamens and stigma. Many are cultivated for their colour and perfume.)

flowers of sulphur *noun* powdered sulphur, which is used to dust on plants to prevent mildew

fluke *noun* a parasitic flatworm (NOTE: Flukes may settle inside the liver, bloodstream or in other parts of the body.)

fluoride *noun* a chemical compound of fluorine, usually with sodium, potassium or tin

COMMENT: Fluorides such as hydrogen fluoride are emitted as pollutants from certain industrial processes and can affect plants, especially citrus fruit, by reducing chlorophyll. They also affect cattle by reducing milk yields. On the other hand, sodium fluoride will reduce decay in teeth and is often added to drinking water or to toothpaste. In some areas, the water contains fluoride naturally and here fluoridation is not carried out. Some people object to fluoridation, although tests have proved that instances of dental decay are fewer in areas where fluoride is present in drinking water.

fluorosis /flɔː'rəʊsɪs/ *noun* a condition caused by excessive fluoride in drinking water or food. It causes discoloration of the teeth and affects the milk yields of cattle. (NOTE: It causes discoloration of teeth and affects the milk yields of cattle.)

flush *noun* a rapid growth of grass

flushing ewes *plural noun* ewes brought into good condition prior to breeding, usually by improving their diet

fly *noun* a general term for a small insect with two wings, of the order *Diptera*. Some flies cause diseases of plants (the frit fly) and some harm animals (the gadfly).

fly blown *adjective* referring to a fleece laden with maggot-fly eggs

flying flock *noun* a flock of sheep imported onto a farm for a time, normally for less than a year, and then sold

fly strike *noun* a serious condition caused by maggots breeding on the animal's hindquarters. It can quickly cause death from shock.

fly-tipping *noun* the dumping of rubbish somewhere other than at an official site

'A quarter of all farms in England and Wales have experienced fly-tipping on their land in recent years. The total quantity of waste tipped in 2001 was estimated at 600000 tons. (Agricultural Waste: Opportunities for Change. Information from the Agricultural Waste Stakeholders' Forum)'

FMA *abbreviation* Fertiliser Manufacturers Association

FMD *abbreviation* foot-and-mouth disease

foal *noun* a young horse of either sex in its first year ○ **mare in-foal** pregnant mare ■ *verb* to produce an offspring

foaling *noun* the act of giving birth to a foal

fodder, fodder crop *noun* plant material or a crop which is grown to give to animals as food, e.g. grass or clover ○ *winter fodder*

fodder beet *noun* a root crop bred from sugar beet and mangolds, usually grown after cereals and used to feed stock

'Cattle are in-wintered on a diet based on fodder beet, maize silage, and distillers' dark grains. Big bale straw is dumped in the pens, the cattle taking what they need for feed and spreading the rest.' [*Farmers Guardian*]

fodder radish *noun* a type of brassica grown primarily for use as a green fodder crop

fodder storage *noun* the storing of fodder for use in winter. Buildings used for storing fodder may be of the simple Dutch barn type, and can be built cheaply using poles and a galvanised iron roof.

FoE *abbreviation* Friends of the Earth

foetal /ˈfiːt(ə)l/ *adjective* referring to a foetus

foetus *noun* an unborn animal in the womb at the stage when all the structural features are visible, i.e. after eight weeks in humans (NOTE: The usual scientific spelling is 'fetus', although 'foetus' is common in non-technical British English usage.)

foggage /ˈfɒɡɪdʒ/ *noun* **1.** a winter grazing of cattle on non-ryegrass swards **2.** grass which has been left standing to provide winter grazing for sheep and cattle

fold *noun* a moveable enclosure, made of hurdles or of electric wire fencing, used to keep cattle or sheep in a certain place ■ *verb* to put sheep into a fold

folded sheep *noun* sheep kept in movable folds, as a means of controlling their grazing

foldland /ˈfəʊldlənd/ *noun* formerly, an area of land allotted to each manor for the purpose of grazing the manor's sheep

foliage *noun* the leaves on plants ○ *In a forest, animals are hard to see through the thick foliage on the trees.*

foliar /ˈfəʊliə/ *adjective* referring to leaves

foliar spray *noun* a method of applying pesticides or liquid nutrients as droplets to plant leaves ○ *needs weekly foliar sprays*

'Other features include effective activity in wet or dry soils, highly systemic activity making it suitable for application as a foliar spray, drench, or in drip irrigation, a favourable safety profile and rapid degradation under field conditions.' [*Arable Farming*]

folic acid /ˌfəʊlɪk ˈæsɪd/ *noun* a vitamin in the vitamin B complex found in milk, liver, yeast and green plants such as spinach

follicle /ˈfɒlɪk(ə)l/ *noun* **1.** the small structure in the skin from which each hair develops **2.** one of many small structures in the ovaries where egg cells develop

followers /ˈfɒləʊəz/ *plural noun* young cows in a dairy herd which are not yet in milk, and are being reared to replace the older stock ■ *noun* cattle put to graze a pasture after another group of animals has used it. ◊ **leader**

following crop *noun* a crop sown by a tenant farmer before leaving the farm at the end of his tenancy. The tenant farmer is permitted to return and harvest the crop and remove it.

food *noun* **1.** the nutrient material eaten by animals for energy and growth **2.** the nutrient material applied to plants as fertiliser

Food Advisory Committee *noun* a UK agency which advises ministers on matters relating to the labelling, composition and safety of food. Abbr **FAC**

Food and Agriculture Organization *noun* an international organisation that is an agency of the United Nations. It was established with the purpose of improving standards of nutrition and eradicating malnutrition and hunger. Abbr **FAO**

Food and Environmental Protection Act, 1986 *noun* legislation which brings the use of agrochemicals under statutory control, as opposed to the previous voluntary arrangement. Abbr **FEPA**

Food and Veterinary Office *noun* a committee working for the European Commission whose job it is to advise on policy in the food safety and quality, veterinary and plant health sectors. Abbr **FVO**

food balance *noun* the balance between food supply and the demand for food from a population

foodborne diseases /ˌfuːdbɔːn dɪˈziːzɪz/ *plural noun* diseases which are transmitted from feedstuff

food chain *noun* a series of organisms that pass energy and minerals from one to another as each provides food for the next (NOTE: The first organism in the food chain is the producer and the rest are consumers.)

COMMENT: Two basic types of food chain exist: the grazing food chain and the detrital food chain, based on plant-eaters and detritus-eaters respectively. In practice, food chains are interconnected, making up food webs.

food colouring *noun* a substance used to colour food

food crop *noun* a plant grown for food

Food From Britain *noun* a trade organisation which specialises in marketing British food abroad. Abbr **FFB**

food grain *noun* a cereal crop used as food for humans, e.g. wheat, barley or rye

food hygiene *noun* the series of actions taken to ensure clean, healthy conditions for the handling, storing and serving of food

Food Industry Sustainability Strategy *noun* a set of policies to promote sustainable development, for producers and suppliers at each point along the food supply chain. Abbr **FISS**

food mile *noun* a measure of the distance that food is transported from its place of origin to the consumer

'Every product carries a label stating how many food miles it has travelled, even if the distance is quite alarming. For example, the fishmonger stocks tuna, because he kept being asked for it. But it comes with a sign pointing out that it has travelled about 4000 miles to the stall.' [*Farmers Weekly*]

food poisoning *noun* an illness caused by eating food that is contaminated with bacteria

food processing industry *noun* the industry involved in the treating of raw materials to produce foodstuffs

food pyramid *noun* a chart of a food chain showing the number of organisms at each level

food safety *noun* the issues surrounding the production, handling, storage and cooking of food that determine whether or not it is safe to eat

Food Safety Act 1990 *noun* legislation which sets hygiene standards for food producers and suppliers

Food Standards Agency *noun* a British government agency set up in 2000 to offer advice on food safety and quality. Abbr **FSA**

foodstuff /ˈfuːdstʌf/ *noun* something that can be used as food ○ *cereals, vegetables and other foodstuffs*

food supply *noun* **1.** the production of food and the way in which it gets to the consumer **2.** a stock of food ○ *The ants will vigorously defend their food supply.*

food value *noun* the amount of energy produced by a specific amount of a type of food

food web *noun* a series of food chains that are linked together in an ecosystem

'Researchers, conservationists and farmers have accepted that pesticides disrupt food webs by removing non-target plants and insects, and that this affects farmland bird populations.' [*Farmers Guardian*]

fool's parsley *noun* a species of hemlock *(Aethusa cynapium)* which looks like parsley. It is not a common source of poisoning in animals.

foot-and-mouth disease *noun* a disease of livestock, especially animals with cloven hooves, characterised by fever and ulcerating cysts. Abbr **FMD**

footbath /ˈfʊtbɑːθ/ *noun* **1.** a trough containing disinfectant through which sheep or cattle are driven to prevent or cure various diseases such as foot rot **2.** a shallow container containing disinfectant in which a person walks to disinfect shoes or boots

footpath *noun* a route along which people walk on foot but along which vehicles are not permitted ○ *Long-distance footpaths have been created through the mountain regions.*

foot rot *noun* a disease of the horny parts and the soft tissue of feet of sheep. It occurs particularly in wet marshy and badly-drained pastures, and is caused primarily by the organism *Fusiformis necrophorus* and sometimes *Fusiformis nodosus*. It makes sheep lame.

forage *noun* a crop planted for animals to eat in the field ■ *verb* to look for food ○ *The woodpecker forages in the forest canopy for insects.*

COMMENT: Forage crops are highly digestible and palatable, and are either very quick growing or very high-yielding. They have the advantage that, whether grazed or harvested and stored, they provide feed at times when grass growth is poor. There are a number of different types of forage harvester. Trailed harvesters can be power take-off or engine-driven. Self-propelled machines are becoming more widely used. After the crop is cut, it is either chopped or lacerated and loaded into a trailer.

forage box *noun* a large movable container used mainly to transport forage from a silo to a trough

forage feeding *noun* the practice of cutting herbage from a sward or foliage from other crops for feeding fresh to animals

forage harvester *noun* a machine which cuts, chops and loads green crops such as lucerne into a trailer, to make silage (NOTE: There are three main groups of **forage harvester**. Precision-chop machines are used for short cut material which gives better clamp fermentation. Double-chop machines cut the crop twice. Single-chop machines use a flail to cut the crop and produce a chop length of 150mm and above. Two different types of this machine are in use: the in-line, directly behind the tractor, and the off-set which allows a field to be cut round and round.)

forage maize *noun* maize grown for ensilage

forage wagon *noun* a mobile container with a pick-up attachment used for collecting and carrying cut forage

foreleg *noun* one of the two front legs of an animal, as opposed to the hind legs

forest *noun* **1.** an area of land greater than 0.5 ha, 10% of which is occupied by trees ○ *The whole river basin is covered with tropical forest.* **2.** same as **plantation** ■ *verb* to manage a forest, by cutting wood as necessary, and planting new trees

forested *adjective* referring to land which is covered with forest

forester /ˈfɒrɪstə/ *noun* a person who manages woodland and plantations of trees

forest floor *noun* the ground at the base of the trees in a forest

forestry *noun* the management of forests, woodlands and plantations of trees

Forestry Commission *noun* a UK government agency responsible for the management of state-owned forests

forget-me-not *noun* a widespread weed (*Myosotis arvensis*) found in all soils, especially near woodland

fork *noun* a common hand implement for turning over soil, and lifting out weeds (NOTE: Forks have four square prongs each sharpened to a point. Forks with flat prongs are used for lifting potatoes. Larger, five-pronged forks with round curved prongs, are used for spreading manure.) ■ *verb* to dig ground with a fork □ **to fork a bed over** to dig a whole bed using a fork □ **to fork manure in** to spread manure with a fork, turning the soil over to cover it

formulation /ˌfɔːmjʊˈleɪʃ(ə)n/ *noun* the form in which a pesticide is sold for use. ◊ **diet formulation** (NOTE: Most pesticides are not soluble in water and have to be formulated so they can be mixed and applied as liquids. They are supplied as emulsions, wettable powders etc. which can be mixed with water.)

fortified *adjective* with something added to make stronger

fortified food *noun* food with vitamins or proteins added to make it more nutritional

fortified wine *noun* wine such as sherry or port with extra alcohol added

forward *adjective* earlier than usual or too early ■ *verb* to improve or send on something

forward creep grazing *noun* a grazing method in which grassland which has been allocated to ewes and lambs during the fattening period, is divided into paddocks, each separated by portable fencing. As one area is finished, the fencing is moved to allow the animals to move to the next.

foul of the foot *noun* a disease of cattle caused by damage to the cleft of the hoof and invasion by a germ (*Fusiformis necrophorus*)

founder crop *noun* a crop that was one of the earliest to be used and developed by humans, e.g. wheat, barley, lentils and chickpeas

four-course rotation *noun* same as **Norfolk rotation**

four tooth sheep *noun* a sheep which is 18–21 months of age

four-wheel drive vehicle *noun* a vehicle in which the power is transmitted to all four wheels, as opposed to only one pair of wheels as is usual in cars

fowl *noun* a bird, especially a hen, raised on a farm for food. ◊ **waterfowl**

fowl pest *noun* a viral disease of chickens

fowl pox *noun* a viral disease in which wart-like nodules appear on the comb, wattles, eyelids and openings of the nostrils of fowls

fowl sick *adjective* referring to land which has become infested with parasites as a result of having been used for free-range hens for too long a period of time

fox *noun* a carnivorous canine predator (*Vulpes vulpes*) with red fur and a large bushy tail

foxglove /ˈfɒksɡlʌv/ *noun* a common weed *(Digitalis purpurea)*. The plant is poisonous and harmful to animals.

foxtail millet /ˈfɒksteɪl ˌmɪlɪt/ *noun* the first cereal to be cultivated in China. It is used for silage, hay, brewing and flour in many parts of the world, and in Britain it is used as birdseed.

frame *noun* the main part of a plough, to which the ploughshare and mouldboard are attached

Fraxinus /ˈfræksɪnəs/ *noun* the Latin name for the ash tree

free *adjective* not attached, confined or controlled ■ *verb* to release something or someone from constraint

Freedom Food *noun* an RSPCA scheme which sets out guidelines for the welfare of livestock, and labels food which comes from participating suppliers

freehold /ˈfriːhəʊld/ *noun* the absolute right to hold land or property for an unlimited time without paying rent. Compare **leasehold**

freeholder /ˈfriːhəʊldə/ *noun* a person who holds a freehold property. Compare **leaseholder**

freehold property *noun* property which the owner holds in freehold

freemartin /ˈfriːˌmɑːtɪn/ *noun* a female calf produced when a male and female embryo share a uterus. The production of testosterone from the testes of the male embryo causes the reproductive system of the female embryo to be effectively masculinised.

free-range eggs *plural noun* eggs from hens that are allowed to run about in the open and eat more natural food

free-running sleeve *noun* a loose sleeve fitted over shafts to stop clothing becoming entangled by riding on the shaft if contact is made, e.g., on manure spreader beater drive shafts

freestone /ˈfriːstəʊn/ *adjective* referring to varieties of peach where the flesh does not cling to the stone. Compare **clingstone**

freeze drying *noun* a method of preserving food or tissue specimens by freezing rapidly and drying in a vacuum

freezer *noun* an appliance for preserving perishable items by keeping them at a very low temperature

French bean *noun* a common green vegetable *(Phaseolus vulgaris)*. The beans grow on dwarf bush plants, and are grown for sale fresh or for processing as canned, frozen or dried vegetables. Some are harvested as a dried seed crop, for sale dried as haricot beans, or for processing (e.g. into baked beans).

French nettle *noun* same as **red dead-nettle**

frequently asked questions *noun* a document that contains common questions and their answers related to a particular subject ○ *Consult the Defra website for a list of FAQs on the new payment scheme.* Abbr **FAQ**

fresh *adjective* not tinned or frozen ○ *Fresh fish is less fatty than tinned fish.*

friable /ˈfraɪəb(ə)l/ *adjective* referring to something such as soil which is light and crumbles easily into fragments

Friends of the Earth *noun* a pressure group formed to influence local and central governments on environmental matters. Abbr **FoE**

Friesian /ˈfriːziən/ *noun* a breed of black and white dairy cattle, famous for its very high milk yields

COMMENT: There are three main types of Friesian recognised today: the Dutch Friesian, the British Friesian and the Holstein-Friesian, which is the North American type. The Friesian is the most important breed in British dairy herds.

Friesland /ˈfriːzlənd/ *noun* a breed of sheep whose milk is used in the production of soft cheese and yoghurt

frit fly /ˈfrɪt flaɪ/ *noun* a small black fly *(Oscinella frit)* that attacks wheat, maize and oats

frits /frɪts/ *plural noun* trace elements fused with silica to form glass. This is crushed into small pieces for distribution on the soil.

frog *noun* **1.** the part of a plough to which the mouldboard and share are attached **2.** a tough flexible pad in the middle of the sole of a horse's hoof

frond *noun* a large compound leaf, divided into many sections, such as that found on ferns and palm trees

frost *noun* **1.** a deposit of crystals of ice on surfaces **2.** freezing weather when the temperature is below the freezing point of water. Frost may lead to a deposit of crystals of ice on surfaces. ○ *There was a frost last night.* □ **frost-hardy plant** a plant which is able to withstand frost □ **frost-**

tender plant a plant which is killed by frost

frost pocket, frost hollow *noun* a low-lying area where cold air collects. Crops which are susceptible to frost should not be planted in such areas.

fructose /ˈfrʌktəʊs/ *noun* fruit sugar, found in honey as well as in fruit

frugivore /ˈfruːdʒɪvɔː/ *noun* an animal that mainly eats fruit (NOTE: Many bats and birds are frugivores.)

fruit *noun* **1.** the structure of a plant formed after flowering and usually containing seeds. Many fruits are eaten as food. ○ *a diet of fresh fruit and vegetables* ○ *A peach is a fleshy fruit.* **2.** the fleshy material round the fruit which is eaten as food ■ *verb* (*of a plant*) to produce fruit ○ *Some varieties of apple fruit very early.*

> COMMENT: Fruit contains fructose which is a good source of vitamin C and some dietary fibre. Dried fruit has a higher sugar content but less vitamin C than fresh fruit.

fruiting season *noun* the time of year when a particular tree has fruit

fruitwaste /ˈfruːtweɪst/ *noun* a residue left after juice has been extracted from fruit, used as an animal feed

fruitwood /ˈfruːtwʊd/ *noun* the wood from a fruit tree such as apple or cherry, which may be used to make furniture

FSA *abbreviation* **1.** *US* Farm Service Agency **2.** Food Standards Agency

ft *abbreviation* foot

fuel *noun* a substance that can be burnt to provide heat or power, e.g. wood, coal, gas or oil ■ *verb* to use a fuel to power something ○ *The boilers are fuelled by natural gas.*

fuelwood /ˈfjuːlwʊd/ *noun* wood that is grown to be used as fuel

fullering /ˈfʊlərɪŋ/ *noun* making a groove on the lower surface of a horse's shoes, into which the heads of the nails will fit

full-mouthed *adjective* referring to an animal which has a complete set of permanent teeth

full-time farmer *noun* a farmer who derives his or her living from agriculture, as distinct from a part-time farmer

full-time worker *noun* a farmworker engaged full-time in work on a farm

fumigant /ˈfjuːmɪgənt/ *noun* a chemical compound that becomes a gas or smoke when heated and is used to kill insects

fumigate *verb* to kill microorganisms or insects by using a fumigant

fumigation /ˌfjuːmɪˈgeɪʃ(ə)n/ *noun* disinfection by means of gas or fumes which penetrate into cracks and holes, a process that is probably more efficient that spraying or scrubbing

fumitory /ˈfjuːmɪtəri/ *noun* a common weed (*Fumaria officinalis*) affecting cereal and clover crops

functional food *noun* a food designed to be medically beneficial, helping to protect against serious conditions such as diabetes, cancer or heart disease. Also called **nutraceutical, neutraceutical**

> 'Thinking laterally it might be possible to feed cows in such a way that they can produce milk that is naturally high in specific oils, minerals, or value and so can be directed to the functional food markets.' [*Farmers Guardian*]

fungal /ˈfʌŋgəl/ *adjective* referring to fungi ○ *Powdery mildew is a fungal disease.*

fungicidal /ˌfʌŋgɪˈsaɪd(ə)l/ *adjective* referring to a substance which kills fungi ○ *fungicidal properties*

fungicide /ˈfʌŋgɪsaɪd/ *noun* a substance used to kill fungi

fungoid /ˈfʌŋgɔɪd/ *adjective* referring to something shaped like a fungus ○ *a fungoid growth on the skin*

fungus *noun* a simple plant organism such as yeast, mushrooms or mould with thread-like cells and without green chlorophyll

> COMMENT: Some fungi can become parasites of animals and cause diseases such as aspergillosis. Other fungi cause plant diseases, such as blight. Others, such as yeast, react with sugar to form alcohol. Fungicides are available as sprays or dusts for use on crops.

funicle /ˈfjuːnɪk(ə)l/ *noun* a short stalk attaching a seed to the inside of the pod

fur *noun* **1.** a coat of hair covering an animal ○ *The rabbit has a thick coat of winter fur.* **2.** skin and hair removed from an animal, used to make clothes ○ *She wore a fur coat and fur gloves.*

furlong *noun* one eighth of a mile, or 220 yards. Originally a furlong was the length of a furrow in the common field.

furrow *noun* a long trench and ridge cut in the soil by the mouldboard of a plough

furrow irrigation *noun* irrigation technique where water is allowed to flow along furrows

furrow press *noun* a special type of very heavy ring roller attached to the plough, used to press the furrow slices

furrow slice *noun* the soil which is displaced by the mouldboard of a plough when it creates a furrow

furze /fɜːz/ *noun* a common shrub (*Ulex europaeus*), found in wasteland and formerly often cut and used as fodder after chaffing. It contains a small amount of a poisonous alkaloid called ulexine, which is seldom present in dangerous quantities.

Fusarium ear blight /fjuːˌzeəriəm ˈɪə ˌblaɪt/ *noun* a serious fungal disease of wheat, that can cause significant loss in yield and quailty

futures *plural noun* stocks of produce which are bought or sold for shipping at some later date, and which may not even have been produced when they are on the market. Compare **actuals**

FUW *abbreviation* Farmers' Union of Wales

FVO *abbreviation* Food and Veterinary Office

FWAG *abbreviation* Farm and Wildlife Advisory Group

FYM *abbreviation* farmyard manure

G

g *symbol* gram

gadfly /ˈɡædflaɪ/ *noun* a fly that bites cattle, of the genera *Tabanus*, the horsefly, or *Oestrus*, the bot fly, most common from late May onwards and causing considerable trouble to cattle

GAEC *abbreviation* Good Agricultural and Environmental Condition

gage /ɡeɪdʒ/ *noun* a variety of plum, especially the greengage

GAI *abbreviation* green area index

Galician blond /ɡəˌlɪʃ(ə)n ˈblɒnd/ *noun* a breed of cattle from northern Spain. It is a triple-purpose breed, red in colour, with yellow horns.

gall *noun* a hard growth on a plant caused by a parasitic insect

gallon *noun* a unit of liquid volume in the Imperial System, approximately equal to 4.5 litres

Galloway /ˈɡæləweɪ/ *noun* a hardy breed of completely black hornless cattle, mainly reared for beef. The coat is distinctive, being formed of long wavy hairs covering a soft undercoat.

Gallus /ˈɡæləs/ *noun* the Latin name for the domestic chicken

galvanised iron /ˌɡælvənaɪzd ˈaɪən/ *noun* iron that has been coated with zinc to prevent it from rusting (NOTE: Sheets of galvanised iron are widely used for roofs.)

Galway /ˈɡɔːlweɪ/ *noun* a breed of sheep found in the Irish Republic. The white-faced Galway is the only native Irish breed and is used to produce store lambs.

game *noun* animals that are hunted and killed for sport or food or both

COMMENT: Game, such as pheasants and partridges, is an important asset on some farms, and letting land for sport shooting is a source of high income.

game birds *plural noun* wild birds which are classified as game, and which can be shot only during certain seasons. The most important in the UK are pheasant, partridge and grouse.

Game Conservancy Trust *noun* an organisation concerned with the conservation of game species, which advises on shoots and woodland management. Abbr **GCT**

gamekeeper *noun* a person working on a private estate who manages it to provide wild birds and animals for shooting and hunting

gamete /ˈɡæmiːt/ *noun* a sex cell

gander *noun* a male goose

gang *noun* a group of workers working together, e.g. a gang of sheep shearers

gangmaster /ˈɡæŋˌmɑːstə/ *noun* a person who gathers together and organises or leads a group of casual and often travelling workers

'The Gangmasters Licensing Authority will develop and operate a licensing scheme, set licensing conditions and maintain a register of licensed labour providers in the agricultural, shellfish and related processing and packaging industries. Once the licensing scheme is running in 2006, it will be an offence for anyone acting as a gangmaster to operate without a licence. It will also be illegal for anyone to use an unlicensed gangmaster.' [*Farmers Guardian*]

gangrene /ˈɡæŋɡriːn/ a condition in which tissues die and decay, as a result of bacterial action, because the animal has lost blood supply to the affected part of the body through injury ■ *noun* serious rot affecting potato tubers. Caused by fungi, it spreads in storage.

gangrenous mastitis /ˌɡæŋɡrɪnəs mæsˈtaɪtɪs/ *noun* a form of the mastitis disease affecting cattle. It may begin as staphylococcal mastitis. The udder becomes blue and cold.

gantry /ˈgæntri/ *noun* a type of farm machine consisting of a long steel beam with implement carriers. The engine and cab are at one end of the beam, and the drive wheel is at the other end.

GAP *abbreviation* Good Agricultural Practice

gapes /geɪps/ *noun* a disease affecting the breathing function of poultry, caused by small worms in the windpipe

garden *noun* an area of land cultivated as a hobby or for pleasure, rather than to produce an income. ◊ **market garden**

garden implements *plural noun* implements such as forks and spades which are used in the garden

garlic *noun* a plant (*Allium sativum*) with a strong-smelling pungent root used as a flavouring in cooking. The bulb consists of a series of wedge-shaped cloves, surrounded by a white fibrous skin.

garrigue /gəˈriːg/ *noun* a dense undergrowth of aromatic shrubs found in Mediterranean regions accompanying evergreen and cork oak

Gasconne /ˈgæskɒn/ *noun* a breed of beef cattle from the Gascony area of southwest France. The animals are silver-grey in colour with medium-length horns.

gastric *adjective* referring to the stomach

gastric juices *plural noun* mixture of hydrochloric acid, pepsin, intrinsic factor and mucus secreted by the cells of the lining membrane of the stomach to help the digestion of food

gastro- /gæstrəʊ/ *prefix* the stomach

gastroenteritis /ˌgæstrəʊentəˈraɪtɪs/ *noun* an inflammation of the membrane lining the intestines and the stomach, caused by a viral infection and resulting in diarrhoea and vomiting

gastrointestinal tract /ˌgæstrəʊɪntestɪn(ə)l ˈtrækt/ *noun* same as **alimentary canal**

GATT /gæt/ *noun* an international organisation aiming to reduce restrictions on trade between countries. It was replaced in 1995 by the World Trade Organization (WTO). Full form **General Agreement on Tariffs and Trade**

GCT *abbreviation* Game Conservancy Trust

GE *abbreviation* genetic engineering

geese plural of **goose**

geest /geɪst/ *noun* an infertile sandy lowland region of North and East Germany, covered with heath

gelatin /ˈdʒelətɪn/ *noun* a protein which is soluble in water, made from collagen

Gelbvieh /ˈgelbviː/ *noun* a breed of dairy cattle from Bavaria in south Germany. The colour varies from cream to yellow. Also called **German Yellow**

geld /geld/ *verb* to castrate and animal, especially a horse

gelding /ˈgeldɪŋ/ *noun* a castrated horse

gene *noun* a unit of DNA on a chromosome which governs the synthesis of one protein and may combine with other genes to determine a particular characteristic

COMMENT: Genes exist in different forms, called alleles. They are either dominant, in which case the characteristic is always passed on to the offspring, or recessive, where the characteristic only appears if both parents have contributed a copy of the same allele.

genera /ˈdʒenərə/ plural of **genus**

General Agreement on Tariffs and Trade *noun* full form of **GATT**

generic *adjective* **1.** relating to or suitable for a broad range of things or situations **2.** referring to a genus

COMMENT: Organisms are usually identified by using their generic and specific names, e.g. *Homo sapiens* (human) and *Felis catus* (domestic cat). The generic name is written or printed with a capital letter. Both names are usually given in italics or are underlined if written or typed.

genetic *adjective* referring to genes or genetics ○ *Breeders of new crop plants are dependent on genetic materials from wild forms of maize and wheat.*

genetically modified *adjective* referring to an organism that has received genetic material from another in a laboratory procedure, leading to a permanent change in one or more of its characteristics. Abbr **GM**

genetically modified organism *noun* a plant or animal produced by the technique of genetic modification. Abbr **GMO**

genetic code *noun* the information carried by an organism's DNA which determines the synthesis of proteins by cells and which is passed on when the cell divides. Also called **genetic information**

genetic engineering *noun* same as **genetic modification**. Abbr **GE**

genetic improvement *noun* the improvement of an animal or plant by breeding

genetic information *noun* same as **genetic code**

genetic manipulation *noun* same as **genetic modification**

genetic material *noun* the parts of a cell that carry information that can be inherited, e.g. DNA, genes or chromosomes

genetic modification /dʒə,netɪk ˌmɒdɪfɪ'keɪʃ(ə)n/ *noun* the alteration and recombination of genetic material under laboratory conditions, resulting in transgenic organisms. Abbr **GM**. Also called **genetic manipulation**, **genetic engineering**

'Cotton is one of Australia's most controversial crops, stirring up big issues such as genetic modification, pesticides and water use. 95% of Australian farmers plant a third of their cotton acreage with the Monsanto GM variety Ingard, genetically modified to protect it from insect attack.' [*Arable Farming*]

genetic resources *plural noun* the genes found in plants and animals that have value to humans ◊ *Modern plant varieties have been developed from genetic resources from South America.*

genetics *noun* the study of the way in which the characteristics of an organism are inherited

COMMENT: Comparisons of today's farm animals with those of the past show considerable differences in appearance and productivity. Today's dairy cattle have no horns, and produce two or three times as much milk as their ancestors in the 19th century. This is in part due to genetic improvement of livestock by selection of superior animals for breeding.

genetic variation *noun* the inherited differences between the members of a species

genome *noun* **1.** the set of all the genes in an individual **2.** the set of genes which are inherited from one parent

genomic /dʒɪ'nəʊmɪk/ *adjective* relating to a genome

genotype /'dʒenətaɪp/ *noun* **1.** the genetic constitution of an organism. ◊ **phenotype 2.** an individual organism

'Once electronic ID becomes the norm in 2008, individual sheep ID numbers will have to be logged on movement documents, and the breed and genotype included in the farm register.' [*Farmers Weekly*]

genotypic /ˌdʒenə'tɪpɪk/ *adjective* relating to a genotype

Gentile di Puglia /ʒen,tiːleɪ di 'pʊliə/ *noun* a breed of Italian sheep found in the Foggia region. A fine-wool merino breed used in a transhumance system.

genus *noun* a group of closely related species (NOTE: The plural is **genera**.)

Gerber test /'dʒɜːbə test/ *noun* a test to determine the butterfat content of milk

germ *noun* **1.** a microorganism that causes a disease, e.g. a virus or bacterium (*informal*) **2.** a part of an organism that develops into a new organism **3.** the central part of a seed, formed of the embryo. It contains valuable nutrients. ◊ **wheatgerm**

German Red Pied /ˌdʒɜːmən red 'paɪd/ *noun* a breed of cattle from northwest Germany. Mainly raised for meat, the animals are red and white in colour.

German Yellow *noun* same as **Gelbvieh**

germicide /'dʒɜːmɪsaɪd/ *adjective*, *noun* a substance that can kill germs

germinate *verb* (*of a seed or spore*) to start to grow

germination /ˌdʒɜːmɪ'neɪʃ(ə)n/ *noun* the beginning of the growth of a seed, resulting from moisture and a high enough temperature

germination percentage *noun* the number of seeds which germinate, taken from a representative sample of 100 seeds

gestation period /dʒe'steɪʃ(ə)n/, **gestation** *noun* the period from conception to birth, when a female mammal has live young in her womb

GH *abbreviation* growth hormone

gherkin /'gɜːkɪn/ *noun* a small cucumber grown for pickling

GHG *abbreviation* greenhouse gas

gibberellin /ˌdʒɪbə'relɪn/ *noun* a plant hormone that stimulates growth and seed germination

gid /gɪd/ *noun* a brain disease of young sheep which also occurs in cattle. Caused by ingestion of tapeworm eggs voided by dogs and foxes. Blindness is an early symptom.

gilt *noun* a young female pig

gimmer /ˈgɪmə/ *noun* a female sheep after its first shearing

gizzard /ˈgɪzəd/ *noun* a thick-walled muscular part of the gut of many birds where food is mechanically crushed. Also called **proventriculus** (NOTE: A gizzard is also present in some insects, fish and crustaceans.)

GLA *abbreviation* Gangmasters Licensing Authority

glanders /ˈglændəz/ *noun* a serious contagious disease of horses, no longer present in Britain, but still found in Asia and Africa

glasshouse *noun* a large structure made of glass inside which plants are grown, especially commercially or for scientific purposes

gley /gleɪ/ *noun* a thick rich soil found in waterlogged ground

gleying /ˈgleɪɪŋ/ *noun* a set of properties of soil which indicate poor drainage and lack of oxygen (NOTE: The signs are a blue-grey colour, rusty patches and standing surface water.)

glidewort /ˈglaɪdwət/ *noun* same as **common hemp nettle**

global distillation *noun* the movement of persistent organic pollutants from warm tropical and subtropical regions to cooler higher latitudes via evaporation and condensation

Global Environment Facility *noun* an organisation set up in 1991 to tackle environmental problems that go beyond country boundaries. It is funded by the World Bank.

globe *noun* a ball-shaped vegetable such as the globe artichoke or a variety of mangel

globe artichoke *noun* ♦ **artichoke**

Gloucester /ˈglɒstə/ *noun* **1.** a hard British cheese **2.** a rare breed of cattle, mahogany in colour, with a white strip passing down the back, over the tail, down the hind quarters and along the belly. Its milk was originally used in the production of Double Gloucester cheese.

Gloucester Old Spot *noun* a breed of pig from the Southwest of England, Wiltshire, Somerset and Gloucester. It is large, with clearly defined black spots on a white coat, and is now a rare breed.

glucose *noun* a simple sugar found in some fruit

glucosinolate /ˌgluːkəʊˈsɪnəʊleɪt/ *noun* a compound left in rape meal after the oil has been extracted. Also called **glucos**

COMMENT: The animals convert the compound to toxin after eating it. Although glucosinolates can be removed by processing, plant breeders are trying to breed new varieties of rape that are low in glucos, and therefore avoid the extra production cost.

glufosinate ammonium /glu ˌfɒsɪneɪt əˈməʊniəm/ *noun* a systemic herbicide acting against a wide range of species. Some crops have been genetically modified to tolerate it.

glume /gluːm/ *noun* a small leaf or scale enclosing a grass spikelet. Most grasses have two glumes.

glume blotch *noun* a fungal disease of wheat

gluten /ˈgluːt(ə)n/ *noun* a protein found in some cereals which makes a sticky paste when water is added (NOTE: The gluten content of flour affects the quality of the bread made from it.)

COMMENT: The gluten is what makes dough elastic and bread soft. Millet and rice do not contain gluten and so cannot be used for making bread.

glyphosate /ˈglaɪfəseɪt/ *noun* a systemic herbicide acting against a wide range of species. Some crops have been genetically modified to tolerate it.

GM *abbreviation* **1.** genetically modified **2.** genetic modification

GMO *abbreviation* genetically modified organism

goad *noun* a spiked stick used to prod cattle

goat *noun* a small animal with horns, kept for its milk and meat

COMMENT: In Europe goats are important for milk production. Goat's milk has a higher protein and butterfat content than cow's milk, and is used especially for making cheese. Elsewhere goats are reared for meat. They are useful as browsers and will eat materials which are not normally eaten by cattle.

goatling /ˈgəʊtlɪŋ/ *noun* a female goat between the ages of one and two years, which has not yet borne a kid

Golden Guernsey *noun* a breed of goat

Good Agricultural and Environmental Condition *noun* one of the Statutory Management Requirements which a farmer must fulfil, which covers the proper maintenance of soil, pastureland, stone walls and hedgerows. Abbr **GAEC**

Good Agricultural Practice *noun* a set of codes which provide practical guidance for farmers on the proper maintenance of soil, water and air. Abbr **GAP**

goose *noun* a large heavy bird, between a duck and a swan in size. Possibly this was one of the first wild birds to be domesticated. Geese are raised especially for table birds at Christmas. In France, goose livers are used to make pâté de foie gras. (NOTE: The males are **ganders,** the young are **goslings.**)

gooseberry *noun* a soft fruit, usually green in colour, from a small prickly bush

goosegrass /'guːsgrɑːs/ *noun* same as **cleavers**

gosling *noun* a young goose

Gossypium /gɒ'sɪpiəm/ *noun* the Latin name for cotton

gourd *noun* the fruit of a trailing or climbing plant. Many varieties are cultivated either as ornamental plants or to provide dried bottle-like containers which can be used as utensils such as water carriers.

gout fly /'gaʊt flaɪ/ *noun* a small fly whose larvae hatch and feed on shoots and ears of cereals, especially barley

government agencies *plural noun* organisations which provide specialist advice for farmers, e.g. ADAS, set up by the British Department for Environment, Food and Rural Affairs

government assistance *noun* financial aid in the form of grants and subsidies provided by governments to help farmers

gr *abbreviation* grain

grade *noun* a category of something which is classified according to quality or size ■ *verb* to divide produce into different categories, according to its quality or size ○ *Eggs are graded into classes A, B, and C.*

COMMENT: Agricultural land is classified into five grades. Grade 1 is land with very minor or no physical limitations to agricultural use. Grade 2 has some minor limitations in soil texture, depth or drainage. Grade 3 has moderate limitations due to soil, relief or climate, it has no potential for horticulture, but can produce good crops of cereals, roots and grass. Grade 4 has severe limitations and is basically used for pasture. Grade 5 is of little agricultural value, mainly for rough grazing.

graded seed *noun* a seed such as sugar beet which is formed of a cluster of seeds and can be separated out by rubbing. Also called **rubbed seed**

grader /'greɪdə/ *noun* a machine which grades fruit or vegetables, according to size

grading up *noun* a selective breeding process, using the males of one breed to mate with females of another breed for at least four generations. The result will be that the female breed will disappear and be replaced by that of the males.

graft *noun* a piece of plant or animal tissue transferred onto another plant or animal and growing there ■ *verb* to transfer a piece of tissue from one plant or animal to another

COMMENT: Many cultivated plants are grafted. The piece of tissue from the original plant (the scion) is placed on a cut made in the outer bark of the host plant (the stock) so that a bond takes place. The aim is usually to ensure that the hardy qualities of the stock are able to benefit the weaker cultivated scion.

grain *noun* **1.** the seed, which is technically a fruit, of a cereal crop such as wheat or maize **2.** a cereal crop such as wheat of which the seeds are dried and eaten ○ *grain farmers* (NOTE: In this sense, **grain** does not have a plural.) **3.** a measure of weight equal to 0.0648 grams. Abbr **gr**

grain aphid *noun* an insect which lives on crops such as barley and can destroy them by feeding on their sap

'Further reports of increasing aphid numbers in unsprayed crops of late October/early November emerging winter barley have been received from central and southern areas, and there are also unconfirmed reports of grain aphids overwintering in crops which received an aphicide last autumn.' [*Farming News*]

grain crop *noun* a cereal crop such as wheat of which the seeds are dried and eaten

grain drier *noun* a machine which dries moist grain before storage. The grain is dried under a blast of hot or warm air.

grain drill *noun* a machine used for sowing cereals in rows

COMMENT: Grain drill feed mechanisms may be internal force feed, external force feed or studded roller.

grain lifters *plural noun* attachments to the cutter bar of a combine harvester, which lift the stems of cereal crops which have been beaten down by bad weather, and so allow the crop to be cut and gathered

grain pan *noun* the part of a combine where the threshed grain collects and is shaken through to the bottom of the machine

grain reserves *plural noun* the amount of cereal grain held in a store by a country which is estimated to be above the country's requirements for one year

grain rolled *noun* cereal rolled or crushed between two rotating cylinders for feeding to livestock

grain spear *noun* an instrument for measuring the temperature and moisture of stored grain. It consists of a thermometer and hygrometer at the end of a long rod which is pushed into the grain.

grain storage *noun* the practice or means of keeping grain until it is sold or used (NOTE: Most grain is stored on the farm until it is sold, and is kept in bins or in bulk on the floor of the granary. The system of storage depends on whether the grain is to be used for feeding animals on the farm or is to be sold.)

grain tank *noun* a storage area at the top of a combine, in which threshed grain is kept. When the tank is full, the grain is transferred to a trailer.

grain weevil *noun* a reddish-brown weevil which lays eggs in stored grain. The larvae feed inside the grain, where they also pupate.

gram *noun* **1.** a metric measure of weight equal to one thousandth of a kilogram. Abbr **g 2.** same as **chickpea**

Gramineae /græˈmɪnɪiː/ *plural noun* former name for **Poaceae**

graminicide /græˈmɪnɪsaɪd/ *noun* a herbicide which kills grasses

Granadilla /ˌɡrænəˈdɪlə/ *noun* the passion fruit, a climbing plant with purple juicy fruit. It is native to Brazil.

granary *noun* a place where threshed grain is stored

granular /ˈɡrænjʊlə/ *adjective* in the form of granules

granule *noun* a small artificially made particle. Fertilisers are produced in granule form, which is easier to handle and distribute than powder.

grape *noun* the fruit of woody perennial vines (*Vitis*)

COMMENT: Grapes are grown in most areas of the world that have a Mediterranean climate, and even in temperate areas like southern England and central Germany. They are eaten as fruit, dried to make currants and raisins, or crushed to make grape juice and wine.

grapefruit *noun* a citrus fruit of a tree (*Citrus paradisi*) similar to the orange. The fruit is lemon-yellow or pink when ripe, about twice the size of an orange, and very juicy.

grapevine *noun* the vine on which grapes grow

grass *noun* a flowering monocotyledon of which there are a great many genera, including wheat, barley, rice, oats. Grasses are an important food for herbivores and humans. □ **cows at grass** cows which are grazing in a field

COMMENT: Grass is the most important crop in the UK. It occupies about two-thirds of the total crop area.

grassland *noun* land covered mainly by grasses. ◊ **acid grassland, calcareous grassland** ■ *plural noun* **grasslands** wide areas of land covered mainly by grasses, e.g. the prairies of North America and the pampas of South America

COMMENT: Grasslands can be divided into the following types. **Rough mountain and hill grazing**: not of great value, the plants being mainly fescues, bents, nardus and molinia grasses. **Permanent pastures**: these are never ploughed, and the quality depends on the percentage of perennial ryegrass. **Leys**: these are temporary grasslands which are sown to grass for a limited period (usually one to five years). The year in which the seed mixture is sown is known as the 'seeding year'. At the end of the first year there is the first year harvest. Sowing the seeds mixture with a cover crop is known as 'undersowing'. 'Direct sowing' is sowing on bare ground without a cover crop. The main species used in grasslands are the following. **Grasses**: perennial ryegrass, cocksfoot, Timothy, Italian ryegrass and meadow fescue; **clovers**: red clover, white clover; **other legumes**: lucerne, sainfoin; **herbs**: yarrow, chicory, rib grass, burnet. Farmers depend on reliable seed firms to supply them with standard seed mixtures. Varieties and strains of herbage plants have different growth characteristics and the choice of mixtures will depend on the purposes of the ley.

grass sickness *noun* a sudden and usually fatal illness affecting sheep and cattle. Symptoms include depression, inflamed membranes, discharge from nostrils. No effective treatment.

grass staggers *plural noun* same as **hypomagnesaemia**

gravity feed *noun* a system where pellets, seeds or granules fall from a hopper into a distribution channel

graze *verb* 1. (*of animals*) to feed on low-growing plants 2. to put animals in a field to eat grass

grazier /'greɪziə/ *noun* a farmer who looks after grazing animals

grazing *noun* 1. the action of animals feeding on growing grass, legumes or other plants ○ *Spine on plants may be a protection against grazing.* 2. an area of land covered with low-growing plants suitable for animals to feed on ○ *There is good grazing on the mountain pastures.*

grazing cycle *noun* the length of time between the beginning of one grazing period and the next

'Cows are now on their second grazing cycle and, while the herd is still split into high and low yielding groups, are managed on a leader/follower basis.' [*Farmers Guardian*]

grazing food chain *noun* a cycle in which vegetation is eaten by animals, digested, then passed into the soil as dung and so taken up again by plants which are eaten by animals

grazing management *noun* looking at the way in which land is grazed and seeing how it can be done most efficiently

grazing pressure *noun* the number of animals of a specified class per unit weight of herbage at a point of time

grazing season *noun* the time of year when animals can feed outside on grass

grazing systems *plural noun* different methods of pasture management

greaseband /'griːsbænd/ *noun* a strip of paper covered with a sticky substance, wrapped round the trunk of a tree to prevent pests from climbing up the tree

greasy pig disease *noun* a bacterial disease which causes skin abrasions and can rapidly affect an entire litter

green *adjective* 1. referring to a colour like that of grass ○ *The green colour in plants is provided by chlorophyll.* 2. immature ○ *green shoots* 3. referring to an interest in ecological and environmental problems ○ *green policies* ■ *noun* 1. a colour like that of grass 2. *also* **Green** a person with a concern for ecological and environmental problems

green area index *noun* the total area of leaves, green fruits and green stems per unit of ground area covered by a plant. Abbr **GAI**

Green Belt *noun* an area of agricultural land, woodland or parkland which surrounds an urban area

COMMENT: Green Belt land is protected, and building is restricted and often completely forbidden. The aim of setting up a Green Belt is to prevent urban sprawl and reduce city pollution.

Green Chemistry Network *noun* a Royal Society of Chemistry initiative designed to foster the development of environmentally benign chemical products that prevent pollution and reduce environmental and human health risks

green claim *noun* any text, symbols or graphics on food packaging which tell the consumer something about its environmental impact, e.g. whether the packaging is recycled or biodegradable

green currencies, green rates *plural noun* fixed exchange rates for currencies used for agricultural payments in the EU

greenfield site *noun* a place in the countryside, not previously built on, that is chosen as the site for a new housing development or factory ○ *Urban fringe sites are less attractive to developers than greenfield sites.* Compare **brownfield site**

greenfly /'griːnflaɪ/ *noun* a type of aphid, a small insect which sucks sap from plants and can multiply very rapidly

COMMENT: Greenfly attack young shoots which have a softer texture. Various species of greenfly feed on cereal crops in May and June. Greenfly can carry virus diseases from infected plants to clean ones.

greengage /'griːngeɪdʒ/ *noun* a variety of cooking plum, which is hard and green

greenhouse *noun* a structure made of glass inside which plants are grown

COMMENT: Greenhouses are used in temperate areas to grow plants which cannot be grown out of doors, either to bring the plants on early (raising seedlings to be planted out later) or to grow plants out of season (tomatoes can be grown in greenhouses during the winter months). A cold greenhouse (i. e. a greenhouse without any heating) can be used for protection of more or less hardy plants during the winter or for growing plants in late spring and summer. A heated greenhouse will be necessary to raise tender plants during the winter.

greenhouse effect *noun* the effect produced by the accumulation of carbon dioxide crystals and water vapour in the upper atmosphere, which insulates the Earth and raises the atmospheric temperature by preventing heat loss

COMMENT: Carbon dioxide particles allow solar radiation to pass through and reach the Earth, but prevent heat from radiating back into the atmosphere. This results in a rise in the Earth's atmospheric temperature, as if the atmosphere were a greenhouse protecting the Earth. Even a small rise of less than 1°C in the atmospheric temperature could have serious effects on the climate of the Earth as a whole. The polar ice caps would melt, causing sea levels to rise everywhere with consequent flooding. Temperate areas in Asia and America would experience hotter and drier conditions, causing crop failures. Carbon dioxide is largely formed from burning fossil fuels. Other gases contribute to the greenhouse effect, for instance methane is increasingly produced by rotting vegetation in swamps, from paddy fields, from termites' excreta and even from the stomachs of cows. Chlorofluorocarbons also help create the greenhouse effect.

greenhouse gas *noun* a gas that occurs naturally in the atmosphere or is produced by burning fossil fuels and rises into the atmosphere, forming a barrier which prevents heat loss ○ *The government is planning to introduce a tax to inhibit greenhouse gas emissions.* Abbr **GHG**

COMMENT: The six greenhouse gases with a direct effect are carbon dioxide, methane, nitrous oxide (all of which occur naturally), hydrofluorocarbons and perfluorocarbons, and sulphur hexafluoride. Indirect greenhouses gases are nitrogen oxides, which produce ozone during their breakdown in the atmosphere, carbon monoxide and non-methane volatile compounds.

greenhouse mealy bug *noun* a horticultural pest, a distant relative of the aphid. It may spoil the appearance of some glasshouse crops, particularly orchids.

greening /ˈɡriːnɪŋ/ *noun* **1.** the process of planting trees and other vegetation in an area **2.** the process of becoming more aware, or of increasing others' awareness, of the environment and environmental issues **3.** the process of turning green, which can occur, e.g., when potatoes are left too long in the light

green manure *noun* fast-growing green vegetation such as mustard or rape which is grown and ploughed into the soil to rot and act as manure

green manuring *noun* the process of growing green crops and ploughing them in to increase the organic content of the soil

'Research also confirmed that what grandfather knew was right could be measured in scientific terms – that some crops were more capable than others at putting organic matter back into the soil. Hence the interest in green manuring and, of late, ploughing straw back into the soil.' [*Arable Farming*]

green pound *noun* the fixed sterling exchange rate as used for agricultural payments in sterling between the UK and other members of the EU

Green Revolution *noun* the development in the 1960s of new forms of widely grown cereal plants such as wheat and rice, which gave high yields and increased food production especially in tropical countries

greens *plural noun* green vegetables such as cabbages

green tea *noun* tea where the leaves are heated to prevent fermentation

green top milk *noun* untreated milk, identified by the green tops to the bottles. Sales to the public are banned in the UK.

Greyface /ˈɡreɪfeɪs/ *noun* a crossbred sheep resulting from a Border Leicester ram and a Blackface ewe. The ewes are mated with Suffolk rams to produce good-quality lambs.

grey leaf *noun* a disease of cereals caused by manganese deficiency

grey water, greywater /ˈɡreɪˌwɔːtə/ *noun* the relatively clean waste water from sinks, baths, and kitchen appliances

grid *noun* a pattern of equally spaced vertical and horizontal metal rods or bars

grind *verb* **1.** to reduce a substance to fine particles by crushing **2.** to move or work noisily and with difficulty

grist /ɡrɪst/ *noun* **1.** corn for grinding **2.** malt crushed for brewing

grit *noun* small particles of various substances fed to poultry

COMMENT: There are two different kinds of grit: hard insoluble grit, such as flint and gravel which the fowl has to take into its gizzard to do the grinding of its feed; and the soluble grit, such as oyster-shell and limestone, which contains lime and which the birds need for bone formation and, later, for the formation of egg shells.

Groningen   Whiteheaded
/ˌgrəʊnɪŋən ˈwaɪthedɪd/ *noun* a dual-purpose breed of cattle developed in the Netherlands. The body is black but the head is white.

groom *noun* a person who looks after horses ■ *verb* to look after animals, especially horses, by brushing cleaning and combing

gross value added *noun* the annual value of goods sold and services paid for inside a country, less tax and government subsidies. Abbr **GVA**

ground *noun* **1.** a surface layer of soil or earth ○ *stony ground* **2.** an area of land, especially one used for a particular purpose

ground cover *noun* plants that grow densely close to the ground, either in natural conditions or planted to prevent soil erosion or the spread of weeds

groundnut /ˈgraʊndnʌt/ *noun* the peanut, a grain legume, and one of the main oilseeds

COMMENT: Groundnuts (or peanuts) are used in the production of vegetable oil for cooking, in salad dressings and in the making of margarine. Poorer quality oils are used to make soap. In the USA, much of the crop is made into peanut butter. The USA, Argentina, Nigeria, Sudan and Indonesia are major exporters of groundnuts, while Canada and Western Europe are the main importing countries.

groundnut cake *noun* the residue left after oil extraction from groundnuts, a valuable protein concentrate for livestock

groundsel /ˈgraʊndsəl/ *noun* a common weed (*Senecio vulgaris*) which affects most crops. Also called **birdseed**

ground water *noun* water that stays in the top layers of soil or in porous rocks and can collect pollution. Compare **surface water**

grouse *noun* a small game bird. There are two main species in Europe: the rare **black grouse** *Lyrurus tetrix* and the Scottish **red grouse** *Lagopus scoticus*.

grow *verb* **1.** (*of plants*) to exist and develop well ○ *Bananas grow only in warm humid conditions.* **2.** (*of plants and animals*) to increase in size ○ *The tree grows slowly.* ○ *A sunflower can grow 3 cm in one day.* **3.** to cultivate plants ○ *Farmers here grow two crops in a year.* ○ *He grows peas for the local canning factory.*

growing point *noun* a point on the stem of a plant where growth occurs, often at the tip of the stem or branch

growing season *noun* the time of year when a plant grows ○ *Alpine plants have a short growing season.*

growth *noun* **1.** an increase in size ○ *the growth in the population since 1960* ○ *The disease stunts the conifers' growth.* **2.** the amount by which something increases in size ○ *The rings show the annual growth of the tree.* **3.** a shoot which has grown from a plant ○ *The cordon should be pruned by cutting back all growths over one metre long.* **4.** a type of plant which grows in a certain area, e.g. vines growing in different areas of France, coffee growing in different areas of Colombia, etc.

growth hormone *noun* a natural or artificial chemical that makes an animal grow more quickly. Abbr **GH**

'The European Parliament has approved moves by the European Commission to ban the use of six growth hormones in the EU meat and poultry production industry, because of concerns that they may harm consumers.' [*Farmers Guardian*]

growth regulator *noun* a chemical used to control the growth of plants, mainly used for weed control in cereals and grassland

growth ring *noun* same as **annual ring**

growth stages *noun* the different stages of development of a crop, measured as an increase in weight or area. Also called **stages of growth**

grub *noun* a small caterpillar or larva ■ *verb* □ **to grub up, to grub out** to dig up a plant with its roots ○ *Miles of hedgerows have been grubbed up to make larger fields.*

grunt *noun* a sound made by a pig ■ *verb* (*of a pig*) to make the sound characteristic of a pig. Compare **bleat, low, neigh**

guano /ˈgwɑːnəʊ/ *noun* a mass of accumulated bird droppings, found especially on small islands and used as organic fertiliser

guaranteed prices *plural noun* a feature of national agricultural policy in which the producers of a commodity are guaranteed a minimum price for their produce

guard cell *noun* either of a pair of cells that border a leaf pore and control its size (NOTE: The guard cells and pore are called a stoma, and are most common on the underside of leaves.)

Guernsey /ˈɡɜːnzi/ *noun* a breed of dairy cattle that has a fawn coat with distinct patches of white

guinea /ˈɡɪni/ *noun* a former British coin, equivalent to the present £1.08, which is still used in quoting prices at livestock sales (NOTE: It is abbreviated in prices to **gn: 3,400gns were paid for the Longhorn bull**.)

guinea corn *noun* sorghum

guinea fowl *noun* a table bird, found wild in savanna regions of Africa. They are now raised for their meat which has a delicate flavour similar to that of game birds.

gully *noun* **1.** a deep channel formed by soil erosion and unable to be filled in by cultivation **2.** a small channel for water, e.g. an artificial channel dug at the edge of a field or a natural channel in rock

Gunter's chain /ˈɡʌntəz tʃeɪn/ *noun* a chain originally used by surveyors to measure land

gut *noun* same as **alimentary canal**

Guzerat /ˈɡuːzəræt/ *noun* an American Brahman breed of cattle

GVA *abbreviation* gross value added

gymnosperm /ˈdʒɪmnəʊspɜːm/ *noun* a seed-bearing plant in which the seeds are carried naked on the scales of a cone rather than being inside a fruit. ◊ **angiosperm**

gypsum /ˈdʒɪpsəm/ *noun* a soft white or colourless mineral consisting of hydrated calcium sulfate, used in cement, plaster and fertilisers

H

ha *symbol* hectare

habit *noun* the characteristic way in which a specific plant grows ○ *a bush with an erect habit* ○ *a plant with a creeping habit*

habitat *noun* the type of environment in which a specific organism lives

habitat action plan *noun* a detailed description of a specific habitat together with the detailed actions and targets proposed for conserving it. Abbr **HAP**

habitat management *noun* same as **nature management**

habitat restoration *noun* activity carried out to return an area to a former more favourable condition for wildlife

HACCP /'hæsəp/ *noun* a process for identifying and controlling hazards within a process, e.g. in the food industry. Full form **Hazard Analysis Critical Control Points**

hack *noun* 1. a riding horse 2. a horse let out to hire ■ *verb* to ride a horse, especially to ride a horse to a show, as opposed to taking the horse in a box

hackles *plural noun* the long feathers on the neck of a domestic cock

hackney /'hækni/ *noun* a type of horse used both for riding and as a draught animal

haemoglobin /ˌhiːməˈɡləʊbɪn/ *noun* a red protein in red blood cells that combines reversibly with oxygen and transports it round the body. Abbr **Hb**

COMMENT: Haemoglobin absorbs oxygen in the lungs and carries it in the blood to the tissues. Haemoglobin is also attracted to carbon monoxide and readily absorbs it instead of oxygen, causing carbon monoxide poisoning.

Hagberg /'hæɡbɜːɡ/, **Hagberg test** *noun* a test used to determine the milling quality of wheat

COMMENT: The test measures the falling time of wheat, using ground wheat in suspension in water. A good milling wheat has a high falling time, and wheat with low falling times is not normally used in milling. Hagberg test kits are available for farmers to make their own tests on samples of wheat.

Hagberg falling number *noun* the falling time in seconds in the Hagberg test ○ *a top quality wheat with a specific weight of 79–80, Hagberg 350 and protein of 12%*

hair balls *noun* balls of hair which collect in the stomach of animals making digestion difficult. They can cause fits and convulsions in very young calves, and sight may be slightly impaired.

hairworm /'heəwɜːm/ *noun* a very thin worm of the species *Capellaria,* which infests poultry

hake bar /'heɪk bɑː/ *noun* an attachment which links a trailed plough to the tractor

half-breed *noun* an animal of mixed breed, mainly applied to crossbred sheep

half-hardy *adjective* referring to a plant that is able to tolerate cold weather down to about 5C. ◊ **hardy**

Half long *noun* a sheep produced by crossing a Cheviot ram with a Blackface ewe

half-standard *noun* a type of fruit tree with a trunk shorter than that of a full standard, about 1.2m from the ground to the first branches

halo- /hæləʊ/ *prefix* salt

halo blight *noun* a disease which affects the pods of peas and beans, making them brown and withered

halomorphic soil /ˌhæləʊmɔːfɪk ˈsɔɪl/ *noun* soil that contains large amounts of salt

halophyte /'hæləfaɪt/ *noun* a plant that is able to grow in salty soil, as in estuaries

halothane gene /ˈhæləʊθeɪn dʒiːn/ *noun* a recessive gene found in some breeds of pigs which affects the animal's susceptibility to stress and can lead to porcine stress syndrome (NOTE: The gene is called the **halothane gene** because it can be tested for by exposing the pigs to the anaesthetic **halothane**.)

'Molecular biology has enable tremendous strides. Eliminating the halothane gene is one example and there will be more equally significant advances to come.' [*Pig Farming*]

halter /ˈhɔːltə/ *noun* a rope with a noose for holding and leading horses or cattle

ham *noun* **1.** the thigh of the back leg of a pig **2.** meat from this part of the pig, usually cured in brine and dried in smoke

hammer mill *noun* a machine used in the preparation of animal feed by grinding cereals into meal

COMMENT: A typical hammer mill has a high-speed shaft with a grinding rotor at one end and a fan at the other. Eight flails are attached to the rotor which beat the grain into meal.

Hampshire /ˈhæmpʃə/ *noun* an American breed of black-haired pig with white markings. It is similar to the British Saddleback which has black skin and a white saddle.

Hampshire Down *noun* a short stocky early-maturing sheep, originating from Berkshire ewe flocks and Southdown rams

hand *noun* a measure used to show the height of a horse. One hand is 10.16cm, and the measurement is taken from the ground to the withers of the horse.

hand collection, hand picking *noun* the picking of fruit such as bananas or peaches by hand

hand feeding, hand rearing *noun* the process of bringing up orphaned animals by feeding them with a bottle

hand hoe *noun* a garden implement with a small sharp blade, used to break up the surface of the soil or to cut off weeds

COMMENT: There are several types of hand hoe, including the Dutch hoe, where the blade is more or less straight and is pushed by the operator, the draw hoe has the blade set at right angles to the handle and is used for drawing drills at seed-sowing time. The Canterbury hoe does not have a blade, but is like a three-pronged fork, with the prongs set a right angles to the handle.

handle *noun* a term used to describe the texture or feel of wool

hand pulling *noun* the act of pulling weeds or plants out of the ground by hand

hank /hæŋk/ *noun* wool which has been spun into a thread and coiled into a loop for convenience. A hank is 560 yards long.

HAP *abbreviation* habitat action plan

harden off *verb* to make plants which have been raised in a greenhouse become gradually more used to the natural temperature outdoors ○ *After seedlings have been grown in the greenhouse, they need to be hardened off before planting outside in the open ground.*

hardjo /ˈhɑːdjəʊ/ ♦ **Leptospira hardjo**

hardpan /ˈhɑːdpæn/ *noun* a hard cement-like layer in the soil or subsoil, which can be very harmful as it prevents good drainage and stops root development

hard wheat *noun* wheat with a hard grain rich in gluten

hardwood *noun* a slow-growing broad-leaved tree, e.g. oak, teak or mahogany

hardy *adjective* referring to a plant able to tolerate cold weather, especially below 5 °C. ◊ **half-hardy**

hare *noun* a long-eared furry animal, similar to but larger than a rabbit, with hind legs longer than forelegs

haricot bean /ˈhærɪkəʊ biːn/ *noun* the dry ripe seed of the French bean

harrow /ˈhærəʊ/ *noun* a piece of equipment with teeth or discs, used for breaking up soil or levelling the surface of ploughed soil ■ *verb* to level the surface of ploughed soil with a harrow, covering seeds that have been sown in furrows

harvest *noun* **1.** the time when a crop is gathered **2.** a crop that is gathered ○ *We think this year's wheat harvest will be a good one.* ■ *verb* **1.** to gather a crop that is ripe ○ *They are harvesting the barley.* **2.** to gather a natural resource

harvester /ˈhɑːvɪstə/ *noun* a machine which harvests a crop ○ *Most crops are now harvested by machines such as combine harvesters or sugar beet harvesters.*

HASSOP /ˈhæsəp/ *noun* ♦ **HACCP** (NOTE: HACCP is pronounced as if it were spelt HASSOP.)

hatch *verb* to become mature and break out of the egg ■ *noun* a brood of chicks

hatchery /ˈhætʃəri/ *noun* a place where eggs are kept warm artificially until the animal inside becomes mature enough to break out

hatchery waste *noun* surplus chicks or embryos produced in a hatchery (NOTE: The animal welfare code lays down rules for the humane slaughter of hatchery waste, and there are also strict rules governing how this waste may be disposed of.)

haulm /hɔːm/ *noun* the stalks and stems of peas, beans and potatoes

haulm roller *noun* a roller found on potato harvesters and grading machinery

haulm silage *noun* silage made from the stems and leaves of peas and beans left after harvest

haunch /hɔːnʃ/ *noun* the hind leg of an animal, especially a deer

hawthorn /ˈhɔːθɔːn/ *noun* a small tree (*Crataegus monogyna*) with spiny shoots, used for hedges round grazing areas

hay *noun* grass mowed and dried before it has flowered, used for feeding animals

COMMENT: Hay is cut before the grass flowers and at this stage in its growth it is a nutritious fodder. If it is mowed after it has flowered it is called straw, and is of less use as a food and so is used for bedding.

hay bale *noun* hay which has been compressed into a square, rectangular or round bale, so that it can be handled and stored more easily

hay baler *noun* a machine which gathers cut hay and makes it into bales

haycock /ˈheɪkɒk/ *noun* formerly, a conical heap of raked hay

hay fever *noun* same as **pollinosis**

haylage /ˈheɪlɪdʒ/ *noun* hay for silage, cut and compressed in plastic bags so that it stays green without any fungus being able to spread

hayloader /ˈheɪləʊdə/ *noun* an implement for loading hay from the field into a trailer

haymaking /ˈheɪmeɪkɪŋ/ *noun* the cutting of grass in fields to make hay

COMMENT: Haymaking normally needs three to four days of fine weather in early season and two to three days when the humidity falls as temperatures rise. The critical period for hay occurs when the crop is partly dried in the field. The object should be to dry the crop as much as possible without too much exposure to sun, and the least possible movement after the crop is partially dry. Field-dried

hay is normally baled, and further barn drying is common.

hay net *noun* a coarse meshed net bag which is filled with hay and hung up for horses to feed from

hay quality *noun* the nutritional value of hay, which can depend on the weather conditions and the time taken to dry

'The farm is now mainly permanent pasture, or grass and legumes for grazing and hay. Some of this pasture has been renovated with clover, and rye is direct drilled into hay fields in the autumn for late season grazing and to improve hay quality.' [*Farmers Weekly*]

hay rack *noun* a wooden frame containing hay, which is placed where livestock can feed from it

hay rake *noun* an implement used for raking hay prior to collection or baling

hay seed *noun* grass seed obtained from hay

haystack /ˈheɪstæk/ *noun* a heap of hay, built in the open air and protected by thatching; not used very often nowadays

hay-sweep *noun* an implement used to collect hay from swaths and carry it to a stack

hazard *noun* something with the potential to cause injury, damage or loss ○ *a fire hazard* ○ *a health hazard* Compare **risk**

Hazard Analysis Critical Control Points *noun* full form of **HACCP**

hazel /ˈheɪz(ə)l/ *noun* a nut-bearing tree (*Corylus avellana*)

Hb *abbreviation* haemoglobin

HCC *abbreviation* Hybu Cig Cymru (NOTE: The English name is 'Meat Promotion Wales'.)

HDC *abbreviation* Horticultural Development Council

headage /ˈhedɪdʒ/ *noun* the number of animals of a specified type, such as cattle, used as a basis for calculating subsidy payments

head corn *noun* the largest grains in a sample of cereal

header *noun* a machine which removes the seed heads from plants. ◊ **stripper-header**

headfly /ˈhedflaɪ/, **head fly** *noun* a parasitic insect *Hydrotaea irritans* which mainly affects sheep and can transmit summer mastitis

heading date *noun* the average date by which a certain percentage of a crop has formed seedheads. This is used by farmers to make decisions on which variety of a crop is more suitable for the environment in which they are planning to grow it.

headland /'hedlənd/ *noun* an uncultivated area of soil at the edge of a field, where a tractor turns when ploughing. ◊ **conservation headland**

headrail /'hedreɪl/ *noun* a rail across the front of a cubicle, to which a halter can be attached

Health and Safety Executive *noun* a UK government organisation responsible for checking people's working environment. Abbr **HSE**

health and welfare plan *noun* a written report made by a farmer in consultation with a vet, describing how livestock will be cared for

'The guidance suggested that farmers should prepare a herd health and welfare plan with their vet which should include measures to control the disease such as early removal of diseased cattle, not breeding from their offspring and making sure that calves only receive colostrum from their own mother where possible. This culminated in last September's launch of the Johne's Initiative with backing from various industry bodies.' [*Dairy Farmer*]

heart *noun* **1.** a muscular organ that pumps blood round an animal's body **2.** the compact central part of a vegetable such as lettuce, cabbage or celery, where new leaves or stalks form **3.** the innermost part of something ○ *This tree grows only in the heart of the forest.* **4.** □ **soil which is in good heart** soil which is fertile and produces large yields of crops

heart rot *noun* a disease of sugar beet and mangolds, caused by boron deficiency. A dry rot spreads from the crown downwards and attacks the roots. The growing point is killed, and replaced by a mass of small deformed leaves.

heartwood /'hɑːtwʊd/ *noun* the hard dead wood in the centre of a tree trunk which helps support the tree. Compare **sapwood**. Also called **duramen**

heat *noun* **1.** □ **to sow lettuces under heat** to sow lettuce seed in a heated greenhouse **2.** the period when a female animal will allow mating. Technical name **oestrus** □ **an animal on heat** a female animal in the period when she will accept a mate

heath *noun* an area of acid soil where low shrubs such as heather and gorse grow and which are treeless as a result of grazing by animals

COMMENT: Lowland heaths are found on dry sandy soils or gravel below 300 m. Upland heaths are found on mineral soils or shallow peat and may be dry or wet, with mosses growing in wetter conditions.

heather *noun* a plant (*Calluna vulgaris*) found on acid soils, common in upland areas. It is used by game birds such as grouse for cover and food.

COMMENT: The main competing uses for heather moorland in the UK are grouse shooting, sheep grazing, afforestation, recreational use and landscape conservation.

heathland /'hiːθlænd/ *noun* a wide area of heath

heat stress *noun* distress and discomfort suffered by an animal because it is too hot

heat treatment *noun* the use of high temperatures, typically 45°C, to disinfest storage areas or containers

heaves /hiːvz/ *noun* a condition of horses, where spores from mouldy hay block the animals' lungs, making breathing difficult

heavy cropper *noun* a tree or plant that produces a large crop of fruit

heavy grains *plural noun* cereal crops such as maize, rye and wheat. Abbr **HG**. Compare **light grains**

heavy soils *plural noun* soils with a high clay content, which need more tractor power when ploughing and cultivating

Hebridean sheep /ˌhebrɪˈdiːən ˌʃiːp/ *noun* a rare breed of small black sheep of Scandinavian origin. The fleece is jet-black in colour, and the animals have one pair of horns curling downwards and another pair almost upright.

hectare *noun* an area of land measuring 100 by 100 metres, i.e. 10000 square metres or 2.47 acres. Symbol **ha**

hecto- /hektəʊ/ *prefix* one hundred, 10^2. Symbol **H**

hedge *noun* a row of bushes planted to provide a barrier around a field or garden

COMMENT: It is said that you can judge the age of a hedge by counting the woody plant species in it and multiplying by 100; the more species there are, the older the hedge is. About three-quarters of the farms in England and Wales have hedges, with an estimated total length of 500,000km. Since 1997, there have been regulations in place to restrict the removal of hedges and it is illegal to remove a hedge without permission from the local planning authority.

hedgebank /'hedʒbæŋk/ *noun* a raised strip of earth on which a hedge is planted ○ *primroses growing on the hedgebank*

hedgecutter /'hedʒˌkʌtə/, **hedgetrimmer** *noun* an implement attached to a tractor, used to trim hedges. Smaller handheld units are available.

hedgelaying /'hedʒleɪɪŋ/ *noun* a traditional method of cultivating hedges, where tall saplings are cut through halfway and then bent over so that they lie horizontally and make a thick barrier

hedgerow *noun* a line of bushes forming a hedge (NOTE: Under the Hedgerow Regulations, it is now forbidden to remove a hedgerow without permission from the local planning authority.)

hedging /'hedʒɪŋ/ *noun* the skill of cultivating hedges

heel in *verb* to place plants in a trench and cover with soil until needed for permanent planting

heft /heft/ *noun* a group of mountain sheep which graze the same area in which they were born, although not kept in by fences

heifer *noun* a female cow which has not calved or has calved for the first time

hemlock /'hemlɒk/ *noun* a poisonous plant. Latin name: *Conium maculatum.*

hemlock poisoning *noun* poisoning of young cattle by eating fresh shoots of the hemlock. Sheep and goats are believed to be resistant to the poison.

hemp *noun* a plant used to make rope and that also produces an addictive drug. Latin name: *Cannabis sativa.*

hemp nettle *noun* a common weed (*Galeopsis tetrahit*) which affects spring cereals and vegetables. Also called **glidewort, holyrope**

hen *noun* **1.** a female of the common domestic fowl **2.** any female bird, e.g. a hen pheasant. ◊ **fat hen**

henhouse /'henhaʊs/ *noun* a small wooden building for keeping hens

hen in lay *noun* a bird which is laying eggs

herb *noun* **1.** a plant that is used to add flavour in cooking **2.** a plant that has medicinal properties **3.** a non-woody flowering plant that has no perennial stem above the ground in winter

herb- /hɜːb/ *prefix* referring to plants or vegetation

herbaceous /hə'beɪʃəs/ *adjective* referring to plants with soft non-woody tissue that die down above ground to survive through the winter

herbage /'hɜːbɪdʒ/ *noun* the green plants, especially grass, eaten by grazing animals

herbage allowance *noun* the weight of herbage per unit of live weight at a point in time

herbage consumed *noun* the mass of herbage once it has been consumed by grazing animals

herbage mass *noun* the weight of herbage produced in a specified area

herbage residual *noun* herbage remaining after defoliation

herbarium /hɜː'beəriəm/ *noun* a collection of preserved plant or fungal specimens, especially one that is used for scientific study and classification

herbicide *noun* a chemical that kills plants, especially used to control weeds

herbivore *noun* an animal that feeds only on plants. ◊ **carnivore, detritivore, frugivore, omnivore**

herbivorous /hɜː'bɪvərəs/ *adjective* referring to an animal that feeds only on plants

herd *noun* a number of animals such as cattle kept together on a farm or looked after by a farmer ○ *They have a herd of beef cattle.* ○ *Dairy herds have been reduced since the introduction of the milk quota system.* (NOTE: The word 'herd' is usually used with cattle; for sheep, goats, and birds such as hens or geese, the word to use is 'flock'.) ■ *verb* **1.** to tend a herd of animals **2.** to gather animals together ○ *herding the cows into the yard*

herd book *noun* the record of animals kept by breeding societies in which only the offspring of registered animals can be recorded

herd health *noun* the welfare of a herd of cattle taken as a whole, particularly regarding the spread of infectious diseases

herdmark *noun* a unique marker assigned to each herd of pigs by Defra and used to identify the animals when being moved from the farm

herd register *noun* an official record of a herd's movements, medical history and birth and death figures

'The bovine herd register is available as a printed report and all subsidy claims are catered for, while all programs record movements, calvings and drying off as a standard.' [*Farmers Guardian*]

herdsman /'hɜːdzmən/ *noun* someone who looks after a herd of animals

herdsperson /'hɜːdzpɜːsən/ *noun* a farm worker who looks after a herd of livestock

herd tester *noun* a person who tests a dairy herd for butterfat content

Herdwick /'hɜːdwɪk/ *noun* a mountain breed of sheep, native to the Lake District, which are able to survive in bitter winter conditions. The rams have horns.

hereditary *adjective* referring to a genetically controlled characteristic that is passed from parent to offspring

heredity *noun* the transfer of genetically controlled characteristics from parent to offspring

Hereford /'herɪfəd/ *noun* a breed of large, hardy cattle that are deep red in colour, with a white head and chest. Herefords are an early-maturing breed, and are important for beef production.

herringbone parlour /'herɪŋbəʊn ˌpɑːlə/ *noun* a type of milking parlour with no stalls and where the operator works from a central pit. The cattle stand side by side at an angle on each side of the central pit.

heterologous /ˌhetə'rɒləgəs/ *adjective* differing in structural features or origin

heterosis /ˌhetə'rəʊsɪs/ *noun* an increase in size or rate of growth, fertility or resistance to disease found in offspring of a cross between organisms with different genotypes. Also called **hybrid vigour**. Compare **inbreeding depression**

heterotroph /'hetərəʊtrɒf/ *noun* an organism that requires carbon in organic form and cannot manufacture it for (NOTE: Animals, fungi and some algae and bacteria are heterotrophs.)

heterotrophic /ˌhetərəʊ'trɒfɪk/ *adjective* referring to a heterotroph ○ *a heterotrophic organism*

heterozygous /ˌhetərəʊ'zaɪgəs/ *adjective* relating to a cell or organism that has

two or more variant forms (**alleles**) of at least one of its genes (NOTE: The offspring of such an organism may differ with regard to the characteristics determined by the gene or genes involved, depending on which version of the gene they inherit.)

Hevea /'hiːviə/ *noun* the Latin name for the rubber tree

HFA *abbreviation* Hill Farm Allowance

HFCS *abbreviation* high fructose corn syrup

Hg *symbol* mercury

HG *abbreviation* heavy grains

HGCA *noun* an organisation established to improve the production and marketing of UK cereal crops and oilseeds, and to promote research. Full form **Home Grown Cereals Authority**

hide *noun* the skin of an animal, which is important commercially both in its raw state and as leather

hide-bound *noun* a condition where dehydration makes it difficult for the animal's skin to move over the underlying tissues

HIE *abbreviation* Highlands and Islands Enterprise

Higher Level Stewardship *noun* one of the categories under the Environmental Stewardship scheme, where farmers can apply for funding in return for implementing complex environmental management schemes on their land. Abbr **HLS**

high fructose corn syrup *noun* a sweetener used in the soft drinks industry, extracted from maize. Abbr **HFCS**. Also called **isoglucose**

high-input farming *noun* intensive agriculture which uses fertiliser, pesticides and modern machinery to guarantee a large crop output. Compare **low-input farming**

highland /'haɪlənd/ *adjective* referring to a hilly or mountainous area ○ *Highland vegetation is mainly grass, heather and herbs.*

Highland *noun* a hardy breed of beef cattle, with long shaggy hair hiding a dense undercoat. The breed is small, has very long horns, matures slowly and is native to the Highlands and Western Isles of Scotland.

highlands *noun* an area of mountains ○ *Farmers in the highlands mostly raise sheep.*

Highlands and Islands Enterprise *noun* a business and community develop-

high mortality rate *noun* a high percentage of animals in a group which die

high-performance *adjective* designed to operate very efficiently

high-tech *adjective* technologically advanced (*informal*)

high temperature short time method *noun* the usual method of pasteurising milk, where the milk is heated to 72°C for 15 seconds and then rapidly cooled. Abbr **HTST**

high-yielding *adjective* producing a large crop ○ *They have started to grow high-yielding varieties of wheat.*

hill *noun* an area of ground higher than the surrounding areas but not as high as a mountain

hill drainage *noun* a small open channels about ten to sixty metres apart, dug to drain hilly grazing areas

hill farm *noun* a farm in mountainous country, with 95% or more of its land classified as rough grazing, mainly for breeding ewe flocks

Hill Farm Allowance *noun* a support payment available for owners of hill farms to help with running costs. Abbr **HFA**

hill grazing *noun* grassland used for sheep and cattle grazing in hilly and mountainous areas

hill land *noun* land on hills, mountains or moors (NOTE: The Hill Livestock Compensatory allowance order classifies such marginal land in upland and hilly areas and makes payments to compensate for farming in these more difficult conditions.)

Hill Radnor *noun* ♦ **Radnor**

hilum /ˈhaɪləm/ *noun* the point where a seed is attached to a pod. When the seed is ripe and has been separated from the pod, a black scar can be seen on the seed.

hind /haɪnd/ *adjective* referring to the back part of an animal ■ *noun* a female deer

hind legs *plural noun* the back legs of an animal

hindquarters /ˈhaɪndkwɔːtəz/ *noun* the back part of an animal, including the haunches and hind legs

hinge *noun* soil which is left uncut by a plough when it has failed to cut a full furrow

hirsel /ˈhɜːsəl/ *noun* 1. a heft of sheep 2. a piece of ground and flock looked after by one shepherd

Hisex Brown /ˌhaɪseks ˈbraʊn/, **Hisex White** *noun* a hybrid breed of laying fowl

histidine /ˈhɪstədiːn/ *noun* an amino acid which is considered essential in infants and children

hitch *noun* the mechanism for connecting implements to tractors. Also called **drawbar** ■ *verb* to attach an implement or trailer to a tractor

hive *noun* a box in which bees are kept

hive-bee *noun* the domesticated bee (*Apis mellifera*)

hl *abbreviation* hectolitre

HLS *abbreviation* Higher Level Stewardship

hock /hɒk/ *noun* the hind leg joint of mammals, between the knee and the fetlock

hoe *noun* an implement pulled by a tractor to turn the soil between rows of crops, and so to control weeds ■ *verb* to cultivate land with a hoe

hog *noun* 1. a castrated male pig 2. US any pig

hog cholera *noun* same as **swine fever**

hogg /hɒg/ *noun* a young sheep before the first shearing

hogget /ˈhɒgɪt/ *noun* a sheep roughly six to twelve months old

hold *verb* to conceive after artificial insemination ○ *Thirty ewes were AI'd and nineteen of them held.*

holding *noun* land and buildings held by a freehold or leasehold occupier

holdover /ˈhəʊldəʊvə/ *noun* a situation where a tenant farmer uses buildings and crops on a farm, after leaving the farm at the end of a tenancy. It may, e.g., involve harvesting crops later in the year after the tenancy has expired.

holly *noun* an evergreen tree (*Ilex aquifolium*) producing hard white timber

Holstein /ˈhɒlsteɪn/ *noun* a Friesian cattle imported into Canada from Holland at the end of the 19th century, now a breed of dairy cattle, black and white in colour. Also called **Canadian Holstein**, **Holstein-Friesian**

holyrope /ˈhəʊlɪrəʊp/ *noun* same as **hemp nettle**

home farm *noun* a farm on a large estate, usually farmed by the owner

Home Grown Cereals Authority *noun* full form of **HGCA**

home-saved seed *noun* same as **farm-saved seed**

homestead *noun* a farmhouse with dependent buildings and the land which surrounds it

homogenised milk /hə,mɒdʒənaɪzd 'mɪlk/ *noun* milk which is made more digestible by breaking up the fat droplets into smaller particles which are evenly distributed through the liquid

homograft /'hɒməgrɑːft/ *noun* same as **allograft**

homologous pair /hɒ,mɒləgəs 'peə/ *noun* a pair of chromosomes in a diploid organism that are structurally similar and have the same arrangement of genes, although they may carry different alleles (NOTE: One member of each pair is inherited from each parent.)

honey *noun* a sweet yellow fluid collected by bees from nectar in flowers, and stored in a beehive

honey bee *noun* a common hive-bee

honeycomb /'hʌnikəʊm/ *noun* a construction of wax, made by bees for storing honey

honey fungus *noun* a fungus which primarily attacks trees and shrubs. Roots become infected by rhizomorphs. In the spring the foliage wilts and turns yellow.

hoof *noun* a horny casing of the foot of a horse, a cow or other animals (NOTE: The plural is either **hoofs** or **hooves**.)

hoof and horn meal *noun* a fertiliser made from animal hooves and horns

hookworm /'hʊkwɜːm/ *noun* a parasitic worm in the intestine which holds onto the wall of the intestine with its teeth and lives on the blood and protein of the carrier

hoose /huːs/ *noun* a popular name for a lungworm

hop *noun* a climbing plant that has long thin groups of green flowers which are used dried in brewing to add flavour to beer. Latin name: *Humulus lupulus*.

hop bine *noun* the new shoot of a hop plant, which has to be made to coil round climbing strings

hop mildew *noun* a fungal disease of hops

hopper /'hɒpə/ *noun* a container with a hole at the bottom, for holding seed or fertiliser granules. The seed drops from the hole onto the ground or into channels taking it to drills.

Hordeum /'hɔːdiəm/ *noun* the Latin name for barley

horizon *noun* a layer of soil which is of a different colour or texture from the rest

COMMENT: In general, the topsoil is called the 'A' horizon, the subsoil the 'B' horizon, and the underlying rock the 'C' horizon.

hormonal /hɔː'məʊn(ə)l/ *adjective* referring to hormones

hormone *noun* **1.** a substance produced in animals in one part of the body which has a particular effect in another part of the body **2.** a plant growth factor

horn *noun* a hard growth which is formed on the tops of the heads of animals such as cattle, deer, goats and sheep

hornbeam /'hɔːnbiːm/ *noun* a tree which produces a very hard wood, formerly used in making wheels for farm carts

horned /hɔːnd/ *adjective* with horns ○ *a horned variety of sheep*

hornless /'hɔːnləs/ *adjective* without horns ○ *a hornless breed of cattle*

horse *noun* a hoofed animal with a flowing mane and tail, used on farms as a working animal, now mainly replaced by tractors

COMMENT: The main groups of horses are: the Heavy Draught Class, such as the **shire horse**; the Light Draught Class, such as the **Cleveland Bay**; the Saddle and Harness Class, such as the **hackney**; and the Pony Class, such as the **Shetland**.

horse bean *noun* a broad bean used as a fodder

horsebox /'hɔːsbɒks/ *noun* a closed vehicle used for transporting horses

horsebreaker /'hɔːsbreɪkə/ *noun* a person who trains a horse

horsefly /'hɔːsflaɪ/ *noun* a general name for many bloodsucking Tabanid flies. Also called **cleg**

horseshoe *noun* an iron shoe nailed to the hard part of a horse's hoof

horsetail /'hɔːsteɪl/ *noun* a poisonous weed found in grassland

horticultural /,hɔːtɪ'kʌltʃərəl/ *adjective* referring to horticulture

Horticultural Development Council *noun* a non-departmental government body providing information and support to people in the horticulture industry, including producers of soft fruit, mushrooms and tree fruit. Abbr **HDC**

Horticultural Trades Association *noun* a body which represents the interests of growers, workers and suppliers in the garden industry. Abbr **HTA**

horticulture *noun* the cultivation of flowers, fruit and vegetables in gardens, nurseries or glasshouses, as a science, occupation or leisure activity. ◊ **botanical horticulture**

Horticulture Research International *noun* the horticultural research department of the University of Warwick. Abbr **HRI**

horticulturist /ˌhɔːtɪˈkʌltʃərɪst/ *noun* a person who specialises in horticulture

Hosier system /ˈhəʊziə ˌsɪstəm/ *noun* a system of dairy cattle management, where the milking of cows is done in the field using a milking pail

host *noun* a plant or animal on which a parasite lives ■ *adjective* referring to a plant or animal on which a parasite lives

hothouse /ˈhɒthaʊs/ *noun* a heated greenhouse

house *noun* a structure where animals or machinery are kept ○ *the reptile house* ○ *the engine house* ■ *verb* to keep livestock in a building ○ *The animals are housed in clean cubicles.*

housing *noun* a series of buildings for livestock

HRI *abbreviation* Horticulture Research International

HSA *abbreviation* Humane Slaughter Association

HSE *abbreviation* Health and Safety Executive

HTA *abbreviation* Horticultural Trades Association

HTST method *abbreviation* high temperature short time method

hull *noun* **1.** the outer covering of a cereal seed. Hulls form bran. **2.** the pod of peas or beans

huller /ˈhʌlə/ *noun* a kind of threshing machine which removes seeds from their husks

human-caused *adjective* referring to a disaster or event which has been brought about by human beings

Humane Slaughter Association *noun* a charity which campaigns for animal

suffering to be minimised during the slaughter process. Abbr **HSA**

humate /ˈhjuːmeɪt/ *noun* a salt that is derived from humus

humid *adjective* relating to air that contains moisture vapour ○ *Decomposition of organic matter is rapid in hot and humid conditions.*

humidify /hjuːˈmɪdɪfaɪ/ *noun* to add water vapour to air to make it more humid

humidity *noun* a measurement of how much water vapour is contained in the air

humification /ˌhjuːmɪfɪˈkeɪʃ(ə)n/ *noun* the breakdown of rotting organic waste to form humus

humify /ˈhjuːmɪfaɪ/ *verb* to break down rotting organic waste to form humus

hump *noun* the rounded flesh on the back or shoulders of an animal, such as a camel, or certain breeds of cattle

humus /ˈhjuːməs/ *noun* the fibrous organic matter in soil, formed from decomposed plants and animal remains, which makes the soil dark and binds it together

hundredweight *noun* a measure of weight of dry goods such as grain (NOTE: Abbreviated after numbers to **cwt: 5cwt**. The British hundredweight is equivalent to 50.8kg, and the US hundredweight is equivalent to 45.4kg.)

hungry soil *noun* soil which lacks nutrients, and so needs large amounts of fertiliser to produce good crops

hurdle *noun* a portable rectangular wooden frame used for temporary fencing for sheep

husband *verb* to use a resource carefully

husbanding /ˈhʌzbəndɪŋ/ *noun* the activity of using a resource carefully ○ *a policy of husbanding scarce natural resources*

husbandry /ˈhʌzbəndri/ *noun* the activity of looking after farm animals and crops ○ *a new system of intensive cattle husbandry*

husbandry system *noun* a written plan for looking after a group of farm animals, looking at considerations such as their habitat, diet, medical care, production rates and general welfare

husk /hʌsk/ *noun* **1.** the dry outer covering of cereal grains, which has little food value, and which is removed during threshing **2.** a parasitic form of bronchitis which is caused by lungworms and is found mainly in cattle ■ *verb* to remove the husk from seeds

hybrid *noun* a new form of plant or animal resulting from a cross between organisms that have different genotypes ○ *high-yielding maize hybrids* ■ *adjective* being the result of a cross between organisms that have different genotypes

hybridisation /ˌhaɪbrɪdaɪˈzeɪʃ(ə)n/, **hybridization** *noun* the production of hybrids

'Another area of interest was the hybridisation of Italian and perennial ryegrass varieties. Italian ryegrasses were valued for their fast growth and bulk, but Italian ryegrass swards tended to last only two years. By cross-breeding perennial ryegrass and Italian ryegrass, it was expected to produce hybrid varieties retaining many of the advantages of Italian ryegrass but lasting for four or five years, said Mr Johnston.' [*Dairy Farmer*]

hybridise /ˈhaɪbrɪdaɪz/, **hybridize** *verb* to produce hybrids by crossing varieties of plants or animals

hybrid vigour *noun* same as **heterosis**

Hybu Cig Cymru *noun* the red meat promotion board for Wales. Abbr **HCC**

hydrated /haɪˈdreɪtɪd/ *adjective* referring to a chemical compound in which water is bound

hydrated lime *noun* a substance produced when burnt lime is wetted. It is a mixture of calcium oxide and water, used to improve soil quality. The lime is in powder form, having been burnt to break it down from large lumps. Also called **slaked lime** (NOTE: The chemical formula is $Ca(OH)_2$.)

hydraulic conductivity /haɪˌdrɒlɪk ˌkɒndʌkˈtɪvɪti/ *noun* the rate at which water can move through soil, used as a factor in soil management

hydrocarbon *noun* a compound formed of hydrogen and carbon

hydrochloric acid /ˌhaɪdrəklɒrɪk ˈæsɪd/ *noun* an inorganic acid which forms in the stomach and is part of the gastric juices

hydrocool /ˌhaɪdrəˈkuːl/ *verb* to cool fresh fruit to prevent it from deteriorating during transport to the retail shop or market

hydroelectricity /ˌhaɪdrəʊɪlekˈtrɪsɪti/ *noun* the electricity produced by water power

hydroelectric power /ˌhaɪdrəʊɪlektrɪk ˈpaʊə/ *noun* the electricity produced by using a flow of water to drive turbines. Also called **hydropower**

hydrogen *noun* a gaseous chemical element that combines with oxygen to form water, with other elements to form acids, and is present in all animal tissue

hydrology *noun* the study of water, its composition and properties and in particular the place of water in the environment

hydromorphic soil *noun* waterlogged soil found in bogs and marshes

hydrophobia /ˌhaɪdrəˈfəʊbiə/ *noun* same as **rabies**

hydrophyte /ˈhaɪdrəfaɪt/ *noun* a plant that lives in water or in marshy conditions

hydroponics /ˌhaɪdrəʊˈpɒnɪks/ *noun* the practice of growing plants in a nutrient liquid with or without sand, vermiculite or other granular material

hydropower /ˈhaɪdrəʊpaʊə/ *noun* same as **hydroelectric power**

hydrops uteri /ˌhaɪdrɒps ˈjuːt(ə)ri/ *noun* a disease of livestock caused by excessive amount of fluid in the pregnant womb. The womb may need to be drained.

hydrostatic pressure /haɪdrəʊˌstætɪk ˈpreʃə/ *noun* the pressure of water that is not moving

hygiene *noun* the state or practice of being clean and keeping healthy conditions

hygrometer /haɪˈgrɒmɪtə/ *noun* an instrument used for the measurement of humidity ○ *The most common type of hygrometer is the wet and dry bulb thermometer arrangement.*

hyperphosphate /ˌhaɪpəˈfɒsfeɪt/ *noun* a soft rock phosphate obtained from North America

hyphae /ˈhaɪfiː/ *plural noun* long thin filaments which make up a typical fungus

hypocalcaemia /ˌhaɪpəʊkælˈsiːmiə/ *noun* same as **milk fever**

hypomagnesaemia /ˌhaɪpəʊmægnɪˈziːmiə/ *noun* a condition caused by lack of magnesium in the bloodstream that makes animals shiver and stagger. Cattle may be affected shortly after being turned out onto spring pastures after having wintered indoors. Also called **grass staggers**

HYV *abbreviation* high-yielding variety

I

I *symbol* iodine

IACR *abbreviation* Institute of Arable Crops Research

IACS *abbreviation* Integrated Administration and Control System

IAH *abbreviation* Institute for Animal Health

IBR *abbreviation* infectious bovine rhinotracheitis

ICA *abbreviation* International Coffee Agreement

ICCA *abbreviation* International Cocoa Agreement

ICCO *abbreviation* International Cocoa Organization

-icide /ɪ'saɪd/ *suffix* substance which destroys a particular organism

ICM *abbreviation* integrated crop management

ICO *abbreviation* International Coffee Organization

ICRISAT /'ɪkrɪsæt/ *abbreviation* International Crops Research Institute for the Semi-Arid Tropics

IFA *abbreviation* Irish Farmers Association

IFR *abbreviation* Institute of Food Research

Ig *abbreviation* immunoglobulin

IGER *abbreviation* Institute of Grassland and Environmental Research

IITA *abbreviation* International Institute of Tropical Agriculture

ILCA *abbreviation* International Livestock Centre for Africa

Ile de France /ˌiːl də 'frɒns/ *noun* a large French breed of sheep, the rams of which are kept to provide crossbred lambs for meat production

illuviation /ɪˌluːviˈeɪʃ(ə)n/ *noun* the movement of particles and chemicals from the topsoil into the subsoil

ILRAD /'ɪlræd/ *abbreviation* International Laboratory for Research on Animal Diseases

imbalance *noun* **1.** a situation where the balance between a set of things is unequal ○ *Lack of vitamins A and E creates hormonal imbalances in farm animals.* **2.** a situation where one species is dominant

immature *adjective* referring to an organism or part that is still developing ○ *an immature duck* ○ *an immature fruit*

immune *adjective* referring to a person, other animal or plant that is not affected by a specific microorganism ○ *This barley strain is not immune to the virus.*

immunisation /ˌɪmjʊnaɪˈzeɪʃ(ə)n/, **immunization** *noun* the production of immunity to a specific disease, either by injecting an antiserum or by giving an individual the disease in such a small dose that the body does not develop the disease, but produces antibodies to counteract it

immunise /'ɪmjʊnaɪz/, **immunize** *verb* to make a person or other animal immune to a specific microorganism by inoculating them

immunity *noun* **1.** the natural or acquired ability of a person or other animal to resist a microorganism and the disease it causes ○ *The vaccine gives immunity to tuberculosis.* **2.** the ability of a plant to resist disease through a protective covering on leaves, through the formation of protoplasts or through the development of inactive forms of viruses

immunoassay /ˌɪmjʊnəʊˈæseɪ/ *noun* a technique for measuring the amount of antigens and antibodies in tissue

'Dr Chambers hopes that a new immunoassay detection technique, developed at CSL, will prove to be commercially viable and a valuable tool for use in the protection of the quality and reputation of UK grain.' [*Arable Farming*]

immunoglobulin /ˌɪmjʊnəʊˈglɒbjʊlɪn/ *noun* a protein produced by specific white blood cells that acts as an antibody in immune responses. Abbr **Ig**

impermeable /ɪmˈpɜːmiəb(ə)l/ *adjective* **1.** referring to a substance which does not allow a liquid or gas to pass through ○ *rocks which are impermeable to water* **2.** referring to a membrane which allows a liquid to pass through, but not solid particles suspended in the liquid

implement *noun* **1.** the process of carrying out a plan **2.** a piece of equipment used for a certain job ■ *verb* to put legislation into action

implementation /ˌɪmplɪmənˈteɪʃ(ə)n/ *noun* the process of carrying out a plan ○ *the rapid implementation of flood defence plans*

import *verb* **1.** to introduce new things from elsewhere **2.** to buy crops or produce in foreign countries and bring them back into the home country ○ *They import dates from North Africa.* ■ *noun* crops or produce which are bought abroad and brought into the country

importer /ɪmˈpɔːtə/ *noun* a person or company which imports produce ○ *a grain importer*

import levy *noun* a tax on farm produce which is imported into the EU

import quotas *plural noun* limits set to the amount of a type of produce which can be imported

impoverish /ɪmˈpɒvərɪʃ/ *verb* to reduce the quality of something □ **to impoverish the soil** to make soil less fertile ○ *Overcultivation has impoverished the soil.*

impoverished *adjective* referring to something with reduced quality ○ *If impoverished soil is left fallow for some years, nutrients may build up in the soil again.*

impregnate *verb* to fertilise a female, by introducing male spermatozoa into the female's body so that they fuse with the female's ova

improved varieties *plural noun* new species of plants which are stronger, or more productive than old species

improvement *noun* the act or an instance of something becoming or being made better ○ *the improvement of crop varieties by selection* ○ *There is still room for improvement in performance.* ○ *We need to achieve improvements in efficiency.*

in- /ɪn/ *prefix* used to refer to a pregnant female animal □ **in-calf, in-pig, in-foal** a cow, sow or mare which is going to have young

inactivate /ɪnˈæktɪveɪt/ *verb* to make something unable to act ○ *The ultraviolet component of sunlight inactivates some herbicides.*

inactive *adjective* **1.** not doing anything **2.** (*of a chemical*) not reacting with other substances **3.** (*of a disease*) not producing symptoms **4.** biologically inert

inactivity /ˌɪnækˈtɪvɪti/ *noun* the state of not being active

inbred /ɪnˈbred/ *adjective* resulting from inbreeding

inbreeding /ˈɪnbriːdɪŋ/ *noun* the process of mating or crossing between closely related individuals, leading to a reduction in variation. Compare **outbreeding** (NOTE: Inbreeding as a result of self-fertilisation occurs naturally in many plants. Inbreeding sorts out some of the best qualities in stock and has been used to establish uniform flocks or herds carrying distinctive traits.)

inbreeding depression *noun* a reduction in variation and vigour arising in a population that is repeatedly inbred. Compare **heterosis**

incisor /ɪnˈsaɪzə/ *noun* a flat sharp-edged tooth in the front of the mouth for cutting and tearing food

incompatible *adjective* unable to cross-fertilise and produce offspring

incorporate *verb* **1.** to apply chemicals such as slug pellets by spreading them in rows at the same time as the seed is sown, as opposed to broadcasting **2.** to plough back straw or green manure into the soil

incorporation /ɪnˌkɔːpəˈreɪʃ(ə)n/ *noun* the process of introducing chopped straw, green manure, etc., into the soil as it is being ploughed

incubation /ˌɪŋkjʊˈbeɪʃ(ə)n/ *noun* the process of keeping eggs warm until the young birds come out, either by an adult bird sitting on them or by artificial means

incubator /ˈɪŋkjʊbeɪtə/ *noun* a special unit providing artificial heat used to hatch eggs. Incubators are available as small trays, or as large rooms for large-scale producers.

indefinite inflorescence *noun* a type of inflorescence in which the stems bearing the flowers continue to grow. Compare **definite inflorescence**

indehiscent /ˌɪndɪˈhes(ə)nt/ *adjective* referring to seed pods, fruit or capsules that do not open to release seeds when ripe. Compare **dehiscent**

Indian corn /ˈɪndiən kɔːn/ *noun* ♦ **maize**

Indian game *noun* a breed of table poultry, often black with yellow legs

indigenous *adjective* native to a place ○ *There are six indigenous species of monkey on the island.* ○ *Bluebells are indigenous to the British Isles.*

indigestible /ˌɪndɪˈdʒestɪb(ə)l/ *adjective* referring to food which cannot be digested, e.g. roughage

indigo /ˈɪndɪɡəʊ/ *noun* a tropical plant of the pea family which is a source of blue dye. Genus: *Indigofera*.

induced twinning *noun* the act of producing twin young after embryo transfer with twin embryos

industrial crop *noun* a crop grown for purposes other than food, e.g. flax grown for fibre

'Farmers who have suffered severe shortfalls in their industrial crops on set-aside should contact the relevant officials to avoid damaging cuts to their arable area payments. Growers in southern Scotland and the north of England have been particularly badly affected by rain and some farmers are ploughing in failed oilseed rape crops.' [*Farmers Weekly*]

infect *verb* **1.** (*of an organism*) to enter a host organism and cause disease ○ *The new strain has infected many people, and the disease is spreading fast.* ○ *All these plants have been infected by a virus.* **2.** to contaminate something with a microorganism that causes disease

infected area *noun* a place where animals must be kept in isolation as a result of a notice issued by the Animals Inspector when an animal is suspected or known to have a notifiable disease

infection *noun* **1.** the process of a microorganism entering a host organism and causing disease **2.** a disease caused by a microorganism ○ *She is susceptible to minor infections.* ○ *West Nile fever is a virus infection transmitted by mosquitoes.*

infectious *adjective* referring to a disease that is caused by microorganisms

and can be transmitted to other individuals by direct means ○ *This strain of flu is highly infectious.* Compare **contagious**

infectious bovine rhinotracheitis /ɪnˌfekʃəs ˌbəʊvaɪn ˌraɪnəʊˌtreɪkiˈaɪtɪs/ *noun* a virus disease of cattle which affects the reproductive, nervous, respiratory or digestive systems. Milk yield is depressed as adults run a high fever. Abbr **IBR**

infective /ɪnˈfektɪv/ *adjective* referring to a disease caused by a microorganism, which can be caught from another person but which cannot always be directly transmitted

infectivity /ˌɪnfekˈtɪvɪti/ *noun* the state of being infective

infertile *adjective* **1.** referring to any organism that is not able to reproduce or produce offspring **2.** referring to trees and plants that are not able to produce fruit or seeds **3.** referring to soil that is not able to produce good crops

COMMENT: An infertile soil is one which is deficient in plant nutrients. The fertility of a soil at any one time is partly due to its natural makeup, and partly to its condition, which is largely dependent on its management in recent times. Application of fertilisers can raise soil fertility and bad management can decrease it.

infertility /ˌɪnfəˈtɪlɪti/ *noun* the inability to reproduce or have offspring

infest *verb* (*of pests*) to be present somewhere in large numbers ○ *Pine forests are infested with these beetles.* ○ *Plants that have been infested should be dug up and burnt.*

infestation /ˌɪnfeˈsteɪʃ(ə)n/ *noun* the presence of large numbers of pests ○ *The crop showed a serious infestation of greenfly.* ○ *The condition is caused by an infestation of lice.*

in-field *noun* formerly, the field nearest the farmstead, regularly manured and cultivated. In-fields are still preserved on some hill farms.

infiltration /ˌɪnfɪlˈtreɪʃ(ə)n/ *noun* **1.** the passing of water into the soil or into a drainage system **2.** an irrigation system in which water passes through many small channels to reach the fields

inflorescence /ˌɪnfləˈresəns/ *noun* a flower or a group of flowers on a stem

COMMENT: There are two types of inflorescence. Indefinite inflorescence is where the branches bearing the flowers continue to grow. Where the main stem ends in a single flower and then stops growing is known as definite inflorescence.

in-going payment *noun* a sum of money paid by a new tenant for the value of the improvements made by the former tenant of a holding

in-ground valuation *noun* the value of tillages or cultivations including direct costs of seed, fertiliser and spray

inherit *verb* to receive a genetically controlled characteristic from a parent ○ *Flower colour is inherited.*

inhibit *verb* to prevent or limit the effect of something ○ *Cloud cover inhibits cooling of the Earth's surface at night.*

inhibitor /ɪn'hɪbɪtə/ *noun* a device or substance which prevents or limits the effect of something

inject *verb* to put a liquid into an animal's body under pressure, by using a hollow needle inserted into the tissues ○ *The cow was injected with antibiotics.*

injection *noun* **1.** the forcing of fluid into something ○ *Power output can be boosted to a value over 100% maximum power, by the injection of a water methanol mixture at the compressor inlet or at the combustion chamber inlet.* **2.** the act of injecting a liquid into a body using a syringe **3.** a preventative measure against a particular disease ○ *a TB injection* **4.** the introduction of something new or stimulating

injurious weed /ɪn'dʒʊəriəs wiːd/ *noun* a weed which causes damage to crops or livestock, e.g. ragwort or creeping thistle

'The Weeds Act does not make it an offence to permit injurious weeds to grow on land but provides Defra with the power to serve a notice on an occupier of any land on which one of the five injurious weeds is growing requiring the occupier to take action to prevent the weeds from spreading.' [*Farmers Guardian*]

in-lay *noun* a period when a hen is laying eggs

in-milk *noun* lactation period of a cow

inocula plural of **inoculum**

inoculate /ɪ'nɒkjʊleɪt/ *verb* **1.** to introduce vaccine into a body in order to stimulate the production of antibodies to a particular organism, giving rise to immunity to the disease ○ *The baby was inoculated against diphtheria.* **2.** to introduce a

microorganism into a plant or a growth medium

inoculation /ɪˌnɒkjʊ'leɪʃ(ə)n/ *noun* **1.** the act of inoculating **2.** an injection against a particular disease ○ *a diphtheria inoculation*

inoculum /ɪ'nɒkjʊləm/ *noun* **1.** material used to inoculate a person or animal against a disease **2.** microscopic airborne spores or other material from an organism that causes disease

inorganic /ˌɪnɔː'gænɪk/ *adjective* **1.** referring to a substance which does not come from an animal or a plant ○ *Inorganic substances include acids, alkalis and metals.* **2.** referring to a substance that does not contain carbon

inorganic acid *noun* an acid which comes from a mineral

inorganic fertiliser *noun* an artificially synthesised fertiliser

inorganic fungicide *noun* a fungicide made from inorganic substances such as sulphur

inorganic herbicide *noun* a herbicide made from inorganic substances such as sulphur

inorganic pesticide *noun* a pesticide made from inorganic substances such as sulphur

inputs /'ɪnpʊts/ *plural noun* substances put into the soil, such as fertilisers which are applied by a farmer

ins /ɪnz/ *noun* a term used to describe the points where the plough enters the ground when leaving the headland (NOTE: The points where the plough is lifted out of the soil are the outs.)

insect *noun* a small animal with six legs and a body in three parts

COMMENT: Insects form the class Insecta. The body of an insect is divided into three distinct parts: the head, the thorax and the abdomen. The six legs are attached to the thorax and two antennae are on the head.

insect bite *noun* a sting caused by an insect which punctures the skin and in so doing introduces irritants

insect-borne *adjective* referring to infection which is carried and transmitted by insects ○ *insect-borne viruses* ○ *Malaria is an insect-borne disease.*

insecticide *noun* a substance which is used to kill insects

COMMENT: Natural insecticides produced from plant extracts are regarded as less harmful to the environment than synthetic insecticides which, though effective, may be persistent and kill not only insects but also other larger animals when they get into the food chain. In agriculture, most pesticides are either chlorinated hydrocarbons, organophosphorus compounds or carbamate compounds. Insecticides may be sprayed or dusted on, or used in granular form as seed dressings. In the form of a gas, insecticides are used to fumigate greenhouses and granaries.

insectivorous /ˌɪnsekˈtɪvərəs/ *adjective* referring to an animal or plant that feeds mainly on insects (NOTE: Pitcher plants and sundews are insectivorous.)

inseminate /ɪnˈsemɪneɪt/ *verb* to impregnate, by introducing male spermatozoa into the female's body so that they link with the female's ova

insemination *noun* the introduction of sperm into the vagina

inspection *noun* a careful check to see if something is in the correct condition or if there are problems (NOTE: Animal welfare codes lay down rules on how closely animals such as laying hens should be inspected in order to ensure that they are healthy.)

inspector *noun* an official whose job is to examine animals, soil, buildings, etc., to see if they conform to government regulations

Institute for Animal Health *noun* a major centre for research into infectious diseases in livestock. Abbr **IAH**

Institute of Arable Crops Research *noun* a research group based in Harpenden, Hertfordshire, which investigates techniques in crop growing. Abbr **IACR**

Institute of Food Research *noun* a not-for-profit scientific institute which advises on food safety, diet and health. Abbr **IFR**

Institute of Grassland and Environmental Research *noun* a research organisation in the field of land use and conservation. Abbr **IGER**

Institute of Terrestrial Ecology *noun* a former ecological research organisation, now merged with the Centre for Ecology and Hydrology. Abbr **ITE**

intake *noun* **1.** an amount of a substance taken into an organism, either eaten or absorbed ○ *a study of food intake among grassland animals* ○ *The bird's daily intake of insects is more than half its own weight.* **2.** hill pasture which has been improved and fenced in

Integrated Administration and Control System *noun* measures intended to combat fraud in aid applications and ensure fair competition in Europe, part of the CAP reforms of 1992. Abbr **IACS**

integrated crop management *noun* an approach to growing crops that combines traditional good farm husbandry with reduction in the use of agrochemicals and takes into consideration the impact of farming practices on the environment

'The farms aim to promote good and profitable agricultural practice, involving integrated crop management to care for the soil, water and the wider environment.' [*Farmers Weekly*]

integrated farm management, integrated farming *noun* an approach to farming that combines the best of traditional methods with modern technology, to achieve high productivity with a low environmental impact

integrated pest management *noun* an appropriate combination of different methods of pest control, involving good cultivation practices, use of chemical pesticides, resistant crop varieties and biological control. Abbr **IPM**

integrated pollution control, integrated pollution prevention and control *noun* an approach which looks at all inputs and outputs from a process that is likely to cause pollution and regulates other factors as well as emissions. Abbr **IPC, IPPC**

intensification /ɪnˌtensɪfɪˈkeɪʃ(ə)n/ *noun* the use of intensive farming methods ○ *Intensification of farming has contributed to soil erosion.* Compare **extensification**

intensify *verb* to use intensive farming methods

intensity /ɪnˈtensɪti/ *noun* the degree to which land is used

intensive *adjective* achieving maximum production from land or animals

intensive agriculture *noun* a method of farming in which as much use is made of the land as possible by growing crops close together, growing several crops in a year or using large amounts of fertiliser. Opposite

extensive agriculture. Also called **productive agriculture**

intensive animal breeding *noun* a system of raising animals in which livestock are kept indoors and fed on concentrated foodstuffs, with frequent use of drugs to control the diseases which tend to occur under these conditions

intensive beef production *noun* the production of a young lean beef animal in a period of less than a year

intensive cultivation, intensive farming *noun* same as **intensive agriculture**

intensive livestock production *noun* a specialised system of livestock production where the livestock are housed indoors. This system can be started up at any time of the year. Disease hazards are those related to diet and permanent housing for the whole of the animal's life.

intensively /ɪn'tensɪvli/ *adverb* using intensive farming methods

inter- /ɪntə/ *prefix* between

interbreed /ˌɪntə'briːd/ *verb* **1.** to mate and have offspring **2.** to cross animals or plants with different characteristics to produce offspring with distinctive features (NOTE: Individuals from the same species can interbreed, those from different species cannot.)

COMMENT: Interbreeding of close relatives can sometimes give a concentration of desirable traits. This was much used by breed pioneers, but it can also increase the frequency of inherited physical defects and breeding plans based on interbreeding are now rare.

intercrop /'ɪntəkrɒp/ *noun* a crop which is grown between the rows of other crops, e.g. barley and mustard or pigeon pea and black gram ■ *verb* to grow crops between the rows of other crops

intercropping /'ɪntəˌkrɒpɪŋ/ *noun* the growing of crops with different characteristics and requirements on the same area of land at the same time ○ *intercropping beans with maize*

'In other work it was shown intercropping cabbages with white clover could contribute to pest control, particularly in circumstances where no insecticide was applied, such as in organic farming…' [*Farmers Guardian*]

interculture /'ɪntəkʌltʃə/ *noun* the practice of mixed cropping, where two or more different crops are grown together on the same area of land

intermuscular /ˌɪntə'mʌskjʊlə/ *adjective* referring to something which is between muscles. Compare **intramuscular**

internal laying *noun* a condition in hens caused by a fault in the oviduct, which results in the yolks not being passed along the oviduct for covering with membranes and shell

international *adjective* referring to more than one country

International Cocoa Agreement *noun* an agreement between countries to stabilise the price of cocoa. Abbr **ICCA**

International Cocoa Organization *noun* an international organisation set up to stabilise the international market in cocoa by holding buffer stocks to offset seasonal differences in production quantities. Abbr **ICCO**

International Coffee Agreement *noun* an agreement between countries to stabilise the price of coffee. Abbr **ICA**

International Coffee Organization *noun* an international organisation set up to stabilise the international market in coffee by holding buffer stocks to offset seasonal differences in production quantities. Abbr **ICO**

International Crops Research Institute for the Semi-Arid Tropics *noun* an organisation established in 1972 at Hyderabad, India. Abbr **ICRISAT**

International Institute of Tropical Agriculture *noun* an organisation established at Ibadan, Nigeria in 1965. Abbr **IITA**

International Laboratory for Research on Animal Diseases *noun* an organisation established at Nairobi, Kenya in 1974. Abbr **ILRAD**

International Livestock Centre for Africa *noun* an organisation established in 1974 at Addis Ababa, Ethiopia. Abbr **ILCA**

International Programme on Chemical Safety *noun* a collaboration between the World Health Organization and the United Nations Environment Programme, to investigate issues relating to chemical safety. Abbr **IPCS**

International Rice Research Institute *noun* an organisation established at Los Banos, Philippines in 1959. Abbr **IRRI**

International Sugar Organization *noun* an international organisation formed of sugar-exporting countries. Abbr **ISO**

International Whaling Commission *noun* an international body set up under an agreement signed in 1946 to control the commercial killing of whales. Abbr **IWC**

International Wheat Council *noun* a group of wheat-exporting countries. Abbr **IWC**

International Wool Secretariat *noun* a group which represents countries which export wool. Abbr **IWS**

internode /ˈɪntənəʊd/ *noun* the part of a plant stem between two adjacent nodes

Interorganisation Programme for the Sound Management of Chemicals *noun* a plan administered by seven participating international organisations, including WHO, to promote the safe use of chemicals. Abbr **IOMC**

interrelay cropping /ˌɪntəˈriːleɪ ˌkrɒpɪŋ/ *noun* a cropping system in which the crops are grown in quick succession, so that the succeeding crop is sown in the standing one, some time before it is harvested

intersow /ˈɪntəsəʊ/ *verb* to sow seed between rows of existing plants

COMMENT: Studies conducted at the Punjab Agricultural University suggest the possibility of intersowing wheat in the furrows between the consecutive potato ridges at the time of earthing up. Summer mungbean (green gram) can then be intersown in the standing wheat crop a few days before harvest, using the space released by the potatoes.

interspecific /ˌɪntəspəˈsɪfɪk/ *adjective* involving two or more species

interveinal /ˌɪntəˈveɪnəl/ *adjective* between the veins

interveinal yellowing *noun* a condition of plants caused by magnesium deficiency, where the surface of the leaves turns yellow and the veins stay green

intervention *noun* 1. the act of making a change in a system 2. □ **to sell into intervention** to sell to a government agency at an intervention price because the market price is too low

Intervention Board *noun* a body set up in 1972 to implement the regulations of the Common Agricultural Policy in the UK. It has now been replaced by the Rural Payments Agency

intervention buying *noun* a feature of the Common Agricultural Policy, whereby governments or their agents offer to buy surplus agricultural produce at a predetermined price. It is subject to a minimum quality standard. Also called **support buying**

intervention price *noun* same as **support price**

COMMENT: The intervention price is the price at which the national intervention agencies are obliged to buy up agricultural commodities offered to them. There are intervention prices on products such as wheat, barley, beef and pigmeat. The application of the system of intervention prices has led to the accumulation of vast stocks of commodities, some of which are sold on the world markets at very low prices.

intestinal /ɪnˈtestɪn(ə)l/ *adjective* referring to the intestine

intestinal diseases *plural noun* diseases and conditions which affect the intestines of animals, e.g. anthrax, dysentery, parasites, enteritis or swine fever

intestine *noun* the digestive canal between the stomach and the anus or cloaca in which food is digested and absorbed (NOTE: In mammals, the small intestine digests and absorbs food from the stomach, and the large intestine then absorbs most of the remaining water.)

intramuscular /ˌɪntrəˈmʌskjʊlə/ *adjective* referring to something which is inside the muscle, as intramuscular fat in meat. Compare **intermuscular**

intraspecific /ˌɪntrəspeˈsɪfɪk/ *adjective* occurring within a species ○ *an intraspecific cross between two cultivars*

intrinsic factor *noun* a protein produced in the gastric glands which reacts with vitamin B12 controls the absorption of extrinsic factor, and which, if lacking, causes pernicious anaemia

introduce *verb* 1. to bring something into being or start to use something new ○ *The lab introduced a new rapid method of testing.* 2. to bring something to a new place ○ *Several of the species of plant now common in Britain were introduced by the Romans.* ○ *Starlings were introduced to the USA in 1891.*

introduction *noun* 1. the process of bringing something into being or using something new ○ *the introduction of a new rapid testing method* ○ *The death rate from* malaria was very high before the introduction of new anti-malarial techniques. **2.** a plant or animal that has been brought to a new place ○ *It is not an indigenous species but a 19th-century introduction.*

in utero /ˌɪn ˈjuːtərəʊ/ *adverb* in the womb

invasion *noun* the arrival of large numbers of unwanted organisms into an area ○ *an invasion of weeds*

invertebrate /ɪnˈvɜːtɪbrət/ *noun* an animal that has no backbone. Compare **vertebrate** ■ *adjective* referring to animals that have no backbone ○ *marine invertebrate animals*

invertebrate pests *plural noun* pests such as grain mites and storage insects such as saw-toothed beetles and the grain weevil, which cause considerable damage to crops in tropical or warm temperate areas

inwintering *noun* the practice of housing cattle and sheep during the winter months rather than keeping them outdoors

iodine *noun* a chemical element. It is essential to the body, especially to the functioning of the thyroid gland, and is found in seaweed.

iodophor /aɪˈɒdəfɔː/ *noun* a disinfectant used to disinfect teats of cows to prevent mastitis

IOMC *abbreviation* Interorganisation Programme for the Sound Management of Chemicals

ion /ˈaɪən/ *noun* an atom or a group of atoms that has obtained an electric charge by gaining or losing one or more electrons (NOTE: Ions with a positive charge are called cations and those with a negative charge are anions.)

IPC ♦ IPPC

IPCS *abbreviation* International Programme on Chemical Safety

IPM *abbreviation* integrated pest management

IPPC *abbreviation* integrated pollution prevention and control

IPU *abbreviation* isoproturon

Irish Moiled /ˌaɪrɪʃ ˈmɔɪld/ *noun* a rare breed of medium-sized dual-purpose cattle. The animals have a distinctive white back strip.

iron *noun* a metallic element that is essential to biological life and is an essential part of human diet. Iron is found in liver, eggs, etc.

COMMENT: Iron is an essential part of the red pigment in red blood cells. Lack of iron in haemoglobin results in iron-deficiency anaemia. Its role in the physiology of plants appears to be associated with specific enzymatic reactions and the production of chlorophyll.

irongrass /ˈaɪəngrɑːs/ *noun* same as **knotgrass**

ironweed /ˈaɪənwiːd/ *noun* same as **knotgrass**

irradiate /ɪˈreɪdieɪt/ *verb* **1.** to subject something to radiation **2.** to treat food with radiation to prevent it going bad

irradiation /ɪˌreɪdɪˈeɪʃ(ə)n/ *noun* **1.** the spread of something from a centre **2.** the use of rays to kill bacteria in food

COMMENT: Food is irradiated with gamma rays from isotopes which kill bacteria. It is not certain, however, that irradiated food is safe for humans to eat, as the effects of irradiation on food are not known. In some countries irradiation is only permitted as a treatment of certain foods.

IRRI *abbreviation* International Rice Research Institute

irrigate *verb* to supply water to land to allow plants to grow, by channels, pipes, sprays or other means

irrigation /ˌɪrɪˈgeɪʃ(ə)n/ *noun* the artificial supplying and application of water to land with growing crops

COMMENT: Irrigation can be carried out using powered rotary sprinklers, rain guns, spray lines or by channelling water along underground pipes or small irrigation canals from reservoirs or rivers. Irrigation water can be more effectively used than the equivalent amount of rainfall, because a regular supply is ensured. Basin or flood irrigation is a primitive form of irrigation, where flood waters from rivers are led to prepared basins. Perennial irrigation allows the land to be irrigated at any time. This may be by primitive means such as shadufs, or by distributing water from barrages by canal and ditches. It is usual to measure irrigation water in millimetres: 1mm on one hectare equals $10m^3$ or ten tonnes. Irrigation is not necessarily always advantageous to the land, as it can cause salinisation of the soil. This happens when the soil becomes waterlogged so that salts in the soil rise to the surface. At the surface, the irrigated water rapidly evaporates, leaving the salts behind in the form of a saline crust. Irrigation also has the further disadvantage of increasing the spread of disease. Water insects easily spread through irrigation canals and reservoirs. In the United Kingdom the greatest need for irrigation is in the east, where the lower rainfall and higher potential evaporation and transpiration means that irrigation is beneficial nine years out of ten. In the UK, potatoes, sugar beet, horticultural crops and grassland are the main irrigated crops.

irrigator /'ɪrɪgeɪtə/ *noun* a device for irrigating, e.g. the Baars irrigator

isinglass /'aɪzɪŋglɑːs/ *noun* a pure soluble gelatin, used to make alcoholic drinks clear; formerly used to preserve eggs

iso- *prefix* equal

isobar /'aɪsəʊbɑː/ *noun* a line on a map linking points which are of equal barometric pressure at a given time

isoglucose /ˌaɪsə'gluːkəʊz/ *noun* same as **high fructose corn syrup**

isohyet /aɪsəʊ'haɪət/ *noun* a line on a map linking points of equal rainfall

isolate *verb* to separate and keep objects or organisms apart from others ○ *isolated the sick animals in a separate enclosure*

isolation *noun* the process of keeping infected animals away from others

isoleucine /ˌaɪsəʊ'luːsiːn/ *noun* an essential amino acid

isoproturon /ˌaɪsəʊ'prɒtjʊrɒn/ *noun* a herbicide used on cereals that is found as a contaminant of surface water (NOTE: It is commonly used in the UK but it is under review for withdrawal from use in the European Union.)

isotherm /'aɪsəʊθɜːm/ *noun* a line on a map linking points of equal temperature

Italian ryegrass /ɪˌtælɪən 'raɪgrɑːs/ *noun* a short lived ryegrass *(Lolium multiflorum)* which is sown in spring and is very quick to establish. It produces good growth in its seeding year and early graze the following year, and is commonly used for short duration leys.

itch *noun* a form of mange

itch mite *noun* an arachnid *(Sarcoptes scabiei)* which burrows into the animal's skin, causing itching

ITE *abbreviation* Institute of Terrestrial Ecology

IUCN – The World Conservation Union *noun* a union of 140 countries that generates scientific knowledge, advice and standards on environmental subjects and monitors the status of species, publishing findings in its Red Lists (NOTE: 'IUCN' stands for the organisation's original name, 'International Union for the Conservation of Nature and Natural Resources', although **World Conservation Union** has largely replaced this title since 1990.)

ivy *noun* a climbing evergreen plant *(Hedera helix)*

ivy-leaved speedwell *noun* a widespread weed *(Veronica hederifolia)* which affects most autumn sown crops. Also called **bird's-eye, eye-bright**

IWC *abbreviation* 1. International Wheat Council 2. International Whaling Commission

IWS *abbreviation* International Wool Secretariat (NOTE: Now called 'The Woolmark Company'.)

J

J *abbreviation* joule

Jack bean *noun* a tropical legume (*Canavalia ensiformis*) grown as a fodder crop

Jacob /ˈdʒeɪkəb/ *noun* a rare breed of sheep with multi-coloured fleece. It is medium-sized and multi-horned, with a white coat and brown or black patches.

Jersey /ˈdʒɜːzi/ *noun* an important breed of dairy cattle, originally from the island of Jersey. Jersey cows are smaller than most other breeds and produce high yields of high butterfat content milk. The cattle are variously coloured from light fawn to red and almost black.

Jerusalem artichoke /dʒəˌruːsələm ˈɑːtɪtʃəʊk/ *noun* ◆ **artichoke**

jetting /ˈdʒetɪŋ/ *noun* **1.** a method of applying insecticide under pressure, used on sheep **2.** a method of cleaning out blocked field drains using high pressure water jets

jetting gun *noun* a gun used to apply insecticide

JNCC *abbreviation* Joint Nature Conservation Committee

Johne's disease /ˈjəʊnəz dɪˌziːz/ *noun* a serious infectious inflammation of the intestines, particularly in cattle. Affected animals rapidly become extremely thin.

joint *noun* **1.** a place at which two bones are connected **2.** a piece of meat ready for cooking, usually containing a bone ■ *adjective* **1.** shared by two or more people ○ *a joint effort* ○ *a joint venture* **2.** referring to a joint in the body ○ *joint pains*

joint-ill *noun* a disease of young livestock, especially newborn calves, kids, and lambs. It causes abscesses at the navel and swellings in some joints. Also known as **navel-ill**

Joint Nature Conservation Committee *noun* a UK government advisory body on conservation. Abbr **JNCC**

Joint Regulatory Authority *noun* the body which processes applications to grow and sell genetically modified crops. Abbr **JRA**

jojoba /həˈhəʊbə/ *noun* a perennial plant, grown in the USA, whose seeds yield an oil which is liquid wax

joule /dʒuːl/ *noun* an SI unit of measurement of energy. Symbol **J**

COMMENT: One joule is the amount of energy used to move one kilogram the distance of one metre, using the force of one newton 4.184 joules equal one calorie.

JRA *abbreviation* Joint Regulatory Authority

juice *noun* **1.** liquid inside a fruit or vegetable **2.** liquid inside cooked meat

juice extractor *noun* a device for extracting juice from a fruit or vegetable

June agricultural census *noun* an annual survey of agricultural activity in Great Britain carried out by Defra (NOTE: Until 2000, the census covered all holdings, regardless of their size. From 2000 onwards, the information on crop areas, numbers of livestock, production and yields, number and size of holdings, numbers of workers, farm machinery, prices and incomes has been derived from a sample set of holdings.)

June drop *noun* a natural fall of small fruit in early summer, which allows the remaining fruit to grow larger

juniper /ˈdʒuːnɪpə/ *noun* a small coniferous tree or shrub of the northern hemisphere, with cones that resemble berries. Genus: *Juniperus*. (NOTE: *Juniperus communis* is native to the British Isles.)

jute /dʒuːt/ *noun* a coarse fibre from a plant (*Corchorus* sp), used to make sacks, coarse cloth and cheap twine (NOTE: The main producers of jute are Bangladesh (which produces over 50% of the total world production), India and Brazil.)

juvenile *noun* a young animal or plant ■ *adjective* referring to an animal, plant, organ or type of behaviour that is not yet adult ○ *The juvenile foliage of eucalyptus is different from its adult foliage.*

K

k *symbol* kilo-

K *symbol* potassium

kainite /'keɪnaɪt/ *noun* potash fertiliser, made of a mixture of potassium and sodium salts, with sometimes magnesium salts added, used mainly on sugar beet and similar crops

kale /keɪl/ *noun* a type of brassica, sometimes used as a green vegetable for human consumption, but mainly grown as animal forage. Also called **bore cole**

COMMENT: Kale can be fed to animals in the field, or made into silage for use during winter. The main types of kale are the marrowstem, which produces heavy crops but is not winter hardy, the thousand-headed, which is hardier, and the dwarf thousand-head, which produces a large number of new shoots late in the winter. Other hybrid varieties are also available. Kale is the commonest of green crops other than grass. The highest feeding value is in the leaf rather than the stem.

karst /kɑːst/ *noun* ground typical of limestone country, with an uneven surface and holes and cracks due to weathering

ked /ked/ *noun* the sheep tick; a blood-sucking fly (*Melophagus ovinus*) which is a parasite of sheep and causes extreme irritation

keep *noun* grass or fodder crops for the grazing of livestock ■ *verb* to remain in good condition after harvest ○ *Conference pears will keep until spring.*

keeper *noun* **1.** a person who looks after deer, pheasants or other animals and birds which are reared to be hunted **2.** a fruit which keeps well

keiserite /'keɪzəraɪt/ *noun* magnesium sulphate powder, used as a fertiliser where magnesium deficiency is evident, especially in light sandy soil

kemp /kemp/ *noun* a very coarse fibre in fleece, covered with a thick sheath and shed each year

kemp-free mohair *noun* mohair which does not have any kemp

kenaf /kə'næf/ *noun* a fibre-producing plant similar to jute

kennel *noun* **1.** a small hut for a dog **2.** □ **kennels** a commercial establishment where dogs are reared or kept for their owners

Kent /kent/ *noun* ♦ **Romney**

Kentish cob /ˌkentɪʃ 'kɒb/ *noun* a commercially grown variety of hazel nut

kernel *noun* **1.** the soft edible part of a nut **2.** the seed and husk of a cereal grain

Kerry /'keri/ *noun* a rare breed of dairy cattle which is native to Ireland. The animals are small and black, sometimes with white patches on the udder, and have upturned horns.

Kerry Hill *noun* a breed of small hill sheep originating in the Kerry hills of Powys in Wales. It has a soft white fleece and speckled face and legs. The ewes are crossed with Down rams for lamb production.

ketosis /kiː'təʊsɪs/ *noun* a wasting disease in livestock caused by low levels of glucose in the blood. The symptoms are a chronic lack of energy, depletion of fat reserves and a sudden drop in milk production.

kg *symbol* kilogram

kg/ha *noun* kilograms per hectare

kibbled /'kɪbæə)ld/ *adjective* coarsely ground, as in kibbled maize

kibbutz /kɪ'bʊts/ *noun* a form of settlement in Israel, based on the collective farming principles (NOTE: The plural is **kibbutzim**.)

kid *noun* a young goat of either sex, up to one year old

kidding pen *noun* a pen in which a doe is kept when giving birth to kids

kidney bean *noun* a climbing French bean, with red seeds, used as a vegetable

kid-snatching *noun* the practice of taking a new-born kid away from its mother to prevent her from licking it and so passing on caprine arthritis-encephalitis

kieserite /ˈkiːzərait/ *noun* another spelling of **keiserite**

killing age *noun* the age of an animal or bird when it is slaughtered

killing out percentage *noun* the deadweight of an animal expressed as a percentage of its live weight

kilo *noun* same as **kilogram**

kilo- *prefix* one thousand, 10^3. Symbol **K**

kilocalorie /ˈkiːləʊkæləri/ *noun* a unit of measurement of heat equal to 1000 calories (NOTE: In scientific use, the SI unit **joule** is now more usual. 1 calorie = 4.186 joules.)

kilogram *noun* the base unit of mass in the SI system, equal to 1000 grams or 2.2046 pounds. Symbol **kg**. Also called **kilo**

kilojoule /ˈkiːləʊdʒuːl/ *noun* an SI unit of measurement of energy or heat equal to 1000 joules. Symbol **kJ**

kilometre *noun* a measure of length equal to 1000 metres or 0.621 miles. Symbol **km** (NOTE: The US spelling is **kilometer**.)

kindling *noun* the birth of a litter, especially rabbits

kip /kɪp/ *noun* a hide of a young animal, used for leather

kitchen garden *noun* a garden with herbs and small vegetables, ready for use in the kitchen

kJ *abbreviation* kilojoule

km *abbreviation* kilometre

knacker /ˈnækə/ *noun* a person who slaughters casualty animals, particularly horses

knapweed /ˈnæpwiːd/ *noun* a perennial weed *(Centaurea nigra)*

knee cap *noun* a felt protector for the knees of horses, used especially when transporting them as a protection against damage caused when slipping. Also used on young horses when jumping.

knife *noun* an attachment on the cutter bar of a mower or combine harvester (NOTE: There are two types of knife, one with smooth sections which need frequent sharpening, and the other with serrated sections which need no sharpening.)

knotgrass /ˈnɒtgrɑːs/ *noun* a common weed *(Polygonum aviculare)* which affects spring cereals, sugar beet and vegetable crops. Its spreading habit prevents other slower-growing plants from growing. Also called **ironweed**, **irongrass**, **pigweed**, **wireweed**

knotter /ˈnɒtə/ *noun* the mechanism on a baler which ties the bales. It has three basic parts: billhook, retainer disc and the knife.

kohlrabi /kəʊlˈrɑːbi/ *noun* a variety of cabbage with a swollen stem, used as a fodder crop, and also sometimes eaten as a vegetable. The leaves may be green or purple. Also called **turnip-rooted cabbage**

Kyloes /ˈkailəʊz/ *noun* a breed of small long-horned shaggy Highland cattle

L

label *noun* a piece of paper attached to produce, showing the price and other details ■ *verb* to identify something by using a label ○ *Parts are labelled with the manufacturer's name.*

COMMENT: Government regulations cover the labelling of food; it should show not only the price and weight, but also where it comes from, the quality grade, and a sell-by date.

labourer *noun* a person who does heavy work

Lacaune /læ'kəʊn/ *noun* a breed of sheep found mainly in Aveyron, France and used mainly for milk production

lactate /læk'teɪt/ *verb* to produce milk as food for young

lactation /læk'teɪʃ(ə)n/ *noun* **1.** the production of milk as food for young **2.** the period during which young are nourished with milk from the mother's mammary glands. ◊ **suckling**

COMMENT: Lactation is stimulated by the production of the hormone prolactin by the pituitary gland. In cows, goats and sheep kept for milk production, the lactation period is made longer by regular milking. For a dairy cow, the period is ten months, followed by a two-month rest before calving again.

lactic acid *noun* a sugar which forms in cells and tissue, and is also present in sour milk, cheese and yoghurt

lactose *noun* a sugar found in milk

ladder farm *noun* a farm with a series of small long narrow fields

ladybird *noun* a beetle of the *Coccinellidae* family, which is useful to the farmer because it feeds on aphids which would damage plants if they were not destroyed (NOTE: The US term is **ladybug**.)

lagoon *noun* a pool of water or other liquid ○ *Slurry can be stored in lagoons.*

LAI *abbreviation* leaf area index

laid crop *noun* a crop which has been flattened by rain and wind

laid hedge *noun* a hedge that has been made by bending over each stem and weaving it between stakes driven into the ground

laid wool *noun* wool discoloured by the use of salves containing tar

lairage /'leərɪdʒ/ *noun* a shed or outdoor enclosure for the temporary housing of animals, as on the way to market, or when they are being transported for export

'A consequence of the introduction of the livestock movement ban will be that auction markets cease to operate. However, it is possible that some livestock will either be in a lairage or on their way to market. In such an eventuality these animals will either be sent back to their original holding or direct to slaughter.' [*Farmers Guardian*]

lake *noun* a large quantity of liquid produce stored because of overproduction (*informal*) ○ *a wine lake* ○ *a milk lake* ◊ **mountain**

lamb *noun* **1.** a young sheep under six months of age **2.** meat from a lamb (NOTE: Meat from an older sheep is called 'mutton', though this is not common.) ■ *verb* to give birth to lambs ○ *Most ewes lamb without difficulty, but some may need help.*

lamb dysentery *noun* a bacterial disease which enters the lamb from the pasture. The bacteria infects the land for a very long time. The disease can be avoided by vaccinating the lambs as soon as possible after birth or by vaccinating the ewes before lambing.

lambing *noun* the action of giving birth to lambs

lambing pen *noun* a pen in which a ewe is kept when giving birth to lambs

lambing percentage *noun* the number of live lambs born per hundred ewes

lambing season *noun* the period of the year when a flock of ewes produces lambs, usually between December and January. The object is to produce lambs for the market when the price is highest, usually between February and May.

lambing sickness *noun* a bacterial disease of sheep picked up from the soil, which can cause rapid death

lambing tunnel *noun* a covered enclosure for ewes and lambs

lamb's foot *noun* same as **parsley piert**

lamina /ˈlæmɪnə/ *noun* **1.** (*in mammals with hooves*) one of several layers of sensitive tissue just inside the hard exterior of the hoof **2.** the blade of a leaf

laminitis /ˌlæmɪˈnaɪtɪs/ *noun* the inflammation of the lamina in a hoof, causing swelling, and often leading to deformed hooves. It is possibly caused by too much grain feed.

LaMIS *abbreviation* Land Management Information Service

land *noun* **1.** the solid part of the Earth's surface □ **back to the land** encouragement given to people who once lived in the country and moved to urban areas to return to the countryside **2.** a section of a field, divided from other sections by a shallow furrow, a term used in systematic ploughing

land agent *noun* a person employed to run a farm or an estate on behalf of the owner

land capability *noun* an estimate of the potential of land for agriculture, made on purely physical environmental factors such as climate and soil. Compare **land suitability**

COMMENT: In 2004 the total area of agricultural land in the UK was 18,436,000 hectares occupying more than 70% of the total land area. On average only 16.1% of this is Grade 1 or 2.

land classification *noun* the classification of land into categories, according to its value for a broad land use type

COMMENT: In England and Wales, the Defra classification map has five main grades, between Grade 1 (completely suitable for agriculture) and Grade 5 (land with severe limitations, because of its soil, relief or climate).

land clearance *noun* the removal of trees or undergrowth in preparation for ploughing or building

land consolidation /ˈlænd kən ˌsɒlɪdeɪʃ(ə)n/ *noun* joining small plots of land together to form larger farms or large fields

land drainage *noun* the process of removing surplus water from land

COMMENT: If surplus water is prevented from moving through the soil and subsoil, it soon fills all the pore spaces in the soil and this will kill or stunt the growing crops. Well-drained land is better aerated, and crops are less likely to be damaged by root-destroying fungi. Aerated soil also warms up more quickly in spring. Plants form deeper and more extensive roots systems, grassland is firmer, and disease risk from parasites is reduced. The main methods of draining land are underground pipe drains, mole drains and ditches.

land erosion control *noun* a method of preventing the soil from being worn away by irrigation, planting or mulching

landfill *noun* **1.** the disposal of waste by putting it into holes in the ground and covering it with earth **2.** same as **landfill site**

landfilling /ˈlændfɪlɪŋ/ *noun* the practice of disposing of waste by putting it into holes in the ground and covering it with earth

landfill site *noun* an area of land where waste is put into holes in the ground and covered with earth ○ *The council has decided to use the old gravel pits as a landfill site.* ○ *Landfill sites can leak pollutants into the ground water.* ○ *Landfill sites, if properly constructed, can be used to provide gas for fuel.* Also called **landfill**

landfill tax *noun* a tax on every metric ton of waste put in a landfill site instead of being recycled that is paid by companies and local councils

land improvement *noun* the process of making the soil more fertile

landlord *noun* the owner of land or building who lets it to a tenant for an agreed rent

land management *noun* the use and maintenance of land according to a set of principles for a particular purpose such as the cultivation of crops or recreational activities

'The skylark and lapwing have all declined in the UK because of loss of habitat and changes in land management. Stone-curlew numbers plummeted after the Second World War but this year hit a national recovery target five years early.' [*Farmers Guardian*]

land manager *noun* someone who is responsible for the condition of land, e.g. a farmer or landowner

landowner *noun* a person who owns land freehold, and may let it to a tenant, or may farm it himself

landrace /'lændreɪs/ *noun* a local variety of plant or animal developed over many thousands of years by farmers selecting for favourable characteristics within a species

land reform *noun* a government policy of splitting up agricultural land and dividing it up between those people who do not own any land

landscape *noun* the scenery, general shape, structure and features of the surface of an area of land

landscape gardener *noun* a gardener who creates a new appearance for a garden

landscape manager *noun* somebody such as a farmer or landowner who is responsible for the way land is used and looked after

COMMENT: Many farmers find themselves as landscape managers, required to maintain the countryside in an aesthetically and environmentally pleasing condition for the predominantly urban population to enjoy.

landside /'lændsaɪd/ *noun* the part of the plough which takes the sideways thrust as the furrow is turned

land suitability *noun* the suitability of land for a certain agricultural purpose

COMMENT: Land suitability is similar to land capability, but defines its usefulness for a particular purpose. Suitability tends to emphasise the positive value of land, while capability emphasises its limitations.

land tenure *noun* the way in which land is owned and possessed. This may be by an individual owning the freehold, by a tenancy agreement between freeholder and tenant, or by a form of community ownership.

land use *noun* the way in which land is used for different purposes such as farming or recreation ○ *a survey of current land use*

COMMENT: In the UK, the main uses of land are classified as: crops and fallow, temporary grass, permanent grass, rough grazing, other land, urban land, forestry and woodland, and miscellaneous.

land use classification *noun* the classification of land according to the way it is used

lanolin /'lænəlɪn/ *noun* a fat extracted from sheep's wool used in making soaps, skin creams and shampoos

Lantra /'læntrə/ *noun* an organisation which works with employers in the agriculture and land industries to ensure that workers in those industries have the necessary skills and training

LAQM *abbreviation* local air quality management

larch /lɑːtʃ/ *noun* a deciduous European softwood tree that has cones. It is fast-growing and used as a timber crop. Genus: *Larix.*

Large Black *noun* a dual-purpose hardy breed of pig that is black with long lop ears

Large White *noun* an important commercial breed of pig that is white with pricked up ears

larva *noun* the form of an insect or other animal in the stage of development after the egg has hatched but before the animal becomes adult (NOTE: The plural is **larvae.**)

larval /'lɑːv(ə)l/ *adjective* referring to larvae ○ *the larval stage*

lasso /lə'suː/ *noun* a rope with a noose at the end, used to catch cattle ■ *verb* to catch cattle, using a lasso

lateral *noun* a shoot which branches off from the leader or main branch of a tree or shrub

latex *noun* **1.** a white fluid from a plant such as poppy, dandelion or rubber tree **2.** a thick white fluid from a rubber tree, which is treated and processed to make rubber

laurics /'laʊrɪks/, **lauric oils** *plural noun* oils from palm seed and coconut

lavender *noun* a shrub (*Lavandula officinalis*) with small lilac-coloured flowers and narrow leaves, cultivated for perfume

laver /'lɑːvə/ *noun* Welsh name for a variety of seaweed which is edible

laxative *adjective, noun* substance which encourage movements of the bowel ○ *Succulent food such as root crops have a laxative effect.*

Laxton's Superb /ˌlækstənz suˈpɜːb/ *noun* a variety of dessert apple formerly grown commercially in the UK

lay *noun* □ **hen in lay** bird which is laying eggs ■ *verb* to produce an egg

layer *noun* **1.** a flat area of a substance under or over another area (NOTE: In geological formations, layers of rock are called **strata**; layers of soil are called **horizons**.) **2.** a stem of a plant which has made roots where it touches the soil **3.** a bird that is in lay ■ *verb* to propagate a plant by bending a stem down until it touches the soil and letting it form roots there

layering *noun* **1.** a method of propagation where the stem of a plant is bent until it touches the soil, and is fixed down on the soil surface until roots form **2.** a process by which the half-cut stems of hedge plants are bent over and wove around stakes set in the ground, to form a new hedge

layers' ailments *plural noun* disorders of fowls in lay, especially birds that are in heavy production. These disorders include egg binding, internal laying and layer's cramp.

layer's cramp *noun* a condition found in pullets after the first few weeks of their laying life. The bird appears weak, but the trouble usually disappears after a few days.

laying *noun* the action of producing eggs

laying cage *noun* a specially built cage for laying hens. The cages are arranged in tiers and each cage should allow the birds to stand comfortably, allow the eggs roll forward and permit access to food and water, easy cleaning and easy handling of the birds.

laying hen *noun* a female domestic fowl which is kept primarily for egg production

laying period *noun* the period during which a hen will continue to lay eggs. This begins at 18 weeks of age and normally lasts for 50 weeks.

lazy-bed *noun* a small arable plot used for growing potatoes, cereals and other crops, found in the West Highlands of Scotland. If the soil is thin, seed potatoes are placed on the surface of the soil and covered with turf.

lea /liː/ *noun* open ground left fallow or under grass

leach /liːtʃ/ *verb* to be washed out of the soil by water ○ *Excess chemical fertilisers on the surface of the soil leach into rivers and cause pollution.* ○ *Nitrates have* leached into ground water and contaminated the water supply.

leachate /ˈliːtʃeɪt/ *noun* **1.** a substance which is washed out of the soil **2.** a liquid which forms at the bottom of a landfill site

leaching *noun* the process by which a substance is washed out of the soil by water passing through it

leader *noun* an animal which goes first, which leads the flock or herd

leader-follower system *noun* a system of grazing where priority is given to a group of animals (the leaders) and the crop is later grazed by a second group of animals (the followers). So first-year heifers might be followed by second-year heifers.

leaf *noun* a structure growing from a plant stem, with a stalk and a flat blade. It is usually green and carries out photosynthesis. (NOTE: A leaf stalk is called a **petiole**, and a leaf blade is called a **lamina**.)

LEAF *noun* an independent organisation that promotes better understanding of farming by the public and helps farmers improve the environment by combining the best traditional farming methods with modern technology. Full form **Linking Environment and Farming**

leaf area index *noun* the area of green leaf per unit area of ground. Abbr **LAI**

leaf blotch *noun* a disease of cereals (*Rhynchosporium secalis*) where dark grey lesions with dark brown margins occur on the leaves

leaf burn *noun* same as **leaf scorch**

leaf cutting *noun* a piece of a leaf, root or stem cut from a living plant and put in soil where it will sprout

leaf joint *noun* a point on the stem of a plant where a new shoot may grow

leaf roll *noun* a viral disease of potatoes, transmitted by aphids. The leaves roll up and become dry, and the crop yield is affected.

leaf scorch *noun* damage done to leaves by severe weather conditions or herbicides. Also called **leaf burn**

leaf spot *noun* a fungal disease of brassicas, where the leaves develop brown and black patches

leaf stripe *noun* a disease of barley and oats (*Pyrenophora graminea*) where the young leaves show pale stripes and seedlings often die

lean meat *noun* meat with very little fat ○ *Animals are bred to produce lean meat.*

lease *noun* a written contract for letting or renting a piece of equipment for a period against payment of a fee ■ *verb* **1.** to let or rent land or equipment for a period ○ *The company has a policy of only using leased equipment.* **2.** to use land or equipment for a time and pay a fee ○ *All the farm's tractors are owned, but the combines are leased.*

leaseback /'liːsbæk/ *noun* arrangement where property is sold and then taken back by the former owner on a lease

leasehold /'liːshəʊld/ *noun, adjective, adverb* possessing property on a lease, for a fixed time ○ *to purchase a property leasehold* ○ *The property is for sale leasehold.*

leaseholder /'liːshəʊldə/ *noun* a person who holds a property on a lease

leather *noun* the skin of an animal, tanned and prepared for use

leatherjacket /'leðədʒækɪt/ *noun* the larva of the cranefly *(Tipuda paludosa)* which hatches from eggs laid on the ground and feeds on the young crop in spring. When grass is ploughed for cereal crops, the larvae feed on the seedling wheat, damaging the plants at or just below ground level.

leek *noun* a hardy winter vegetable *(Allium ameloprasum)* with a mild onion taste (NOTE: To produce high-quality leeks, the lower parts of the stems need to be blanched. The stems are used in soups and stews.)

Leghorn /leg'hɔːn/ *noun* an excellent laying breed of hen. It is a hardy bird, coloured black, brown and white. Leghorns produce good-sized white eggs.

legume /'legjuːm/ *noun* **1.** a member of the plant family that produces seeds in pods, e.g. peas and beans. Family: Leguminosae. **2.** a dry seed from a single carpel, which splits into two halves, e.g. a pea

COMMENT: There are many species of legume, including trees, and some are particularly valuable because they have root nodules that contain nitrogen-fixing bacteria. Such legumes have special value in maintaining soil fertility and are used in crop rotation. Peas, beans, clover and vetch are all legumes.

Leguminosae /le,guːmɪn'əʊsi/ *noun* a family of plants including peas and beans, that produce seeds in pods

leguminous /lɪ'gjuːmɪnəs/ *adjective* referring to a legume

Lehmann system /'leɪmən ˌsɪstəm/ *noun* a system of pig breeding developed in Germany, where bulk food such as potatoes and fodder beet are fed after a basic ration of meal

Leicester longwool /ˌlestə 'lɒŋwʊl/ *noun* a breed of large hornless white-faced sheep, used a lot by Robert Bakewell, but now rare. ◊ **Border Leicester, Blue-faced Leicester**

lemma /'lemə/ *noun* the outer bract which encloses the flowers of grass

lemon *noun* a yellow edible fruit of an evergreen citrus tree *(Citrus limon)*. Lemons have a very tart flavour and are used in flavouring and in making drinks.

Leptospira hardjo /ˌleptəʊspaɪrə 'haːdjəʊ/ *noun* a bacterium which infects cattle and humans, causing leptospirosis and Weil's disease. Abbr **LH, lep hardjo**

leptospirosis /ˌleptəʊspaɪ'rəʊsɪs/ *noun* a disease of cattle caused by bacteria, which causes abortions and low milk yields. It can be carried by sheep or in running water.

LERAP *abbreviation* Local Environment Risk Assessment for Pesticides

lesion *noun* an open wound on the surface of a plant or on the skin of an animal, caused by disease or physical damage

less favoured area *noun* a former name for land in mountainous and hilly areas, which is capable of improvement and use as breeding and rearing land for sheep and cattle. It is now called Disadvantaged or Severely Disadvantaged Areas. The EU now recognises such areas and gives financial help to farmers in them. Abbr **LFA**

Less Favoured Areas Support Scheme *noun* formerly, a government scheme providing financial support to ensure that farming continued in mountainous or hilly areas

let-down *noun* □ **let-down of milk** the release of milk from the mammary gland

COMMENT: The hormone oxytoxin activates the release of milk. The let-down lasts between seven and ten minutes, when the extraction of milk from the udder is easiest.

lettuce *noun* a salad vegetable *(Lactuca sativa)* which comes in a variety of forms and leaf textures. The commonest are cos lettuce, cabbage lettuce, crisphead and loose-leaved lettuces. Cabbage lettuces have roundish heads, while cos lettuces have longer leaves and are more upright.

leucine /ˈluːsiːn/ *noun* an essential amino acid

leucocyte *noun* a white blood cell

level *adjective* having a flat, smooth horizontal surface ■ *noun* **1.** a relative amount, intensity or concentration ○ *an unsafe level of contamination* ○ *reduced noise levels* **2.** a flat low-lying area of usually marshy land, often reclaimed by artificial drainage in parts of Fen Country in Eastern England round the Wash

levy *noun* money which is demanded and collected by a government

ley /leɪ/ *noun* **1.** a field in which crops are grown in rotation with periods when the field is sown with grass for pasture (NOTE: Leys are an essential part of organic farming.) **2.** land which has been sown to grass for a time

ley farming *noun* farming system in which fields are left to pasture in rotation

COMMENT: Strictly speaking, ley farming is a system where a farm or group of fields is cropped completely with leys which are reseeded at regular intervals; alternatively, any cropping system which involves the use of leys is called ley farming. Ley farming is an essential part of organic farming. Pasture land is fertilised by the animals which graze on it, and then is ploughed for crop growing. When the land has been exhausted by the crops, it is put back to pasture to recover.

LFA *abbreviation* less favoured area

LH *abbreviation* Leptospira hardjo

lice plural of **louse**

licence *noun* an official document which allows someone to do something

lie *noun* a place where an animal lies down ○ *Livestock benefit from a dry lie at pasture.*

lifestyle farmland buyer *noun* somebody who purchases farmland for leisure or investment purposes rather than as a working farmer

lifetime *noun* **1.** the time during which an organism is alive ○ *Humans consume tons of sugar in a lifetime.* **2.** the approximate time it would take for the part of an atmospheric pollutant concentration created by humans to return to its natural level assuming emissions cease ○ *Average life-*

times range from about a week for products such as sulphate aerosols to more than a century for CFCs and carbon dioxide. Also called **atmospheric lifetime**

lift *verb* to harvest root crops such as potatoes by digging them out of the ground. Potatoes can be lifted from the soil and, using a spinner or an elevator digger, left in rows for hand-picking.

lifter /ˈlɪftə/ *noun* same as **lifting unit**

liftings /ˈlɪftɪŋz/ *plural noun* crops which have been lifted

lifting unit *noun* a pair of wheels or a triangular-shaped share, used on a harvester to lift the roots and pass them to the main elevator. Also called **lifter**

COMMENT: The roots are lifted by being squeezed out of the ground in between the two wheels. The distance between the two wheels or shares can be adjusted to suit the size of the crop. The wheels should be set quite close together at the bottom when harvesting small roots. The wheels run at an angle to each other so that their rims lie close together when in the soil and farther apart at the top.

light grains *plural noun* cereals such as barley and oats. Compare **heavy grains**

light leaf spot *noun* a disease *Pyrenopeziza brassicae* which affects oilseed rape, causing light green or pale areas to appear on the leaves

light soil *noun* soil consisting mainly of large particles which are loosely held together because of the relatively large pore space. Light soil is usually easier to cultivate than heavy soil, but may dry out too quickly.

Light Sussex *noun* a dual-purpose breed of poultry, one of the several varieties of the Sussex breed. The birds are white, with black stripes to the feathers of the neck and black feathers on the wings and tail.

lignify /ˈlɪɡnɪfaɪ/ *verb* to become hard and woody ○ *Plants are less digestible as they become lignified.*

lignin /ˈlɪɡnɪn/ *noun* the material in plant cell walls that makes plants woody and gives them rigidity and strength

Lim *abbreviation* Limousin (*informal*)

Lima bean /ˈliːmə biːn/ *noun* same as **butter bean**

limb *noun* the leg of an animal

lime *noun* **1.** calcium oxide made from burnt limestone, used to spread on soil to reduce acidity and add calcium **2.** a hardwood tree. Genus: *Tilia*. **3.** a citrus fruit tree, with green fruit similar to, but smaller than, lemons. Latin name: *Citrus aurantifolia*. ■ *verb* to treat acid soil by spreading lime on it

limestone /ˈlaɪmstəʊn/ *noun* a common sedimentary rock, formed of calcium minerals and often containing fossilised shells of sea animals. It is porous in its natural state and may form large caves by being weathered by water. It is used in agriculture and building. ○ *carboniferous limestone*

> COMMENT: Limestone is formed of calcium minerals and often contains fossilised shells of sea animals. It is an important source of various types of lime.

liming /ˈlaɪmɪŋ/ *noun* the spreading of lime on soil to reduce acidity and add calcium. Also called **lime treatment**

Limousin /ˌlɪmuˈzæn/ *noun* a relatively hardy French breed of beef cattle, developed on the uplands around Limoges in central France. The cattle are red, with large bodies. Limousin bulls are used on dairy cattle producing a good crossbred calf.

linch pin /ˈlɪntʃ pɪn/ *noun* a pin used to lock an implement onto the three-point linkage at the rear of a tractor

Lincoln longwool /ˌlɪŋkən ˈlɒŋhɔːn/ *noun* a rare breed of sheep now found mainly in Lincolnshire, with white faces and long shiny wool. The animals are very large and slow to mature.

Lincoln red *noun* a breed of cattle bred from the shorthorn. The animals are deep red in colour. The breed was originally dual-purpose, but now is mainly used for crossing with dairy cows to produce beef calves.

lindane /ˈlɪndeɪn/ *noun* an organochlorine pesticide. It is a persistent organic pollutant and has been banned for all agricultural uses in the European Union. Formula: $C_6H_6Cl_6$.

line *verb* to cover the inside of a container to prevent the contents escaping ○ *Landfill sites may be lined with nylon to prevent leaks of dangerous liquids.*

line breeding *noun* the deliberate crossing or mating of closely related individuals in order to retain characteristics of a common ancestor

> COMMENT: The purpose of line breeding is to try to preserve in succeeding generations the mix of genes responsible for a particularly excellent individual specimen.

liner *noun* the rubber inner tube of a teat cup in a milking machine

ling /lɪŋ/ *noun* a variety of heather (*Calluna vulgaris*)

link *noun* a measurement, forming one loop of a chain (one-hundredth of a surveying chain, or 7.92 inches) ■ *verb* to be related to or associated with something ○ *Health is linked to diet* or *Health and diet are linked.*

linkage /ˈlɪŋkɪdʒ/ *noun* the process of two or more genes situated close together on a chromosome being inherited together

Linking Environment and Farming *noun* full form of **LEAF**

Linnaean system *noun* the scientific system of naming organisms devised by the Swedish scientist Carolus Linnaeus (1707–78) (NOTE: Carl von Linné is another form of his name.)

> COMMENT: The Linnaean system (or binomial classification) gives each organism a name made up of two main parts. The first is a generic name referring to the genus to which the organism belongs, and the second is a specific name which refers to the particular species. Organisms are usually identified by using both their generic and specific names, e.g. *Homo sapiens* (man) and *Felix catus* (domestic cat). The generic name is written or printed with a capital letter. Both names are usually given in italics, or are underlined if written or typed.

linoil /ˈlɪnɔɪl/ *noun* linseed oil

linseed /ˈlɪnsiːd/ *noun* a variety of flax (*Linum usitatissimum*) with a short straw. It produces a good yield of seed used for producing oil.

Linum /ˈliːnəm/ *noun* the Latin name for flax

linuron /ˈlɪnjuːrɒn/ *noun* a residual herbicide which acts in the soil

Lion Quality *noun* the symbol used on eggs in the UK to show that they come from a British Egg Industry Council-approved supplier

lipase /ˈlɪpeɪz/ *noun* an enzyme that breaks down fats

liquefied petroleum gas *noun* propane or butane or a combination of both produced by refining crude petroleum oil.

Abbr **LPG** (NOTE: Liquefied petroleum gas is used for domestic heating and cooking and for powering vehicles.)

liquid *adjective* having a consistency like that of water ○ *Liquid oxygen is stored in cylinders.* ■ *noun* a substance with a consistency like water ○ *Water is a liquid, ice is a solid.*

liquid fertiliser *noun* a simple solution, not kept under pressure, of the normal raw materials of solid fertilisers, as opposed to pressurised solutions such as aqueous ammonia

liquid manure *noun* a manure consisting of dung and urine in a liquid form (NOTE: Manure in semi-liquid form is **slurry**.)

liquor *noun* a concentrated liquid substance. ◊ **rumen liquor, silage liquor**

liquorice *noun* the root of a plant (*Glycyrrhiza glabra*) used in making sweets and soft drinks. It also has medicinal properties.

listeria /lɪˈstɪəriə/ *noun* a bacterium found in human and animal faeces, one species of which can cause meningitis if ingested in contaminated food. Genus: *Listeria.*

List of Chemicals of Concern *noun* a list of chemicals believed to be produced or used in the UK in substantial amounts and which meet specific criteria for concern relating to risks to the environment and human health. The list is drawn up by the UK Chemicals Stakeholder Forum and is intended for discussion and input.

litre *noun* a measure of capacity equal to 1000cc or 1.76 pints. Symbol **l**, **L**

litter *noun* **1.** a group of young mammals born to one mother at the same time ○ *The sow had a litter of ten piglets.* **2.** bedding for livestock (NOTE: Straw is the best type of litter, although bracken, peat moss, sawdust and wood shavings can be used.) ■ *verb* **1.** to leave rubbish in a place **2.** to give birth ○ *Bears litter in early spring.*

Little Red Tractor *noun* the symbol used on food packaging in the UK to show that it comes from a British Farm Standard-approved supplier

liver *noun* a large gland in the upper part of the abdomen, the main organ for removing harmful substances from the blood

liver fluke *noun* a parasitic trematode which lives in the liver and bile ducts of animals, e.g., *Fasciola hepatica* which infests sheep and cattle, causing loss of condition

livery /ˈlɪvəri/ *noun* a stable which keeps a horse for the owner and usually feeds, grooms, and exercises the horse

livestock *noun* cattle and other farm animals which are reared to produce meat, milk or other products ○ *Livestock production has increased by 5%.*

livestock auction *noun* an auction sale where livestock are shown in a ring and sold to the highest bidder

livestock records *plural noun* simple records of all livestock, which each farm has to make and which are then available for the Defra returns which are compiled each year

livestock register *noun* same as **herd register**

livestock unit *noun* the part of a farm where livestock are reared

liveweight /ˈlaɪvweɪt/ *noun* the weight of a live animal. Compare **deadweight**

liveweight marketing *noun* the marketing of live animals

llama /ˈlɑːmə/ *noun* a pack animal in the Andes of South America. It is a ruminant, and belongs to the camel family.

Llanwenog /lænˈwenɒg/ *noun* a breed of sheep found in many parts of West Wales. The fleeces are considered to be the finest produced in the UK. The wool has a very soft handle.

Lleyn /leɪn/ *noun* a breed of sheep native to the Lleyn peninsula in North Wales. The animals are small, hornless and hardy, good milkers, and very productive, often producing triplets.

loader /ˈləʊdə/ *noun* a machine used to load crops, manure, etc., into trailers or spreaders. The front-end tractor-mounted loader is the most common.

loam /ləʊm/ *noun* **1.** dark soil, with medium-sized grains of sand, which crumbles easily and is very fertile **2.** a mixture of clay, sand and humus, used as a potting compost

loamy /ˈləʊmi/ *adjective* referring to soil that is dark, crumbly and fertile

local air quality management *noun* the process of taking steps to improve the air quality in an area where it does not meet accepted standards. Abbr **LAQM**

lockjaw /'lɒkjɔː/ *noun* same as **tetanus**

locks *plural noun* small tufts of wool separated from the fleece during shearing

locomotor disorder /ˌləʊkə'məʊtə dɪsˌɔːdə/ *noun* a disease caused by skeletal growth not keeping pace with the rate of muscle development

locus /'ləʊkəs/ *noun* the position of a gene on a chromosome (NOTE: The plural is **loci**.)

locust /'ləʊkəst/ *noun* a flying insect which occurs in subtropical areas, flies in large groups and eats large amounts of vegetation

locust bean *noun* the broken-down pods of the carob tree, used as animal feed

lode /ləʊd/ *noun* a deposit of metallic ore

lodgepole pine /'lɒdʒpəʊl paɪn/ *noun* a slow growing tree which thrives on poor soil and is used as a pioneer crop (*Pinus contorta*)

lodging *noun* the tendency of cereal crops to bend over, so that they lie more or less flat on the ground

loess /'ləʊɪs/ *noun* a fine fertile soil formed of tiny clay and silt particles deposited by the wind

loganberry /'ləʊɡənb(ə)ri/ *noun* a soft fruit, a cross between a raspberry and a blackberry

Loghtan /'lɒhtən/ *noun* ♦ **Manx Loghtan**

Lomé Convention /'ləʊmeɪ kən ˌvenʃ(ə)n/ *noun* an agreement reached in 1975 between the European Community and 66 developing nations (the ACP states). It gives entry into the EU for certain agricultural produce without duty, with sections on guaranteed prices. ◊ **Cotonou Agreement**

long-day plant *noun* a plant that flowers as the days get longer in the spring. Compare **short-day plant**

long-grain rice *noun* varieties of rice with long grains, grown in tropical climates such as India. See Comment at **rice**

Longhorn /'lɒnhɔːn/ *noun* a dual-purpose hardy breed of cattle, with long down-curving horns. The animals are usually red or brown in colour, with white markings. The breed is now rare.

long ton *noun* same as **ton 1**

longwool /'lɒnwʊl/, **longwoolled sheep** *noun* the name of several breeds of sheep with long wool

Lonk /lɒŋk/ *noun* a breed of moorland sheep, found in the Pennines of Lancashire and Yorkshire. It is one of the Swaledale group, although larger than other varieties, and produces finer wool than most hill sheep. The face and legs are white with dark markings.

loose-box *noun* a stable for animals that are kept untied (NOTE: A loose-box should have a hay rack, manger, water bowl and tying rings. It should also have a grooved floor to make cleaning and drainage easier. Loose-boxes are also useful for housing sick animals.)

loose-leaved *adjective* referring to a plant such as a lettuce with a loose collection of leaves and no heart

loose silky bent *noun* a plant with thin green or purple stems which affects winter cereals (*Apera spica-venti*)

loose smut *noun* a fungus (*Ustilago nuda*) affecting wheat and barley. Masses of black spores collect on the diseased heads; the spores are dispersed in the wind, and only a bare stalk is left.

lop /lɒp/ *verb* to cut the branches of a tree

lop ears *plural noun* long ears which hang down on either side of the animal's head

louping-ill /'luːpɪŋ ɪl/ *noun* an infective parasitic disease of sheep (*Ixodes ricinus*), carried by ticks in hill pastures. Animals suffer acute fever and nervous twitch and staggers. Also called **staggers**, **twitch**, **trembles**

louse *noun* a small wingless insect that sucks blood and lives on the skin as a parasite on animals and humans. There are several types, the commonest being body louse, crab louse and head louse. Some diseases can be transmitted by lice. Genus: *Pediculus*. (NOTE: The plural is **lice**.)

louse disease *noun* an external parasitic disease of cattle. Severe infection leads to loss of condition, wasting and anaemia.

lovage /'lʌvɪdʒ/ *noun* a herb used as a vegetable and for making herbal teas

love-in-idleness /ˌlʌv ɪn 'aɪd(ə)lnəs/ *noun* same as **field pansy**

low *noun* **1.** same as **depression** ○ *A series of lows are crossing the North Atlantic towards Ireland.* **2.** a sound made by a cow ■ *verb* (*of a cow*) to make a sound. Compare **bleat**, **grunt**, **neigh**

low-input farming, lower input farming *noun* a system of farming based on restricted use of chemical fertilisers, pesticides and herbicides

low-intensity land *noun* land on which crops are not intensively cultivated

lowlands *noun* an area of low land where conditions are usually good for farming, as opposed to hills and mountains, or highlands ○ *Vegetation in the lowlands is sparse.*

low loader *noun* a farm trailer with its flat floor near the ground to make loading easier

LPG *abbreviation* liquefied petroleum gas

LSU *abbreviation* livestock unit

lucerne /lu'sɜːn/ *noun* a perennial, drought-resistant, leguminous plant that is rich in protein. It is mainly used either for green feed for animals or for hay or silage. Latin name: *Medicago sativa.* (NOTE: The US name is **alfalfa.**)

lugs /lʌgz/ *plural noun* projections from the tyres of tractor wheels. They increase traction by digging into the soil and by keeping the tyre in contact with solid surfaces in muddy conditions.

Luing /lɪŋ/ *noun* a hardy breed of beef cattle, found mainly in North-West Scotland

lump lime *noun* another name for burnt lime or quicklime

lumpy jaw *noun* same as **actinomycosis**

lungworm /'lʌŋwɜːm/ *noun* a parasitic worm which infests the lungs. Infestation can be cured with anthelmintics. Also called **hoose**

lupin *noun* a leguminous plant (*Lupinus polyphyllus*) grown as a crop for protein and seed oil

COMMENT: Lupins were originally grown in the UK as green manure on acid sandy soils, and for some sheep folding. They are now grown for grain production. The seeds of lupin contain 30–40% protein and 10–12% edible oil. The white lupin is an early-ripening sweet type, but is difficult to harvest and must be combined carefully.

lush *adjective* referring to vegetation which is thick and green ○ *The cattle were put to graze on the lush grass by the river.* ○ *Lush tropical vegetation rapidly covered the clearing.*

lymph /lɪmf/ *noun* a colourless liquid containing white blood cells, which circulates in the body, carrying waste matter away from tissues to the veins (NOTE: It is an essential part of the body's defence against infection.)

lynchet /'lɪntʃɪt/ *noun* **1.** a strip of land formed as the result of a movement of soil down a slope as a result of cultivation. Negative lynchets form at the top of the slope and positive lynchets at the bottom. **2.** an unploughed strip of land forming a temporary boundary between fields

COMMENT: Lynchets on former prehistoric fields can still be seen in the form of steps on the sides of hills.

lyophilise /laɪ'ɒfɪlaɪz/, **lyophilize** *verb* to preserve food by freezing it rapidly and drying in a vacuum

lysimeter /laɪ'sɪmɪtə/ *noun* a device used for measuring the rate of drainage of water through soil and the soluble particles removed in the process

lysine /'laɪsiːn/ *noun* an essential amino acid in protein foodstuffs, essential for animal growth

M

MA *abbreviation* modified atmosphere

machinery syndicate *noun* a group of farmers who join together to buy very large items of equipment, which they can use in turn

macroclimate /ˈmækrəʊklaɪmət/ *noun* the climate over a large area such as a region or country. ◊ **mesoclimate, microclimate**

macronutrient /ˈmækrəʊˌnjuːtriənt/ *noun* a nutrient that an organism uses in very large quantities, e.g. oxygen, carbon, hydrogen, nitrogen, phosphorus, potassium, calcium, magnesium or iron

mad cow disease *noun* same as **BSE** (*informal*)

Maedi-Visna /ˌmaɪdi ˈvɪznə/ *noun* a virus disease of sheep, which causes breathing difficulties. Abbr **MV**

‘All exported Shropshires are selected from flocks that are scrapie monitored and Maedi Visna accredited. Animals are also scrapie genotyped.’ [*Farmers Guardian*]

maggot *noun* a soft-bodied, legless larva of a fly such as a bluebottle, warble fly or frit fly. Maggots may attack crops and livestock.

magnesium *noun* a light, silvery-white metallic element that burns with a brilliant white flame

COMMENT: The addition of magnesium to soil may prevent deficiency diseases in crops or in livestock, such as interveinal yellowing of leaves in potatoes and sugar beet, and hypomagnesaemia or ‘grass staggers’ in grazing animals. Heavy spring applications of potash (potassium) fertilisers will increase the chance of grass staggers occurring.

magnetic resonance imaging /mæg ˌnetɪk ˈrezənəns ˌɪmɪdʒɪŋ/ *noun* a technique that uses electromagnetic radiation to obtain images of invisible parts of a structure or the body's soft tissues. Abbr **MRI** (NOTE: The object is subjected to a powerful magnetic field which allows signals from atomic nuclei to be detected and converted into images by computer.)

mahogany *noun* a tropical hardwood tree producing a dark timber, now becoming rare. Genus: especially: *Swietenia*.

maiden, maiden tree *noun* a tree in its first year after grafting or budding, when it is formed of a single stem

maincrop potatoes /ˌmeɪnkrɒp pə ˈteɪtəʊz/ *plural noun* varieties of potato grown as a main crop. Compare **earlies**

Maine /meɪn/ *noun* ◆ **Bleu du Maine**

Maine-Anjou /ˌmeɪn ˈɒnʒuː/ *noun* a breed of dual-purpose cattle developed in Brittany, now imported into the UK from France, and exported to many other countries. The animals are roan or red and white in colour.

maintenance ration *noun* the quantity of food needed to keep a farm animal healthy but not productive

‘Head herdsman Reginald Green believed that most people didn't know how to feed cows: You have got to know the proper maintenance ration for a cow according to her size and weight before you can start to feed for milk, he said.’ [*Dairy Farmer*]

maize *noun* a tall cereal crop grown in warm climates, that carries its grains on a large solid core (**cob**) of which there are only one or two per plant. Latin name: *Zea mays*. (NOTE: The US term is **corn**.)

COMMENT: Maize is a tall annual grass plant, with a strong solid stem. The male flowers form a tassel on the top of the plant and the females some distance away in the axils of some of the middle stem leaves. After wind pollination of the filament-like styles or silks, the grain develops into long 'cobs' of tightly packed seeds. In Great Britain the crop is grown for making silage, or for harvesting as ripened grain; some is grazed or cut as a forage crop, while a small proportion is sold for human consumption as 'corn on the cob'. Maize needs rich deep well-drained soils and ideally a frost-free growing season with a lot of sunshine before harvest. Maize is the only grain crop which was introduced from the New World into the Old World, and it owes its name of Indian corn to the fact that it was cultivated by American Indians before the arrival of European settlers. It is the principal crop grown in the United States, where it is used as feed for cattle and pigs. In Mexico it is a staple food, being coarsely ground into flour from which tortillas are made. Maize is also a staple food grain in the wetter parts of Africa; in South Africa the cobs are known as 'mealies'.

maize gluten *noun* a type of animal feedingstuff obtained after maize has been milled. It is high in protein.

malathion /ˌmælə'θaɪən/ *noun* an organophosphorus insecticide used to kill aphids, mainly on flowers grown in glasshouses

male *adjective* **1.** referring to an animal that produces sperm **2.** referring to a flower that produces pollen, or a plant that produces such flowers ■ *noun* a male animal or plant

malnutrition *noun* the state of not having enough to eat

malodours /mæl'əʊdəz/ *plural noun* unpleasant smells

malt *noun* best-quality barley grains which have been through the malting process and are used in breweries to make beer and in distilleries to make whisky ■ *verb* to treat grain such as barley by allowing it to sprout and then drying it. The malted grain is used for making beer.

maltase /'mɔːlteɪz/ *noun* an enzyme in the small intestine that converts maltose into glucose

malt culms *plural noun* roots and shoots of partly germinated malting barley. A byproduct of the malting process, the culms are used as a feedingstuff for livestock.

malted meal *noun* a brown wheat flour mixed with flour made from barley

malting *noun* the process by which barley grain is soaked in water, then sprouted on a floor to produce an enzyme.

It is then dried in a kiln and the roots and shoots are removed to leave the malt grains.

malting barley *noun* best-quality barley used for malting

maltose /'mɔːltəʊs/ *noun* a sugar formed by digesting starch or glycogen

maltster /'mɔltstə/ *noun* a person who makes malt for sale to breweries

Malus /'meɪləs/ *noun* the Latin name for the apple tree

mammal *noun* an animal that gives birth to live young, secretes milk to feed them, keeps a constant body temperature and is covered with hair. Class: Mammalia.

mammalian meat and bone meal *noun* same as **meat and bone meal**

mammary glands /'mæməri glændz/ *plural noun* glands in females that produce milk. In cows, sheep and goats, the glands are located in the udder.

manage *verb* **1.** to organise something or control the way in which something happens ○ *The department is in charge of managing land resources.* **2.** to succeed in doing something ○ *We managed to prevent further damage occurring.*

managed woodland *noun* a woodland which is controlled by felling, coppicing, planting, etc.

management *noun* **1.** the organised use of resources or materials **2.** the people who control an organisation or business

management practice *noun* practical ways of using management decisions to organise the use of resources or materials

'Vaccination of ewes and lambs is one of the more common management practices performed by sheep producers each and every year. Some producers will vaccinate their flock for "everything", while others choose a more conservative approach.' [*Farmers Guardian*]

manager *noun* a person who is in charge of an organisation or part of one

mane *noun* a long hair on the neck of a horse

manganese /'mæŋgəniːz/ *noun* a metallic trace element. It is essential for biological life and is also used in making steel.

COMMENT: Manganese deficiency is associated with high pH and soils that are rich in organic matter. It can cause grey leaf of cereals, marsh spot in peas and speckled yellowing of leaves of sugar beet. It is usually cured by applying manganese sulfate as foliar spray.

mange /meɪndʒ/ *noun* a skin disease of hairy or woolly animals, caused by mites, including *Sarcoptes*, the itch mite

mangel /'mæŋgəl/, **mangel wurzel** /'mæŋgəl ˌwɜːzəl/ *noun* a plant similar to sugar beet, but with larger roots. Mangel is mainly grown in southern England as a fodder crop. Also called **mangold**

COMMENT: Varieties of mangels include Globes, Tankards (oblong-shaped), Longs and Intermediates. Mangels contain less than 15% dry matter and are normally harvested before maturity and dried off in a clamp.

mangel fly *noun* a fly whose yellow-white legless larvae cause blistering of the leaves of mangels and sugar beet. This holds back plant growth and in severe cases can kill the plant. Also called **mangold fly**

manger *noun* a trough in a stable, from which horses and cattle feed

mangetout /'mɒnʒtuː/ *noun* a variety of pea, which is picked before the seeds are developed, and of which the whole pod is cooked and eaten

mango /'mæŋgəʊ/ *noun* a tropical tree (*Mangifera indica*) and the fruit it produces (NOTE: The tree originated in India, but is grown widely in tropical countries. The fruit is large, yellow or yellowish-green, with a soft orange pulp surrounding the very large seed. The seeds and bark are also used medicinally.)

mangold /'mæŋgəʊld/ *noun* same as **mangel**

mangold fly *noun* same as **mangel fly**

mangosteen /'mæŋgəstiːn/ *noun* a tree (*Garcinia mangostana*) which is native of Malaysia, but which is now cultivated in the West Indies. The fruit has a dark shiny rind and a soft sweet white flesh.

manioc /'mæniɒk/ *noun* the French name for cassava, used as an animal feedingstuff

manive /'mæniːv/ *noun* cassava meal, used as an animal feedingstuff

manure *noun* animal dung used as fertiliser (NOTE: In liquid or semi-liquid form it is called 'slurry'.) ■ *verb* to spread animal dung on land as fertiliser

COMMENT: All farm manures and slurries are valuable, and should not be regarded as a problem for disposal, but rather as assets to be used in place of expensive fertilisers which would otherwise need to be bought. Manure and slurry have to be spread in a controlled way, or pollution can result from runoffs into streams after rainfall.

manure cycle *noun* the process by which waste materials from plants, animals and humans are returned to the soil to restore nutrients

manure spreader *noun* a trailer with a moving floor conveyor and a combined shredding and spreading mechanism, used to distribute manure over the soil. Also called **muck spreader**

Manx Loghtan *noun* a rare breed of sheep, which is native to the Isle of Man. The wool is mouse-brown and the animals are multi-horned.

maple *noun* a hardwood tree of northern temperate regions, some varieties of which produce sweet sap which is used for making sugar and syrup. Genus: *Acer*.

mapping /'mæpɪŋ/ *noun* the process of collecting information and using it to produce maps

Maran /'mærɒn/ *noun* a heavy continental breed of fowl, which has a greyish-brown barred plumage and produces dark brown eggs

marbling /'mɑːb(ə)lɪŋ/ *noun* the appearance of muscle with intramuscular fat, seen on the cut surface of meat

Marchigiana /ˌmɑːkɪdʒi'ɑːnə/ *noun* a breed of white beef cattle from Italy, now imported into the UK and used for cross-breeding to improve beef-calf quality in dairy cows

mare *noun* a female horse, five years old or more

Marek's disease /'mæreks dɪˌziːz/ *noun* a virus disease of poultry, causing lameness and paralysis

margin *noun* **1.** the edge of a place or thing ○ *unploughed strips at the margins of the fields* ○ *a leaf margin* **2.** the difference between the amount of money received for a product and the money which it cost to produce

marginal /'mɑːdʒɪn(ə)l/ *adjective* **1.** referring to areas of land such as field edges or banks beside roads which are at the edge of cultivated land **2.** referring to a plant which grows at the edge of two types of habitat ○ *marginal pond plants such as irises* **3.** referring to land of poor quality which results from bad physical conditions such as poor soil, high rainfall or steep slopes, and where farming is often difficult ○ *Cultivating marginal areas can lead to erosion.*

margin over purchased feed *noun* the amount of money received for produce such as per litre of milk, shown as a percentage above the amount spent in purchasing feed for the animals. Abbr **MOPF**

mariculture /ˈmærɪkʌltʃə/ *noun* the breeding and keeping of sea fish or shellfish for food in seawater enclosures

COMMENT: Mariculture refers to aquaculture in seawater, such as raising oysters, lobsters and fish in special enclosures.

marine flora *noun* the plants that live in the sea

marjoram /ˈmɑːdʒərəm/ *noun* a Mediterranean aromatic herb (*Origanum*), the dried leaves of which are used as flavouring

market garden *noun* a place for the commercial cultivation of plants, usually vegetables, soft fruit, salad crops and flowers, found near a large urban centre that provides a steady outlet for the sale of its produce

market gardener *noun* a person who runs a market garden

market gardening *noun* the business of growing vegetables, salad crops and fruit for sale

market town *noun* a town with a permanent or regular market, which serves as a trading centre for the surrounding area. Some markets specialise in certain types of livestock or produce.

market weight *noun* the target weight at which livestock will be sold at market or slaughtered

'The emphasis in the UK has been to achieve maximum growth rates from the earliest possible time, using high price, high density diets to achieve the least number of days to reach market weight.' [*Pig Farming*]

marking *noun* **1.** the practice of putting a mark on an animal to identify who it belongs to, e.g. a brand on the skin **2.** (*by an animal*) the act of using urine to identify its territory and warn away competitors

markings *plural noun* coloured patterns on the coat of an animal or in the feathers of a bird

marl /mɑːl/ *noun* a fine soil formed of a mixture of clay and lime, used for making bricks

marram grass /ˈmærəm grɑːs/ *noun* a type of grass planted on sand dunes to prevent them being spread by the wind. Latin name: *Ammophila arenaria.*

marrow *noun* a large vegetable (*Cucurbita pepo*) of the pumpkin family, which may be grown as bush or trailing varieties

marrowstem kale /ˈmærəʊstem ˌkeɪl/ *noun* a variety of kale with a thick stem and large leaves, grown as feed for livestock in the autumn and winter months, though it is not winter hardy

marsh *noun* an area of permanently wet land and the plants that grow on it (NOTE: Marshes may be fresh water or salt water and tidal or non-tidal. A marsh usually has a soil base, as opposed to a bog or fen, which is composed of peat.)

COMMENT: Many former areas of marshland have been reclaimed and have been artificially drained by a system of ditches and sluices which allow surface water to escape to the sea, but prevent salt water entering the area. The drained soils are usually fertile and some of these areas are important for agriculture.

marshland /ˈmɑːʃlænd/ *noun* land that is covered with marsh

marsh spot *noun* a disease affecting peas, caused by manganese deficiency

marshy soil *noun* a very wet soil

martingale /ˈmɑːtɪŋgeɪl/ *noun* a device used to regulate the way a horse's carries its head. It consists of a strap or straps, attached to the girth at one end, and at the other to the reins or to the noseband.

mash *noun* a mixture of feeding meals combined to provide all the necessary elements for a balanced diet

Masham /ˈmæʃəm/ *noun* a crossbred type of sheep which results from a Wensleydale or Teeswater ram mated with a hill ewe of the Swaledale type and has black markings on the face and legs. The breed is an economical ewe with a good lambing average and a useful fleece.

mashlum /ˈmæʃlʌm/ *noun* a mixture of oats and barley (and sometimes wheat), sown to provide grain for feeding to livestock. Also called **maslin, meslen, meslin**

mast *noun* the small hard fruits that have fallen from a beech tree, used as food by pigs and other animals

Master of Foxhounds Association
noun a group which provides information to foxhunters and hunt leaders. Abbr **MFHA**

mastication /ˌmæstɪˈkeɪʃ(ə)n/ *noun* the process of grinding food in the mouth, using the teeth and jaws

mastitis /mæˈstaɪtɪs/ *noun* a common bacterial disease affecting dairy animals in which the udders become inflamed and swollen, and the passage of the milk is blocked

> COMMENT: Common causes are staphylococci such as *Staphylococcus aureus* (staphylococcal mastitis), streptococci (*Streptococcus uberis*) or other bacteria (*E. coli mastitis*). The condition can be treated with antibiotics.

mastitis-metritis-agalactia *noun* full form of **MMA**

mast swine *noun* a German term for a fattening pig

mat *noun* a covering of undecayed grassland vegetation which forms on very acid soil, when the soil lacks the microorganisms necessary to break decaying matter down

mate *noun* an animal that reproduces sexually with another ■ *verb* to reproduce sexually with another of same species

mating likes *noun* same as **assortive mating**

matted /ˈmætɪd/ *adjective* with many fibres woven together

mattock /ˈmætək/ *noun* a heavy hoe

maturity *noun* **1.** the time when a plant's seeds are ripe **2.** the time when an animal has become an adult **3.** the time when an animal is ready for slaughter

maw /mɔː/ *noun* a stomach, especially the last of a ruminant's four stomachs

maximum residue level *noun* the maximum amount of a pesticide that can remain in crops or foodstuffs under European Union regulations. Abbr **MRL**

> 'Amendments to EU legislation governing pesticide maximum residue levels have been voted through with the result that MRL controls will be extended on a range of crops including cereals, potatoes and sugar beet.' [*Farmers Guardian*]

may *noun* a popular name for hawthorn, a common plant for making hedges

mayweed /ˈmaɪwiːd/ *noun* one of a group of weeds which affect cereals (*Chamomilla* spp, *Anthemis* spp, *Matri-*

caria spp). The weeds affect winter crops and vegetables, and are found on headlands. They can cause considerable problems to machinery. Also called **dogdaisy, wild chamomile**

MBM *abbreviation* meat and bone meal

MCPA *noun* a herbicide that kills the most persistent broad-leaved weeds, such as nettles, buttercups, charlock, dock seedlings, plantains and thistles (NOTE: Its full name is **2-methyl-4chloro-phenoxyacetic acid.**)

MCPP *abbreviation* mecoprop

MDC *abbreviation* Milk Development Council

ME *abbreviation* **1.** metabolisable energy ◇ *ME levels in concentrates* **2.** metabolised energy

meadow *noun* a field of grass and wild plants, sometimes grown for fodder. ◊ **water meadow**

meadow fescue *noun* a perennial grass (*Festuca pratensis*) which has considerable importance for hay and grazing. It is a highly productive grass which flourishes when sown with Timothy.

meadowgrass /ˈmedəʊɡrɑːs/ *noun* varieties of grass of the genus *Poa*

meal *noun* a finely ground compound feedingstuff for poultry and pigs, containing all the elements necessary for good health and steady growth

mealworm /ˈmiːlwɜːm/ *noun* the larva of various beetles of the genus *Tenebrio* that infests and pollutes grain products

meat *noun* animal flesh that is eaten as food (NOTE: Meat is formed of the animal's muscle.)

-meat *suffix* showing the flesh of an animal, used in particularly in the EU ◇ *pigmeat* ◇ *sheepmeat*

meat and bone meal *noun* meal made from waste meat and bones, formerly used in animal feed but now banned in the EU because of fears that it was a contributing factor in the spread of BSE. Abbr **MBM**

Meat and Livestock Commission *noun* an organisation which provides services to livestock breeders, including the evaluation of breeding stock potential and carcass grading and classification. The Commission also carries out various research projects. Its staff also provide services for abattoirs and livestock auction markets. The Commission promotes the sale of British meat. Abbr **MLC**

meat chicken *noun* a chicken which is raised for its meat rather than for its eggs or to produce chicks

'While breeding for improved welfare can be economically beneficial when individual animals have high value, like the dairy cow, the meat chicken's value is very low.' [*Farmers Guardian*]

meat-eating animal *noun* same as **carnivore**

meat extender /'miːl ɪkˌstendə/ *noun* any foodstuff or mixture of foodstuffs added to meat preparations to increase their bulk

meat fly *noun* same as **blowfly**

Meat Hygiene Service *noun* a division of the Food Standards Agency which deals with hygiene in slaughterhouses and meat preparation facilities. Abbr **MHS**

Meatlinc /'miːtlɪŋk/ *noun* a new breed of sheep used as a terminal sire. Only the rams are sold.

mechanically recovered meat *noun* the scraps of meat which remain on an animal's carcass after the prime cuts have been removed, which are removed using machinery, ground and used as cheap filler for burgers, pies, sausages, etc. Abbr **MRM**

mechanisation /ˌmekənaɪ'zeɪʃ(ə)n/, **mechanization** *noun* the introduction of machines for agricultural working purposes

COMMENT: Mechanisation has been an important factor in the contraction of the agricultural labour force. Mechanisation has not only involved increases in the number and range of machines, but also dramatic increases in their size and power. This has enabled slopes previously regarded as too steep for ploughing to be cultivated. The increased size of tractors and combines has encouraged enlargement of fields and the removal of hedgerows. This has caused alarm amongst conservationists and led to increased erosion in wet weather in some areas.

mecoprop /'mekəʊprɒp/ *noun* a commonly used herbicide, mostly used to control weeds in cereal and grass crops, that is found as a contaminant of water. Abbr **MCPP, CMPP**

medium *noun* a substance in which an organism lives or is grown

melon *noun* a plant of the cucumber family (*Cucumis melo*) with a sweet fruit.

The flesh of the fruit varies from green to orange or white.

Mendel's laws /'mendəlz lɔːz/ *plural noun* the laws governing heredity

COMMENT: The two laws set out by Gregor Mendel following his experiments growing peas, were (in modern terms): that genes for separate genetic characters assort independently of each other and that the genes for a pair of genetic characters are carried by different gametes. For animal breeders, the main feature of Mendelism is that it is based on simple and clearly-defined traits that are inherited as separate entities: these were traits such as colour, which are controlled by single genes.

merchant *noun* a person who sells a product □ **seed merchant, corn merchant** trader who sells seed or corn, usually wholesale

mercuric chloride *noun* same as **mercury (II) chloride**

mercury /'mɜːkjʊri/ *noun* a metal element that is liquid at room temperature. It is used in thermometers, barometers and electric batteries and is poisonous. Also called **quicksilver**

mercury (I) chloride *noun* a poisonous white compound of mercury and chlorine, used as a moss killer and laxative. Formula: Hg_2Cl_2. Also called **mercurous chloride, calomel**

mercury (II) chloride *noun* a poisonous compound of mercury and chlorine, used as an antiseptic and wood preservative. Formula: $HgCl_2$. Also called **mercuric chloride**

Merino /mə'riːnəʊ/ *noun* a breed of sheep which originated in North Africa and was then introduced into Spain. It is now bred in all parts of the world, especially in Australia, South Africa and New Zealand, for its dense soft fine fleece, with strong and curly fibres.

Merinolandschaf /məˌriːnəʊ'læntʃæf/ *noun* a breed of sheep found in South Germany. Large travelling flocks are common.

meslen /'mezlən/, **meslin** *noun* same as **mashlum**

mesoclimate /ˌmezəʊ'klaɪmət/ *noun* the climate over a specific locality such as a hillside or valley, extending no more than a few kilometres in radius. ◊ **macroclimate, microclimate**

mesotrophic /ˌmezəʊ'trɒfɪk/ *adjective* referring to water that contains a moderate amount of nutrients. Compare **eutrophic**, **oligotrophic**

meta- /metə/ *prefix* **1.** changing **2.** following

metabolic /ˌmetə'bɒlɪk/ *adjective* referring to metabolism

metabolic disease *noun* one of a group of diseases that are caused by animals being called upon to produce an end-product faster than their bodies can process their intake of feed

metabolic size *noun* the size of an animal to which the metabolic rate is proportional

metabolisable protein /məˌtæbəlɪzəb(ə)l 'prəʊtiːn/ *noun* a type of protein which can be metabolised by an animal, used as a feed supplement to improve lactation

'The next and possibly most sophisticated part of the new predictions is that for predicting protein requirements. This involves designating three types of metabolisable protein, namely MPE (metabolisable protein from rumen available energy); MPN (metabolisable protein from N sources); and finally MPB which is effectively metabolisable bypass protein.' [*Dairy Farmer*]

metabolise /mə'tæbəlaɪz/ *verb* to break down or build up organic compounds by metabolism ○ *The liver metabolises proteins and carbohydrates.*

metabolised energy, metabolisable energy *noun* the proportion of energy from feed which is used by an animal through its metabolism. Abbr **ME** (NOTE: **Metabolised energy** is the measure of energy following digestion, after the alimentary gases and urinary losses have been subtracted. Animals cannot be expected to transfer energy from feed with perfect efficiency as there will always be losses through undigested food and as alimentary gases. The energy needs of livestock can be calculated from their size.)

metabolism *noun* the chemical processes of breaking down or building up organic compounds in organisms

COMMENT: Metabolism covers all changes which take place in the body: the building of tissue (anabolism), the breaking down of tissue (catabolism), the conversion of nutrients into tissue,

the elimination of waste matter and the action of hormones.

metaldehyde /met'ældɪhaɪd/ *noun* a substance used in the form of pellets to kill slugs and snails, or in the form of small blocks to light fires

metamorphosis *noun* a process of change into a different form, especially the change of a larva into an adult insect

meter *noun* a device to measure a physical property such as current, rate of flow or air speed ■ *verb* to count or measure with a meter

methane /'miːθeɪn/ *noun* a colourless flammable gas produced naturally from rotting organic waste, as in landfill sites or animal excreta. Formula: CH_4.

COMMENT: Methane is produced naturally from rotting vegetation in marshes, where it can sometimes catch fire, creating the phenomenon called will o' the wisp, a light flickering over a marsh. Large quantities may also be formed in the rumen of cattle. It occurs as the product of animal excretions in livestock farming. Excreta from livestock can be passed into tanks where methane is extracted leaving the slurry which is then used as fertiliser. The methane can be used for heating or as a power source. Methane is also a greenhouse gas, and it has been suggested that methane from rotting vegetation, from cattle excreta, from water in paddy fields, and even from termites' nests, all contribute to the greenhouse effect.

methanol /'meθənɒl/ *noun* an alcohol manufactured from coal, natural gas or waste wood, which is used as a fuel or solvent. Formula: CH_3OH. Also called **methyl alcohol, wood alcohol**

'Biodiesel is made by mixing vegetable oil with methanol, giving glycerine as a by-product. Its source is natural and renewable and it dramatically reduces exhaust emissions of smoke and soot, carbon monoxide and sulphur dioxide, the cause of acid rain.' [*Farming News*]

COMMENT: Methanol can be used as a fuel in any type of burner. Its main disadvantage is that it is less efficient than petrol and can cause pollution if it escapes into the environment, as it mixes easily with water. Production of methanol from coal or natural gas does not help fuel conservation, since it depletes Earth's fossil fuel resources.

methionine /'meθɪəniːn/ *noun* an essential amino acid

methyl alcohol *noun* same as **methanol**

methyl bromide *noun* an effective chemical for sterilising soil and fumigating spaces

methyl phosphine *noun* a compound with specific action against phosphine-resistant strains of storage pests

metre *noun* an SI unit of length ○ *The area is four metres by three.* Symbol **m** (NOTE: The US spelling is **meter.**)

metric ton *noun* same as **tonne**

metritis /meˈtraɪtɪs/ *noun* an infection of the lining of the womb in cattle, the symptoms of which are a white discharge and/or a high temperature. Also called **whites**

Meuse-Rhine-Ijssel *noun* a dual-purpose breed of cattle, originating from the Netherlands. It is used by breeders in Britain to upgrade Dairy Shorthorn. The breed's dairy performance is similar to that of the British Friesian, and it has a fine beef conformation. Cattle are red and white in colour. Abbr **MRI**

mezzadria /meˈtsædriə/ *noun* a system used in Southern Italy, where a vineyard is leased and the landlord is paid a half-share of the wine produced

MFHA *abbreviation* Master of Foxhounds Association

Mg *symbol* magnesium

MGA *abbreviation* **1.** Maize Growers Association **2.** Mushroom Growers' Association

MHC *abbreviation* moisture holding capacity

MHS *abbreviation* Meat Hygiene Service

microbe *noun* a microorganism (NOTE: Viruses, bacteria, protozoa and microscopic fungi are informally referred to as microbes.)

microbial /maɪˈkrəʊbiəl/ *adjective* referring to microbes

microbial ecology *noun* the study of the way in which microbes develop in nature

microbial insecticide *noun* an insecticide based on fungal, bacterial or other microorganisms that are pathogens of insects, or their toxins, e.g. the fungus *Verticillium lecanii* is used to control whitefly in glasshouses

microbial protein *noun* a protein source in ruminants from dead rumen microbes, usually forming 70% to 100% of the ruminant's supply of protein

'The water soluble carbohydrate content of grass is well recognised as a key factor in determining how efficiently ruminants can turn the nitrogen in their diets into microbial protein for milk and meat production.' [*Dairy Farmer*]

microclimate /ˈmaɪkrəʊˌklaɪmət/ *noun* the climate over a very small area such as a pond, tree, field, or even a leaf. ◊ **macroclimate, mesoclimate**

microenvironment *noun* same as **microhabitat**

microhabitat /ˈmaɪkrəʊˌhæbɪtæt/ *noun* a single small area such as the bark of a tree, where fauna and/or flora live. Also called **microenvironment**

micron /ˈmaɪkrɒn/ *noun* a measurement of thickness, one millionth of a metre, used in measuring the fineness of hair or wool

micronutrient /ˈmaɪkrəʊˌnjuːtriənt/ *noun* a nutrient which an organism uses in very small quantities, e.g. iron, zinc or copper

microorganism *noun* an organism that can only be seen with a microscope. Compare **microbe** (NOTE: Viruses, bacteria, protozoa and fungi are all forms of microorganism.)

micropropagation /ˈmaɪkrəʊˌprɒpəɡeɪʃ(ə)n/ *noun* the propagation of plants by cloning a small piece of plant tissue cultured in a growth medium

midden /ˈmɪdən/ *noun* a heap of dung

Middle White *noun* a breed of white pig which comes from a cross between the Large White and the Small White. It is short and compact with long upright ears and a turned-up snout. It is now a rare breed.

mids /mɪdz/ *plural noun* middle-sized potatoes which are graded and sold for human consumption

Midterm Review *noun* the review of the Common Agricultural Policy of the European Union carried out in 2003

migrant *noun* an animal or bird that moves from one place to another according to the season. Compare **nomad**

milch cow /ˈmɪlk kaʊ/ *noun* a cow which gives milk or is kept for milk production

mildew /ˈmɪldjuː/ *noun* a disease caused by a fungus which produces a fine powdery film on the surface of an organism

milfoil /ˈmɪlfɔɪl/ *noun* same as **yarrow**

milk *noun* an opaque white liquid secreted by female mammals during lactation ■ *verb* to extract milk from a cow's udder. Pressure on the teats makes the milk spurt out. Milking can be done by hand, but is usually done by machines in a milking parlour.

COMMENT: In the UK, most milk comes from Friesian cows, and has been heat-treated, pasteurised, sterilised or ultra-heat-treated before it is sold to the public. It may also be calcium-enriched or lactose-reduced. Milk is sold in cartons or plastic bottles, either as homogenised, semi-skimmed or skimmed. In glass bottles it is sold with various coloured metal tops: 'silver top' is pasteurised with an average 3.9% fat and has a noticeable cream line; 'red top' is similar to the silver, but the milk is homogenised to distribute the cream throughout; 'gold top' is pasteurised milk from Guernsey or Jersey breeds of cow, and has an average fat content of 4.9%; 'red and silver striped top': pasteurised semi-skimmed milk with average 1.6% fat content; 'blue and silver checked top': pasteurised skimmed milk, with an average 0.1% fat content; 'green top': untreated whole milk, with an average 3.9% fat (no longer sold in the UK). Sterilised whole milk with fat content of 3.9% is sold in bottles with crown caps or blue foil tops. UHT milk is also available as whole, semi-skimmed or skimmed: it is milk with a shelf life of 6 months, though when opened it should be kept cold and used as ordinary pasteurised milk.

milk composition *noun* the percentages of protein, lactose, fat, minerals and water which make up milk. The composition varies according to the breed of cow, but average percentages are: protein (3.4%), milk sugar (4.75%), fat (3.75%), minerals (0.75%), water (87.35%).

milk cooler *noun* a stainless steel bulk storage tank, in which milk is cooled by running water passing over the outside of the tank

Milk Development Council *noun* a body which collects levies on milk and distributes the money to research and development projects. Abbr **MDC**

milker /ˈmɪlkə/ *noun* **1.** a cow which is giving milk **2.** a cow which is kept for milk **3.** a farmworker who supervises the milking of cows **4.** the part of the milking machine which is attached to the cow's teats with teat cups

milk fever *noun* a disease of milk cows, milk goats and ewes. Technical name **hypocalcaemia** (NOTE: In spite of its name, the disease is not a fever, and may affect a dairy cow just before calving or during the seven days which follow calving. The first symptoms are restlessness, moving the hind feet up and down while standing; these symptoms are followed by loss of balance and later loss of consciousness. The disease is common at the third, fourth or fifth time of calving, and is caused by a metabolic disturbance or imbalance in the system, due to a low calcium content in the blood. The disease is treated by injections of calcium borogluconate.)

'Cows that have had a difficult calving or milk fever prefer a drink of warm water after calving, while Mr Blowey recommends offering good quality hay as part of the feed.' [*Farmers Weekly*]

milk goat *noun* a goat which is reared for its milk

milking machine *noun* a machine which imitates the sucking action of a calf, used to extract milk from the cow's udder (NOTE: It uses a pulsator mechanism to apply pressure to the teats, causing the release of the milk. The milk is then passed into a collecting jar or may pass by pipeline to a large tank.)

milking parlour *noun* a building in which cows are milked, and often are also fed, washed and cleaned

COMMENT: There a four basic designs of parlour: the herringbone parlour, where the cow stands at an angle of 45° to the milker, is commonest for large herds; the abreast parlour, where the cows stand side by side with their backs to the milker; the tandem parlour where they stand in line with their sides to the milker; the most expensive and complex of the four systems is the rotary parlour, where the cows stand on a rotating platform with the milker in the middle.

Milk Marketing Board *noun* until 1994, the board which organised the collection and buying of milk from farmers and its sale to customers

Milk Marque *noun* the name of a national cooperative which replaced the Milk Marketing Board, with the aim of liberalising the milk market. It was split into 3 regional companies in 1999.

milk producer *noun* a farmer who is registered with Defra, and produces milk in compliance with the regulations concerning clean milk production. In 2006 the estimated number of registered producers was just under 19,000 in Great Britain.

milk products *plural noun* milk and other foodstuffs produced from it, which are sold for human consumption. The main milk products are liquid milk (homogenised, pasteurised, sterilised or UHT), butter, cheese, cream, condensed milk and milk powder.

milk quota *noun* a system by which farmers are only allowed to produce certain amounts of milk, introduced to restrict the overproduction of milk in member states of the EU. Abbr **MQ**

COMMENT: Quotas were introduced in 1984, and were based on each state's 1981 production, plus 1%. A further 1% was allowed in the first year. A supplementary levy or superlevy, was introduced to penalise milk production over the quota level. In the UK, milk quotas can be bought and sold, either together with or separate from farmland, and are a valuable asset. The government is responsible for the setting of quotas for milk production, according to the directives of the EU commission.

milk recording *noun* keeping a record of the milk given by each cow at each milking, the quality of the milk is analysed each month

milk ripe stage *noun* a stage in the development of grain such as wheat where the seed has formed but is still soft and white and full of white sap. Also called **milky stage**

milk sheep *noun* a sheep which is reared for its milk

milk sinus *noun* the space in each teat into which the milk is secreted

milk sugar *noun* same as **lactose**

milk yield *noun* the quantity of milk produced each year by a cow

COMMENT: In the UK, the average annual milk yield per dairy type cow increased from 3,989 litres per cow in 1974/5 to 6,530 litres per cow in 2006.

milky stage /'mɪlki steɪdʒ/ *noun* same as **milk ripe stage**

mill *noun* a factory where a substance is crushed to make a powder, especially one for making flour from the dried grains of cereals ■ *verb* to crush a substance to make a powder

millet /'mɪlɪt/ *noun* a cereal crop grown in many of the hot, dry regions of Africa and Asia, where it is a staple food. Genera: especially: *Panicum* or *Eleusine*.

COMMENT: The two most important species are finger millet and bulrush millet. Millet grains are used in various types of food. They can be boiled and eaten like rice, made into flour for porridge, pasta or chapatis, and mixed with wheat flour to make bread. Millets can be malted to make beer. Millets are also grown as forage crops, and the seed is used as a poultry feed.

milling quality *noun* the calculation of how easy it is to separate the white endosperm from the brown seed coat or bran in the milling process. In general hard wheats are of higher milling quality than soft wheats.

'Feed wheat started the season at about £60/t at harvest, but heavy rain delayed progress, and decent premiums were available for anyone with wheat in the barn. As the rain continued, milling quality fell and much was downgraded to feed.' [*Farmers Weekly*]

milling wheat *noun* best-quality wheat used to make flour for making bread

millstone /'mɪlstəʊn/ *noun* a heavy round slab of stone, used to grind corn

milo /'maɪləʊ/ *noun* US sorghum

mineral *noun* an inorganic solid substance with a characteristic chemical composition that occurs naturally (NOTE: The names of many minerals end with the suffix **-ite**.)

COMMENT: The most important minerals required by the body are: calcium (found in cheese, milk and green vegetables) which helps the growth of bones and encourages blood clotting; iron (found in bread and liver) which helps produce red blood cells; phosphorus (found in bread and fish) which helps in the growth of bones and the metabolism of fats; and iodine (found in fish) which is essential to the functioning of the thyroid gland.

mineralisation /ˌmɪnərəlaɪˈzeɪʃ(ə)n/, **mineralization** *noun* the breaking down of organic waste into its inorganic chemical components

mineral matter content *noun* the amount of minerals found in plants

mineral nutrients *plural noun* nutrients except carbon, hydrogen and oxygen which are inorganic and are absorbed by plants from the soil

minimal *adjective* very small in amount or importance ○ *the minimal area for sampling in which specimens of all species can be found*

minimal cultivation *noun* a system of cultivation which subjects the land to shallow working and minimises the number of passes of machinery. No ploughing is needed.

COMMENT: Although suitable for cereal production, minimal cultivation is not suitable for all crops or soil conditions. Crops like sugar beet and potatoes need a deeper tilth than that obtained by minimal cultivation.

minimal disease herd *noun* a herd of livestock with a very low level of infectious diseases

minimum tillage *noun* a method of ploughing in which disturbance of the soil does not affect the deeper layers. The benefits are conservation of organic matter, leading to a better soil structure and less soil erosion, better soil biodiversity and the use of less energy. The disadvantages include the easier germination of grass seeds. Also called **min-till**

'Twenty varieties, including commercially available and yet-to-be released types, are being grown after seedbeds were prepared after ploughing/pressing, a minimum tillage approach involving two passes of a discs/tine combination rig, or the crop direct drilled into wheat stubble.' [*Arable Farming*]

Ministry of Agriculture Fisheries and Food *noun* the former UK government department with responsibility for agricultural and food matters. ◊ **Department for Environment, Food and Rural Affairs**

Minorca /mɪˈnɔːkə/ *noun* a breed of poultry, originating in the Mediterranean. The birds are black or white in colour.

min-till *noun* same as **minimum tillage**

Miranda /mɪˈrændə/ *noun* a breed of cattle found in Portugal. The animals are dark brown in colour, with horns coloured white with black tips. Mirandas are bred for meat and for draught.

miscanthus /mɪsˈkænθəs/ *noun* a plant related to sugar cane that is grown for use as a fuel. Also called **elephant grass**

miscarry *verb* same as **abort 3** (*technical*)

mite *noun* a tiny animal of the spider family which may be free-living in the soil or on stored products, or parasitic on animals or plants

miticide /ˈmɪtɪsaɪd/ *noun* a substance that kills mites

mixed cropping *noun* the practice of growing more than one type of plant on the same piece of land at the same time. Opposite **monocropping**

mixed culture *noun* the process of growing several species of tree together on the same piece of land

mixed farming *noun* the practice of combining arable and dairy farming

mixed fertiliser *noun* same as **compound fertiliser**

mixed grazing *noun* a grazing system where more than one type of animal grazes the same pasture at the same time

'Mixed grazing regimes provide a range of sward lengths which are attractive areas for birds to nest and feed and encourage regeneration of grassland and moorland areas.' [*Farmers Guardian*]

mixed woodland *noun* a wooded area where neither conifers nor broadleaved trees account for more than 75% of the total

MLC *abbreviation* Meat and Livestock Commission

MMA *noun* same as **farrowing fever**

MMBM *abbreviation* mammalian meat and bone meal

Mn *symbol* manganese

mode of action *noun* the way in which a pesticide acts. For example, organophosphorous compounds disrupt the nerve impulses in insects.

moder /ˈməʊdə/ *noun* humus which is partly acid mor and partly neutral mull

modified atmosphere *noun* an oxygen-depleted atmosphere enriched with carbon monoxide or carbon dioxide, used for disinfestation of pests or for increasing the shelf life of food. Abbr **MA**

'After seven weeks lambs were slaughtered, then carcases conditioned at 1°C for seven days, after which steaks were displayed in modified atmosphere packs under retail conditions for measurement of fat oxidation (shelf life). A muscle was taken for fatty acid analysis and eating quality assessed after grilling.' [*Farmers Guardian*]

MOET *abbreviation* multiple ovulation and embryo transfer

mohair *noun* fine wool from a goat, over 30 microns. Compare **cashmere**

Moiled /mɔɪld/ ♦ **Irish Moiled**

moist *adjective* slightly damp, containing a small amount of water

moisture *noun* a slight amount of water as found in soil, grain, etc.

moisture content *noun* the percentage of moisture in something, e.g. harvested crops

COMMENT: The safe moisture content for storage of all grains is 14% or less. In the UK, only fully ripe grain in a very dry period is likely to be harvested at less than 14% moisture content. In a wet season, the moisture content may be as high as 30%, and these grains will have to be dried. The moisture content of hay is 80% when cut, and has to be reduced to below 20% for safe storage.

moisture-holding capacity *noun* the amount of water held by soil between field capacity and the permanent wilting point. The amount of moisture will vary according to the texture and structure of the soil. Abbr **MHC**

molasses *noun* a dark brown syrup, a by-product of sugar production left after sugar has been separated. It is used as a binding agent in compound animal feeds and is also added to silage.

mole *noun* **1.** a small dark-grey mammal which makes tunnels under the ground and eats worms and insects **2.** an SI unit of measurement of the amount of a substance. Symbol **mol 3.** the part of a mole plough that cuts a round channel underground

mole drain *noun* an underground drain formed under the surface of the soil by a mole plough as it is pulled across a field (NOTE: Mole drains are usually made 3 to 4 metres apart, and are used in fields with a clay subsoil.)

molehill /ˈməʊlhɪl/ *noun* a small heap of earth pushed up to the surface by a mole as it makes its tunnel

mole plough *noun* a plough used to make mole drains, pulled by a tractor, and forming a wide round hole underground

Molinia /məˈlɪniə/ *noun* a poor type of grass found in rough mountain or hill grazings. It is of little value as grazing.

mollusc *noun* an invertebrate animal with a soft body, a muscular foot on the underside used for movement and, in many species, a protective shell (NOTE: Molluscs are found on land as well as in fresh and salt water. Slugs, snails and shellfish are molluscs. The US spelling is **mollusk**.)

molluscicide /məˈlʌskɪsaɪd/ *noun* a substance used to kill molluscs such as snails

mono- /mɒnəʊ/ *prefix* single or one. Opposite **multi-**

monocot /ˈmɒnəʊkɒt/ *noun* same as **monocotyledon** (*informal*)

monocotyledon /ˌmɒnəʊˌkɒtɪˈliːd(ə)n/ *noun* a plant with seeds that have a single cotyledon, e.g. a grass or lily. Compare **dicotyledon**. ◊ **cotyledon**

monocropping /ˈmɒnəkrɒpɪŋ/, **monocrop system** /ˈmɒnəʊkrɒp ˌsɪstəm/, **monoculture** /ˈmɒnəkʌltʃə/ *noun* a system of cultivation in which a single crop plant such as wheat is grown over a large area of land often for several years. Opposite **mixed cropping**

COMMENT: Monocropping involving cash crops, groundnuts, cotton, etc., exposes farmers in Africa to price fluctuations on the world market. Diversification is needed to stabilise farm incomes.

monoecious /mɒˈniːʃəs/ *adjective* with male and female flowers on separate plants. Compare **dioecious**

monogastric /ˌmɒnəʊˈɡæstrɪk/ *adjective* referring to animals with only one stomach, e.g. pigs. Compare **ruminant**

monophagous /mɒˈnɒfəɡəs/ *adjective* referring to an organism that feeds on only one kind of food. Compare **polyphagous**

monopitch /ˈmɒnəʊpɪtʃ/ *noun* a type of piggery with artificially controlled natural ventilation

moor *noun* an area of often high land that is not cultivated, and is formed of acid soil covered with grass and low shrubs such as heather

moorland /ˈmʊələnd, ˈmɔːlənd/ *noun* a large area of moor

MOPF *abbreviation* margin over purchased feed

mor /mɔː/ *noun* a type of humus found in coniferous forests, which is acid and contains few nutrients. Compare **mull**

morbidity /mɔːˈbɪdɪti/ *noun* a state of being diseased or sick

'In all, pneumonia was seen and treated in 186 of the 272 calves at risk on the dairy farms in question. The level of morbidity varied widely from 42 to 91% and the signs presented by the calves were variable.' [*Dairy Farmer*]

Moredun /ˈmɔːdʌn/ *noun* a scientific research institute which investigates diseases in animals, their prevention and treatment

Morrey system /ˈmɒri ˌsɪstəm/ *noun* a paddock system of rotational grazing, used in the management of dairy herds

mortality *noun* the occurrence of death ○ *The population count in spring is always lower than that in the autumn because of winter mortality.*

mosaic *noun* a disease of plants that makes yellow patterns on the leaves and can seriously affect some crops. It is often caused by viruses.

moss *noun* a very small plant without roots, which grows in damp places and forms mats of vegetation ○ *Sphagnum is a type of moss.*

mould *noun* **1.** a fungus, especially one that produces a fine powdery layer on the surface of an organism (NOTE: The US spelling is **mold**.) **2.** soft earth

mouldboard /ˈməʊldbɔːd/ *noun* the part of a plough which lifts and turns the slice of earth, so making the furrow (NOTE: Mouldboards are made of steel, and are made in many different styles, each producing a different surface finish. The main ones are general-purpose, digger and semi-digger.)

moult *noun* an occasion of shedding feathers or hair at a specific period of the year ■ *verb* to shed feathers or hair at a specific period of the year ○ *Most animals moult at the beginning of summer.* (NOTE: The US spelling is **molt**.)

mount *verb* to attach an implement to a tractor so that it is held by the tractor and has no other support

COMMENT: If the implement is simply pulled by the tractor and is supported by its own wheels, then it is said to be 'trailed' (a rotary cultivator can be either mounted or trailed). If the implement is supported both by the tractor and by its own wheels it is said to be 'semi-mounted'.

mountain *noun* a surplus or large amount of something, especially something that is being stored ○ *butter mountain* ○ *fridge mountain* ◊ **lake**

mountain sheep *noun* a sheep belonging to a breed which lives in or comes originally from a mountain region

movement licence *noun* a licence which is needed in order to move animals from areas of infectious disease. Restriction on the movement of animals is a measure used to prevent the spread of disease.

movement record *noun* a record kept by a farmer of all movements of animals on and off the farm premises. These records have been compulsory in the UK since 1925.

'As only one of those farms is farm assured, all animals have to have a three-month residency period at the Bradleys to attain assured status. This means all movement records being kept up to date to prove the status of the cattle.' [*Farmers Guardian*]

mow /məʊ/ *verb* to cut grass or a forage crop

mower *noun* a machine used to cut grass and other upright crops

COMMENT: The cut crop is left in a swath behind the cutting mechanism for further treatment in the process of making hay or silage. There are four main types of mower. **Cutter bar mowers** are rear-mounted and consist of a framework with a hinged cutter bar, the cutter bar having a number of fingers with a cutting edge on each side, which support the knife. **Rotary mowers** are made with two or four rotors, each having two, three or four turning blades. **Flail mowers** have a high-speed rotor fitted with swinging flails which cut the grass and leave it bruised in a fluffy swath. **Cylinder mowers** are used for lawns, but rarely used on farms.

MQ *abbreviation* milk quota

MRI *abbreviation* **1.** magnetic resonance imaging **2.** Meuse-Rhine-Ijssel

MRL *abbreviation* maximum residue level

MRM *abbreviation* mechanically recovered meat

MTR *abbreviation* Midterm Review

mucin /ˈmjuːsɪn/ *noun* a glycoprotein that is a constituent of mucus

muck *noun* same as **manure**

muck spreader *noun* same as **manure spreader**

muck weed *noun* same as **fat hen**

mucosal disease /mjuːˈkəʊz(ə)l dɪˌziːz/ *noun* a livestock disease caused by a virus, often fatal

mucus *noun* a slimy solution of mucin secreted by vertebrates onto a mucous membrane to provide lubrication

mudflats /'mʌd 'flts/ *noun* areas of flat mud, usually near the mouths of rivers

mulberry /'mʌlb(ə)ri/ *noun* a tree *(Morus nigra)* with dark fruit, similar to blackberries

mulberry heart disease *noun* a fatal disease of pigs, caused by a lack of vitamin E and selenium

mulch /mʌltʃ/ *noun* an organic material used to spread over the surface of the soil to prevent evaporation or erosion, e.g. dead leaves or straw ∎ *verb* to spread organic material over the surface of the soil to prevent evaporation or erosion

COMMENT: Black plastic sheeting is often used by commercial horticulturists, but the commonest mulches are organic. Apart from preventing evaporation, mulches reduce weed growth and encourage worms.

mule *noun* **1.** a hybrid and usually sterile offspring of a male ass and a mare **2.** a crossbred sheep from a Blue-faced Leicester ram and a Swaledale ewe. Mules have speckled faces and a high lambing rate.

mulesing /'mjuːlzɪŋ/ *noun* an operation to cut away loose skin on sheep to prevent blowfly attacks

mull /mʌl/ *noun* a type of humus found in deciduous forests. Compare **mor** (NOTE: It is formed of rotted leaves, is PH neutral and contains many nutrients.)

multicrop /'mʌltɪ 'krɒpɪŋ/ *verb* to grow more than one crop of something on the same piece of land in one year ○ *Wet rice is often multicropped.*

multicropping /'mʌlti,krɒpɪŋ/ *noun* the cultivation of more than one crop on the same piece of land in one year

'Farming's future lies in pests and weeds controlled by ecological/organic methods: rotations, multicropping, intercropping species and varieties, working with nature' beneficial insects and birds.' [*Farmers Guardian*]

multifactorial   inheritance /,mʌltifæk,tɔːriəl ɪn'herɪt(ə)ns/ *noun* the control of an inherited characteristic by several genes

multigerm seed /'mʌltidʒɜːm ,siːd/ *noun* a seed which occurs as a cluster of seeds fused together and which produces more than one plant when it germinates, a common example of which is the sugar beet seed. The multiple plants must be reduced to one by a process called 'singling'.

multi-horned *adjective* referring to an animal such as a Jacob's sheep which has more than two horns

multiple cropping *noun* the process of growing more than one crop on the same piece of land in one year

multiple ovulation and embryo transfer *noun* a method of insemination where embryos are transferred to cows. Abbr **MOET**. Compare **artificial insemination**

multiple suckling *noun* a system in dairy breeding where nurse cows suckle several beef calves at the same time

Murray Grey *noun* a breed of beef cattle, silver grey in colour. It is a polled early-maturing hardy breed, and the carcass has a high proportion of lean meat.

muscle *noun* an organ that contracts to make part of the body move

Muscovy /'mʌskəvi/ *noun* a utility breed of duck which is large in size, and coloured either black and white, or black, blue and white

mushroom *noun* a common edible fungus, often grown commercially

mushroom compost *noun* a special growing medium for the commercial production of mushrooms

mushroom spawn *noun* a mass of spores of edible mushrooms, used in propagation

musk melon *noun* a variety *(Cucumis melo)* of melon, with large scented fruit

must *noun* grape juice which has been extracted for wine, but which has not started to ferment

mustard *noun* a species of brassica *(Sinapis)*, whose seeds are among the most important spices. Mustard is also used as green manure.

COMMENT: Much of the mustard grown commercially is rape *(Brassica napus)*. The seeds of black mustard *(Brassica nigra)* are ground to produce the yellow spice. White mustard *(Brassica alba)* is grown as a salad crop (used in mustard and cress).

mutagen /'mjuːtədʒən/ *noun* an agent that causes mutation, e.g. a chemical or radiation

mutant *adjective* referring to a gene in which a mutation has occurred, or to an organism carrying such a gene ○ *mutant mice* ■ *noun* an organism carrying a gene in which mutation has occurred ○ *New mutants have appeared.* Also called **mutation**

mutate *verb* (*of a gene or organism*) to undergo a genetic change that can be inherited ○ *Bacteria can mutate suddenly and become increasingly able to infect.*

mutation /mjuː'teɪʃ(ə)n/ *noun* **1.** a heritable change occurring in a gene **2.** same as **mutant**

mutton *noun* the meat of a mature sheep, produced from older sheep such as ewes which are finished for breeding

mutualism /'mjuːtʃuəlɪz(ə)m/ *noun* same as **symbiosis**

muzzle *noun* the projecting part of an animal's head, especially the mouth, jaws and nose

MV *abbreviation* Maedi-Visna

MV Accreditation Scheme *noun* a system which enables farms to set up and maintain sheep known to be free of Maedi-Visna disease after appropriate blood tests, which are continued for member flocks

myc- /maɪk, maɪs/ *prefix* same as **myco-** (NOTE: used before vowels)

mycelium /maɪ'siːliəm/ *noun* a mass of hyphae which forms the main part of a fungus

myco- /maɪkəʊ/ *prefix* fungus or fungal

mycology /maɪ'kɒlədʒi/ *noun* the study of fungi

mycoplasm /'maɪkəʊ,plæz(ə)m/, **mycoplasma** /'maɪkəʊ,plæzmə/ *noun* a microorganism that lacks rigid cell walls. Genus: *Mycoplasma*. (NOTE: Some species cause respiratory diseases.)

mycorrhiza /,maɪkəʊ'raɪzə/ *noun* a mutual association of a fungus with the roots of a plant in which the fungus supplies the plant with water and minerals and feeds on the plant's sugars (NOTE: Many different fungi form mycorrhizas, especially with trees, and many plants such as orchids cannot grow without them.)

'Selective herbicide was applied to remove the ryegrass. It is possible to produce plants without mycorrhiza on the roots and, in this trial, control was significantly better when mycorrhiza were present.' [*Arable Farming*]

mycosis /maɪ'kəʊsɪs/ *noun* an infection with or a disease caused by fungi

mycotic dermatitis /maɪ,kɒtɪk ,dɜːmə'taɪtɪs/ *noun* a disease affecting sheep, caused by fungal-like bacteria which multiply on the skin and cause inflammation. Severe infections cause fleece loss.

mycotoxin /'maɪkəʊ,tɒksɪn/ *noun* a toxic substance produced by a fungus growing on crops in the field or in storage. There are regulations controlling the amount of some mycotoxins such as aflatoxin and ochratoxin permitted in food.

myiasis /'maɪəsɪs/ *noun* an infestation of animals by the larvae of flies

myxomatosis /,mɪksəmə'təʊsɪs/ *noun* a usually fatal virus disease affecting rabbits, transmitted by fleas

N

N *symbol* nitrogen

Na *symbol* sodium

NABIM *abbreviation* National Association of British and Irish Millers

NAC *abbreviation* National Agricultural Centre

naked grain *noun* a grain such as wheat that is easily separated or threshed out from its husk, i.e. in its caryopsis state

nanny goat *noun* a female goat

NASPM *abbreviation* National Association of Seed Potato Merchants

National Agricultural Centre *noun* the site of the annual Royal Show (at Stoneleigh, in Warwickshire), owned by the RASE. Abbr **NAC**

National Animal Welfare Trust *noun* a rescue centre with branches across the UK, which provides short-term care and rehabilitation for unwanted and abused animals. Abbr **NAWT**

National Assembly for Wales Agriculture and Rural Affairs Department *noun* the department of the devolved Welsh Assembly government which deals with farming, the environment, animal welfare and rural development in Wales. Abbr **NAWARAD**

National Canine Defence League *noun* former name for **Dogs Trust**

National Envelope *noun* a source of additional funds to help livestock producers, e.g. a beef national envelope and a sheep national envelope

National Farmers' Union *noun* an organisation representing the interests of British farmers in negotiations with the government and other agencies. Abbr **NFU**

National Institute of Agricultural Botany *noun* an organisation in the UK which tests all new varieties of crops. After successful testing, the varieties are made available to farmers. Abbr **NIAB**

national list *noun* a list of agricultural crop varieties tested by the NIAB and available for sale. Under EU regulations, all seeds sold to farmers or horticulturists must be tested and certified.

National Milk Records *noun* a company which keeps central records for dairy farmers. Abbr **NMR**

National Nature Reserve *noun* a nationally important example of a type of habitat, established as reserve to protect the most important areas of wildlife habitat and geological formations. Abbr **NNR** (NOTE: There are over 200 National Nature Reserves in England, owned or controlled by English Nature or held by approved bodies such as Wildlife Trusts.)

National Office of Animal Health *noun* an organisation which represents the British animal medicines industry. Abbr **NOAH**

national park *noun* a large area of land selected because of its scenic, recreational, scientific, or historical importance for special protection from development, and managed by a local government body for recreational use by the public and the benefit of the local community

National Soil Resources Institute *noun* an association formed in 2001 which provides education and training in the fields of soil and land management practice. Abbr **NSRI**

National Union of Agricultural and Allied Workers *noun* former name for **RAAW**

native *adjective* always having lived, grown or existed in a place ○ *Tigers are native to Asia.*

native breeds *plural noun* breeds which have been developed in a country, and not brought in from other countries

'Three years on the project includes 17 farmers who have received grants to establish herds of eight different native breeds of cattle – and the beef from these cattle is now being marketed as Limestone Beef, generating a premium for the producers involved.' [*Farmers Guardian*]

natural *adjective* referring to nature, or produced by nature not by humans ○ *natural materials* ○ *areas of natural beauty*

natural environment *noun* **1.** same as **natural habitat 2.** the part of the Earth that has not been built or formed by humans. Compare **built environment**

Natural Environment Research Council *noun* a group which carries out research and training in the environmental sciences. Abbr **NERC**

natural habitat *noun* the usual surroundings in which an organism lives in the wild. Also called **natural environment**

natural immunity *noun* immunity from disease inherited by newborn offspring from birth, acquired in the womb or from the mother's milk

natural insecticide *noun* an insecticide produced from plant extracts

naturalise /ˈnætʃ(ə)rəlaɪz/, **naturalize** *verb* to introduce a species into an area where it has not lived or grown before so that it becomes established as part of the ecosystem ○ *Rhododendron ponticum has become naturalised in parts of Britain.*

natural resource *noun* a naturally occurring material that can be put to use by humans, e.g. wood or oil (*often plural*)

natural selection *noun* the process of evolutionary change, by which offspring of organisms with certain characteristics are more able to survive and reproduce than offspring of other organisms, thus gradually changing the composition of a population

natural vegetation *noun* the range of plant communities that exist in the natural environment without being planted or managed by people

nature *noun* **1.** the characteristics that make someone or something what they are ○ *the nature of the task* **2.** all living organisms and the environments in which they live ○ *They try to live in harmony with nature.*

nature conservation *noun* the active management of the Earth's natural resources, plants, animals and environment, to ensure that they survive or are appropriately used

nature management *noun* the activity of managing a natural environment to encourage plant and animal life. Also called **habitat management**

nature reserve *noun* an area where plants, animals and their environment are protected

nature trail *noun* a path through the countryside with signs to draw attention to important and interesting features about plants, animals and the environment

navel-ill *noun* a disease of young livestock, especially newborn calves, kids and lambs. It causes abscesses at the navel and swellings in some joints. Also known as **joint-ill**

navy bean *noun* a dried seed of the common bean (*Phaseolus vulgaris*, used in particular for canning as baked beans. Also called **haricot bean**

NAWARAD *abbreviation* National Assembly for Wales Agriculture and Rural Affairs Department

NAWT *abbreviation* National Animal Welfare Trust

NCDL *abbreviation* National Canine Defence League (NOTE: Now called the 'Dogs Trust'.)

near infrared spectrophotometry *noun* a method of establishing tissue composition, used in agriculture to assess the quality of meat and of grain crops

'Pre-germination in malting barley could become easier to test using near infra red spectroscopy, according to research funded by HGCA. Germination in the ear can lead to poor malting quality and problems in the brewing process.' [*Farmers Weekly*]

neat *noun* an old term meaning a cow or ox

neck collar *noun* a leather band put round the neck of a horse or cow, to hold the animal in a stall

neck rot *noun* a disease affecting bulb onions during storage. The onions become soft and begin to rot from the stem downwards.

necrosis /neˈkrəʊsɪs/ *noun* the death of tissue or cells in an organism

nectar *noun* a sweet sugary liquid produced by flowers, which attracts birds or insects which pollinate the flowers

nectarine *noun* a smooth-skinned variety of peach *(Prunus persica nectarina)*

neigh *noun* a sound made by a horse ■ *verb* *(of a horse)* to make the characteristic sound of a horse. Compare **bleat, grunt, low**

nematicide /ne'mætɪsaɪd/ *noun* a substance which kills nematodes

nematode /'nemətəʊd/ *noun* a type of roundworm, some of which, e.g. hookworms, are parasites of animals while others, e.g. root knot and cyst nematodes, live in the roots or stems of plants

nematode disease *noun* a disease of the alimentary tract and lungs, caused by nematodes. Infection is transmitted from one group of animals to another by means of infective larvae in herbage.

Nematodirus disease *noun* a disease of lambs caused by parasitic roundworms. The animals suffer diarrhoea and loss of condition.

NERC *abbreviation* Natural Environment Research Council

nest *noun* **1.** a construction built by birds and some fish for their eggs **2.** a construction made by some social insects such as ants and bees for the colony to live in ■ *verb* to build a nest

nest box *noun* an open-fronted box in which a hen lays eggs. The box may be a single unit or part of a series of boxes.

net blotch *noun* a fungal disease of barley, with dark brown blotches affecting the leaves

nettle /'net(ə)l/ *noun* a plant, especially one of the genus *Urtica* which possesses stinging hairs. ◊ **hemp nettle, red deadnettle**

net value added *noun* the annual value of goods sold and services paid for inside a country, less tax and Government subsidies and also allowing for the depreciation of capital assets. Abbr **NVA**

neutraceutical /ˌnjuːtrə'sjuːtɪk(ə)l/, **neutriceutical** *noun* same as **functional food**

neutral *adjective* referring to the state of being neither acid nor alkali ○ *pH 7 is neutral.*

neutralise /'njuːtrəlaɪz/, **neutralize** *verb* **1.** to make an acid neutral ○ *Acid in*

drainage water can be neutralised by limestone. **2.** to make a bacterial toxin harmless by combining it with the correct amount of antitoxin **3.** to counteract the effect of something

neutralising value *noun* a measurement of the capability of a lime material to neutralise soil acidity. It is the same as the calcium oxide equivalent.

'Principally a liming agent, slag contains burnt lime, which gives it a neutralising value of more than 50 per cent. In addition, its naturally-occurring minerals and trace elements have generated reports from cereal growers of yield increases in the region of half a tonne an acre.' [*Farming News*]

new blood *noun* genetic variation brought into a breed by, e.g., introducing a new male to a flock or herd

Newcastle disease *noun* an acute febrile contagious disease of fowls. Affected birds suffer loss of appetite, diarrhoea and respiratory problems, and mortality rates are high. It is a notifiable disease.

new chemicals *plural noun* the chemicals that were not listed in the European Inventory of Existing Commercial Chemical Substances between January 1971 and September 1981. Compare **existing chemicals**

New Hampshire Red *noun* a breed of poultry with red plumage, lighter in weight than Rhode Island Red. New Hampshire Reds are mainly kept as layers, producing brownish-tinted eggs.

new variant CJD *noun* ◊ **variant CJD**

new wood *noun* growth made during the current year

NFE *abbreviation* nitrogen-free extract

NFFO *abbreviation* Non-Fossil Fuel Obligation ■ *adjective* referring to technologies which are designed to ensure diversity of power supply, such as hydro power, energy crops and wind power, according to the Non-Fossil Fuel Obligation

NFU *abbreviation* National Farmers' Union

NFYFC *abbreviation* National Federation of Young Farmers' Clubs

NIAB *abbreviation* National Institute for Agricultural Botany

nicotine *noun* a harmful substance in tobacco. It is used as an insecticide.

nightshade /'naɪtʃeɪd/ *noun* a plant of the family *Solanaceae* which, if eaten by stock, are likely to cause sickness or death

night soil *noun* human excreta, collected and used for fertiliser in some parts of the world

nip bar *noun* a bar fitted to moving mechanisms to prevent parts of the body being drawn into the machine

nipplewort /'nɪp(ə)lwɜːt/ *noun* an annual weed, *Lapsana communis*

nitrate *noun* **1.** an ion with the formula NO_3 **2.** a chemical compound containing the nitrate ion, e.g. sodium nitrate **3.** a natural constituent of plants. Beets, cabbage, cauliflower and broccoli can contain up to 1mg/kg.

COMMENT: Nitrates are a source of nitrogen for plants. They are used as fertilisers but can poison babies if they get into drinking water.

nitrate-sensitive area, nitrate-vulnerable zone *noun* a region of the country where nitrate pollution is likely and where the use of nitrate fertilisers is strictly controlled. Abbr **NSA**, **NVZ** (NOTE: Thirty new areas are proposed by a government scheme which will restrict nitrogen use to 150 kg/ha for five years. An EU directive in 1994 was aimed at reducing nitrate pollution on up to 2 million hectares of farmland in the UK.)

nitrification /,naɪtrɪfɪ'keɪʃ(ə)n/ *noun* the process by which bacteria in the soil break down nitrogen compounds and form nitrates which plants can absorb (NOTE: It is part of the nitrogen cycle.)

nitrification inhibitor *noun* a chemical product used to slow down the release of nitrate in organic manure

nitrifier /'naɪtrɪfaɪə/ *noun* a microorganism that is involved in the process of nitrification

nitrify /'naɪtrɪfaɪ/ *verb* to convert nitrogen or nitrogen compounds into nitrates

nitrite /'naɪtraɪt/ *noun* **1.** an ion with the formula NO_2 **2.** a chemical compound containing the nitrite ion, e.g. sodium nitrite

COMMENT: Nitrites are formed by bacteria from nitrogen as an intermediate stage in the formation of nitrates.

nitrogen *noun* a chemical element that is the main component of air and an essential part of protein. It is essential to biological life.

COMMENT: Nitrogen is taken into the body by digesting protein-rich foods. Excess nitrogen is excreted in urine. When the intake of nitrogen and the excretion rate are equal, the body is in nitrogen balance or protein balance. Nitrogen is supplied to the soil by fertilisers, organic matter, nodule bacteria on legumes, and by nitrogen-fixing microorganisms in the soil.

nitrogen compound *noun* a substance such as a fertiliser containing mostly nitrogen with other elements

nitrogen cycle *noun* the set of processes by which nitrogen is converted from a gas in the atmosphere to nitrogen-containing substances in soil and living organisms, then converted back to a gas (NOTE: Nitrogen is absorbed into green plants in the form of nitrates, the plants are then eaten by animals and the nitrates are returned to the ecosystem through animals' excreta or when an animal or a plant dies.)

nitrogen deficiency *noun* a lack of nitrogen in the soil, found where organic matter is low and resulting in thin, weak growth of plants

nitrogen fertiliser *noun* a fertiliser containing mainly nitrogen, e.g. ammonium nitrate

nitrogen fixation *noun* the process by which nitrogen in the air is converted by bacteria in some plant roots into nitrogen compounds (NOTE: When the plants die the nitrogen is released into the soil and acts as a fertiliser.)

'One of the biggest misconceptions currently circulating is a grass ley for grazing and cutting on an organic farm requires a greater range of legumes than in normal practice. In particular, red clover is being prescribed for use in grazing leys in the mistaken belief the nitrogen fixation for the accompanying grass crop will be dramatically increased.' [*Farmers Guardian*]

nitrogen-fixing plant *noun* a leguminous plant which forms an association with bacteria that convert nitrogen from the air into nitrogen compounds in the soil, e.g. a pea plant

nitrogen-free extract *noun* used in the chemical analysis of animal feeding stuffs, the nitrogen-free extract consists mainly of soluble carbohydrates (sugars) and starch. Abbr **NFE**

nitrogen-hungry plants *plural noun* plants which need a lot of nitrogen

nitrogenous fertiliser /naɪˈtrɒdʒənəs ˌfɜːtɪlaɪzə/ *noun* a fertilisers such as sulphate of ammonia which is based on nitrogen

nitrogen oxide *noun* an oxide formed when nitrogen is oxidised, e.g. nitric oxide or nitrogen dioxide. Formula: NO_x.

NMR *abbreviation* National Milk Records

NNR *abbreviation* National Nature Reserve

NOAH *abbreviation* National Office of Animal Health

node /nəʊd/ *noun* a point on the stem of a plant where a leaf is attached

nodule *noun* a small lump found on the roots of leguminous plants such as peas which contains bacteria that can convert nitrogen from the air into nitrogen compounds

nomad *noun* an animal that moves from place to place without having a fixed range. Compare **migrant**

nomadic /nəʊˈmædɪk/ *adjective* referring to nomads

nomadism /ˌnəʊməˈdɪz(ə)m/ *noun* a habit of some animals that move from place to place without having a fixed range

nominated service *noun* artificial insemination with semen from a named and tested male animal

non-centrifugal sugar *noun* a dark semi-solid sugar made by boiling the juices obtained from crushed sugar cane. India is the principal producer.

non-EU *adjective* not in the EU

non-flammable *adjective* referring to a material that is difficult to set on fire

non-food crops *plural noun* crops which are grown for purposes other than producing food, such as to provide renewable energy or chemicals

'Oilseeds are an ideal vector to deliver large industrial volumes to a range of technical industries, and have the potential to be one of the major non-food crops used by industry alongside starch. So says Dr Jeremy Tomkinson of the National Non-Food Crop Centre (NNFCC) at York.' [*Arable Farming*]

Non-Fossil Fuel Obligation *noun* a British government policy to promote the use of energy from renewable sources, such as solar or wind power. Abbr **NFFO**

non-organic *adjective* referring to crops that are not produced according to guidelines restricting the use of fertilisers and other practices

non-persistent pesticide *noun* a pesticide which does not remain toxic for long, and so does not enter the food chain

non-selective herbicide *noun* a chemical herbicide which kills all vegetation

non-till *adjective* same as **no-till agriculture**

noose *noun* a loop in a rope, with a loose knot which allows it to tighten, e.g. in a halter or a lasso

Norfolk horn *noun* a rare breed of sheep adapted to dry heathland. Black-faced and horned.

Norfolk rotation *noun* a system for farming, using arable farming for fodder crops, and involving the temporary sowing of grass and clover (NOTE: The Norfolk rotation system was introduced into England in the early 18th century and involved root crops (turnips or swedes), then cereal (barley), followed by ley (usually red clover), and ended with cereal (usually wheat). The Norfolk rotation provided a well-balanced system for building up and maintaining soil fertility, controlling weeds and pests, providing continuous employment and profitability.)

Normandy /ˈnɔːməndi/ *noun* a breed of cattle from north-west France that have a white coat with red-brown patches. The animals are reared for meat and for milk, from which Camembert cheese is made.

North Country Cheviot *noun* a large-sized breed of sheep with fine good-quality wool. This variety of the Cheviot is found in Caithness and Sutherland.

North Devon ⧫ **Devon**

Northern Dairy Shorthorn *noun* a dairy breed of cattle, which comes from the old Teeswater cattle, with perhaps a little Ayrshire blood. it is now established as a pure breed. The most popular colour is light roan, but red, white and mixtures of shades are found. The animals are thrifty, hardy and suitable for harsh upland conditions.

north-facing *adjective* directed towards the north ○ *a north-facing slope*

North Ronaldsay /ˌnɔːθ 'rɒn(ə)ldsi/ *noun* a rare breed of small sheep, which varies in colour from white through grey, brown and black, and also combinations of these colours. The tail is short, and most of the animals have horns.

Norway rat /'nɔːweɪ ræt/ *noun* same as **brown rat**

noseband /'nəʊzbænd/ *noun* a broad leather band worn around the horse's nose and above the bit, used to prevent a horse from opening its mouth too wide

notifiable disease *noun* a serious infectious disease of plants, animals or people that has to be officially reported so that steps can be taken to stop it spreading

'The safe haven scheme aims to keep Britain free from the damaging bacterial disease ring rot, by creating a supply chain where all seed has been traceably produced from ring rot free stocks. It follows two outbreaks of the notifiable disease in two years.' [*Farmers Weekly*]

COMMENT: The following are notifiable diseases of humans: cholera, diphtheria, dysentery, encephalitis, food poisoning, jaundice, malaria, measles, meningitis, ophthalmia neonatorum, paratyphoid, plague, poliomyelitis, relapsing fever, scarlet fever, smallpox, tuberculosis, typhoid, typhus, whooping cough, yellow fever. The following are some of the notifiable diseases of animals: anthrax, BSE, foot and mouth disease, Newcastle disease, rabies, sheep pox, sheep scab, swine fever.

no-till agriculture, no-till farming *noun* a system of cultivation in which mechanical disturbance of the soil by ploughing is kept to a minimum to reduce soil erosion. Also called **non-till**

novel crop *noun* a non-traditional crop, e.g. miscanthus grown as an energy crop or evening primrose grown for supply to the pharmaceutical industry

nozzle *noun* a projecting part with an opening at the end of a pipe, for regulating and directing a flow of fluid

NPK *noun* nitrogen, phosphorus and potassium, used in different proportions as a fertiliser

NSA *abbreviation* nitrate-sensitive area

NSRI *abbreviation* National Soil Resources Institute

Nubian goat /'njuːbiən gəʊt/ *noun* a breed of goat of mixed Egyptian and Indian origin, now crossed with British goats to produce the Anglo-Nubian breed

nucleus *noun* the central body in a cell, containing DNA and RNA, and controlling the function and characteristics of the cell

nurse cow *noun* a cow used to suckle the calves of others

nurse crop *noun* a crop grown to give protection to young plants of a perennial crop which is being established. Nurse crops provide shade and act as windbreaks.

'Last year two, one acre trial plots of Sitel lucerne were grown, the first sown in early April under a nurse crop of spring barley to protect the emerging crop and help establishment.' [*Farmers Guardian*]

nursery *noun* a place where plants are grown until they are large enough to be planted in their final positions

nursery bed *noun* a bed in which seedlings are planted out from the seedbed until they are large enough to be put in permanent positions

nursery plot *noun* an area of cultivated soil used for growing plants on before they are planted out, or for sowing seed. Also called **seed plot**

nut *noun* **1.** a hard indehiscent fruit with one seed **2.** any hard edible seed contained in a fibrous or woody shell, e.g. groundnuts **3.** a small cube of compressed meal, a convenient form of animal feed

nutraceutical /ˌnjuːtrə'sjuːtɪk(ə)l/, **nutriceutical** *noun* same as **functional food**

nutrient *noun* a substance that an organism needs to allow it to grow, thrive and reproduce, e.g. carbon, hydrogen, oxygen, nitrogen, phosphorus, potassium, calcium, magnesium or sulphur. Plants obtain their nutrients from the soil, while humans and other animals obtain them from their food, including plants.

nutrient leaching *noun* the loss of nutrients from the soil caused by water flowing through it, which deprives the soil of nutrients and may pollute water courses

nutrigenomics /ˌnjuːtrɪdʒɪ'nɒmɪks/ *noun* the study of the way in which genetic and environmental influences act together on an animal, and how this information can be used to boost productivity, health etc.

nutrition *noun* **1.** the process of taking in the necessary food components to grow and remain healthy. ◊ **soil nutrition 2.** nourishment or food which an animal eats

nutritional /njuːˈtrɪʃ(ə)n(ə)l/ *adjective* referring to nutrition ○ *the nutritional quality of meat*

nutritious *adjective* providing the nutrients that are needed for growth and health

nutritive /ˈnjuːtrətɪv/ *adjective* **1.** referring to a substance that provides the necessary components for growth and health ○ *plants grown in a nutritive solution* **2.** referring to nutrition

nutritive value *noun* the degree to which a food is valuable in promoting

health ○ *The nutritive value of white flour is lower than that of wholemeal flour.*

'While enzyme supplements are now widely used to improve the nutritive value of feeds for non-ruminants, the response of ruminants to direct fed fibrolytic enzymes has been both unclear and highly inconsistent, according to a presentation at BSAS in York.' [*Farmers Guardian*]

NVA *abbreviation* net value added

NVZ *abbreviation* nitrate-vulnerable zone

nymph *noun* an insect at the stage in its development between the larval stage and adulthood

O

O *symbol* **1.** oxygen **2.** Below Average (*in the EUROP carcass classification system*)

oak *noun* a deciduous or evergreen hardwood tree of which there are many species. Latin name: *Quercus*.

oarweed /'ɔːwiːd/ *noun* a common seaweed (*Laminaria digitata*) used as food

OAS *abbreviation* Organic Aid Scheme

oasis effect *noun* the loss of water from an irrigated area due to hot dry air coming from an unirrigated area nearby

oasthouse /'əʊsthaʊs/ *noun* a building containing a kiln for drying hops. It is a circular or square building with a characteristic conical roof.

oat /əʊt/ *noun* a hardy cereal crop grown in most types of soil in cool wet northern temperate regions. Latin name: *Avena sativa*. (NOTE: Oats are regarded as environmentally friendly because they require fewer inputs than other cereals.)

oatmeal *noun* a type of feeding stuff produced when the husk is removed from the oats kernel by a rolling process. Oatmeal is particularly good for horses and valuable for cattle and sheep, but not as suitable for pigs because of its high fibre content.

OBF *abbreviation* officially brucellosis free

occupational asthma *noun* asthma caused by materials with which people comes into contact at work, e.g. asthma in farm workers (**farmer's lung**), caused by hay

OCDS *noun* a temporary scheme under which farmers receive payment and support for the disposal of cattle that were born or reared in the United Kingdom before August 1996. The scheme was introduced in 2006 to replace the OTMS and will run until the end of 2008. Full form **Older Cattle Disposal Scheme**

odour nuisance *noun* a smell which is annoying or unpleasant

OECD *abbreviation* Organization for Economic Cooperation and Development

OELS *abbreviation* Organic Entry Level Stewardship

oestrogen *noun* a steroid hormone belonging to a group of hormones that controls the reproductive cycle and the development of secondary sexual characteristics in female primates (NOTE: The US spelling is **estrogen**.)

oestrous cycle *noun* the pattern of reproductive activity shown by most female animals, except most primates

oestrus /'iːstrəs/ *noun* one of the periods of the oestrous cycle that occurs in mature female mammals that are not pregnant. In this period ovulation normally occurs and the female is ready to mate. Also called **heat**

Oestrus /'iːstrəs/ *noun* a family of flies, including the bot fly

offal *noun* the inside parts of an animal, e.g. liver, kidney or intestines, when used as food, as opposed to meat, which is muscle

off-going crop *noun* a crop sown by a tenant farmer before leaving the farm at the end of his tenancy. He is permitted to return and harvest the crop and remove it.

officinalis /ɒ,fɪsɪ'nɑlɪs/, **officinale** /ɒ,fɪsɪ'nɑli/ *adjective* 'used in medicine', often part of the generic name of plants

offspring *noun* a child, the young of an animal, or a descendant of a plant (NOTE: The plural is **offspring**: *The birds usually produce three or four offspring each year.*)

OFS *abbreviation* Organic Farming Scheme

oil *noun* a liquid compound which does not mix with water, occurring as vegetable or animal oils, essential volatile oils and mineral oils

COMMENT: There are three types of oil: fixed vegetable and animal oils, essential volatile oils, and mineral oils. The most important oil-producing crops are the coconut palm, the oil palm, groundnuts, linseed, soya beans, maize and cotton seed. Other sources are olives, rape seed, lupin, sesame and sunflowers. After the nuts or seed have been crushed to extract the oil, the residue may be used as a livestock feed or as a fertiliser.

oilcake /ˈɔɪlkeɪk/ noun same as **oilseed cake**

oil crop noun a crop grown for extraction of the oil in its seeds, e.g. sunflower or oilseed rape

'In the mid-1970s the first 13-acre rape crop was grown alongside 200 acres of wheat and a few acres of oats and barley for the cattle. Now 550 acres of the oil crop is alternated with wheat in a 1:2 rotation to provide a high proportion of first wheat, which yields 10–15cwts/acre more than second wheat at around 3.75t/acre.' [*Arable Farming*]

oilseed cake /ˈɔɪlsiːd ˌkeɪk/ noun a feedingstuff concentrate, high in protein, made from the residue of seeds which have been crushed to produce oil. Also called **oilcake**

oilseed rape noun a plant of the cabbage family with bright yellow flowers, grown to provide an edible oil and animal feed from the processed seeds. Latin name: *Brassica napus*. Also called **rape** (NOTE: Oil produced from oilseed rape is often called 'vegetable oil'.)

oilseeds plural noun crops grown for the oil extracted from their seeds, e.g. oilseed rape or linseed

Old English game noun a breed of poultry, now mainly a fancy breed. The birds are coloured black and white with blue wing tips.

Older Cattle Disposal Scheme noun full form of **OCDS**

old wood noun growth made during previous years

oligotrophic /ˌɒlɪɡəʊˈtrɒfɪk/ adjective (*of water*) referring to water that contains few nutrients. ◊ **dystrophic, eutrophic, mesotrophic**

olive noun a Mediterranean tree with small yellowish-green edible fruit from which an edible oil can be produced. Latin name: *Olea europaea*.

OM abbreviation organic matter

omasum /əʊˈmeɪsəm/ noun the third stomach of a ruminant, which acts as a filter, and where much of the water in food is taken out before the food passes onto the abomasum. ◊ **abomasum, reticulum, rumen** (NOTE: The omasum is also colloquially called **the Bible** or **the Book**.)

omnivore /ˈɒmnɪvɔː/ noun an animal that eats both plant and animal foods. ◊ **carnivore, detritivore, frugivore, herbivore** (NOTE: Humans and pigs are examples of omnivores.)

omnivorous /ɒmˈnɪv(ə)rəs/ adjective referring to an animal that eats both plant and animal foods

once grown seed noun seed obtained from plants grown from a certified seed and intended for use by the farmer on his own farm, and not for resale

onion noun a vegetable crop (*Allium cepa*), grown either for cooking or for eating in salads. The ripe onion consists of the edible swollen leaf bases surrounded by scale leaves. It is harvested when the growing tops have fallen over. It is the dormant bulbs which are harvested and eaten.

onion couch noun a grass weed (*Arrhenatherum elatius*) which grows to 24–48 inches and develops long oat-like hairs like flower heads. Onion couch affects cereals.

onion fly noun an insect pest (*Hylemyia antiqua*) the maggots of which cause damage to onions by eating into the developing bulb

onion set noun a seed onion, a small onion grown from seed, which has been dried, and which is planted the following year so that it will root and grow on to maturity

on-off grazing noun same as **rotational grazing**

on-the-hoof adjective referring to animals which are sold live for slaughter

open countryside noun an area of country without many trees or high mountains

open fields plural noun fields which are not separated by hedges or walls, but by banks of earth. Formerly fields were divided into strips, each worked by a farmer; the system was used originally by the Saxons.

COMMENT: In recent years the removal of many field boundaries as a result of farm consolidation has led to an increase in the size of the average British field, and created large open fields again. Hedges have been removed to allow large farm machinery to be used more economically, and the loss of hedgerows has had a marked effect on the wildlife in the countryside.

open furrow *noun* a furrow shaped like a V, with the furrow slices laid in opposite directions to each other

opening bid *noun* the first bid at an auction

optimise /'ɒptɪmaɪz/, **optimize** *verb* to make something as efficient as possible

optimum *adjective* referring to the point at which the condition or amount of something is the best ○ *optimum height*

orache /'ɒrɪtʃ/ *noun* a common weed (*Atriplex patula*) which affects sugar beet and maize crops, and makes harvesting the crop difficult

orange *noun* the fruit of the *Citrus aurantium*, a native tree of China, whose nutritional value is due mainly to its high vitamin C content. Grown in semi-tropical and Mediterranean regions, it is eaten as fresh fruit or used for juice and for making preserves. The USA, Brazil, Spain, Morocco and Israel are large exporters of oranges.

COMMENT: Blood oranges are coloured by the presence of anthocyanins. Mandarin oranges such as satsumas and tangerines have loose peel. The Seville orange is a bitter orange, grown in Spain and used by marmalade manufacturers.

orchard *noun* an area of land used for growing fruit trees

COMMENT: Orchards were once a common feature of most farms, but now fruit is commercially produced by specialised commercial growers. The modern orchard consists of trees grafted onto dwarfing rootstock, shaped by pruning and closely planted in rows which are separated to allow room for tractors and sprayers to pass. Apples, plums, pears and cherries are the most important tree fruits in Britain, with Kent, Worcestershire and parts of East Anglia being the most important growing regions. In the USA, oranges and other citrus fruits are grown in orchards in the Southern

States, in particular in Florida and California.

Ordnance Survey /ˌɔːdnəns 'sɜːveɪ/ *noun* an agency which generates accurate mapping data for Great Britain. Abbr **OS**

orf /ɔːf/ *noun* a virus disease affecting sheep, cattle and goats and easily passed on to humans. The disease causes scabs and ulcers which affect the mouth, nose and eyes. In its later stages legs, genitals and udders may be affected.

organ *noun* a part of an organism that is distinct from other parts and has a particular function, e.g. an eye or a flower

organelle /ˌɔːgə'nel/ *noun* a specialised structure within a cell, e.g. a mitochondrion or nucleus

organic *adjective* **1.** referring to a compound containing carbon **2.** referring to food produced using only a restricted number of permitted pesticides and fertilisers, or to the production of such food **3.** referring to a substance which comes from an animal or plant

organic agriculture *noun* same as **organic farming**

Organic Aid Scheme *noun* a government-funded scheme that gives one-off support payments to organic farmers, especially to cover their set-up costs. Abbr **OAS**

organically /ɔː'gænɪkli/ *adverb* using only a restricted number of permitted pesticides and fertilisers in growing a crop

organic conversion *noun* the process of converting from conventional agriculture to organic production

'While the Herdman family made the formal move into organic conversion just before the foot-and-mouth outbreak, they had been gradually adopting organic and sustainable farming practices on an informal basis as a practice' to find whether or not they would work successfully at Acton Farm.' [*Farmers Guardian*]

Organic Entry Level Stewardship *noun* one of the categories under the Environmental Stewardship scheme, under which organic farmers can apply for funding in return for implementing certain environmental management schemes on their land. Abbr **OELS**

organic farming *noun* a method of farming which does not involve the use of artificial fertilisers or pesticides ○ *Organic* *farming may become more economic than conventional farming.*

COMMENT: Organic farming uses natural fertilisers and rotates stock farming (i.e. raising of animals) with crop farming. Soil nutrients are maintained by the addition of plant and animal manures, and pest control is achieved by the use of naturally derived pesticides, and by crop rotation, which allows natural predation to take place. Organic farming may produce lower yields than traditional or intensive farming, but the lower yields may be offset by the high cost of the chemical fertilisers used in intensive farming. It may become more economic than conventional farming due to premium prices which are paid for organic products. In areas of overproduction, organic farming has the advantage of reducing crop production without loss of quality and without taking land out of agricultural use. At the present time, Scotland has the UK's largest proportion of organic farmland, at 7%, with an average of 4% across the the rest of the UK. The main factor in controlling conversion to organic farming is the capital cost. A government scheme to encourage farmers to convert to organic agriculture has begun. Payments will be made to farmers in England over a 5-year period to assist with the costs of converting land to organic production. The scheme is also designed to stimulate a form of production which emphasises soil improvement and the control of pests and diseases. In 2004 there were 678,630 hectares registered as organic land in the UK.

Organic Farming Scheme *noun* a former support scheme which gave payments for organic farmers wanting to increase their production, now administered under the Organic Entry Level Stewardship scheme. Abbr **OFS**

organic fertiliser *noun* a fertiliser made from dead or decaying plant matter or animal wastes, e.g. leaf mould, farmyard manure or bone meal

organic material /ɔːˈɡænɪk ˈmætəˌ, **organic matter** *noun* carbon-based material derived from organisms, e.g. decomposed plant material or animal dung

organic matter *noun* **1.** a combination found in soil of plant material that is decomposing, microorganisms such as fungi, and humus. Also called **soil organic matter 2.** same as **organic material**

COMMENT: Organic matter is acted on by bacteria, fungi, earthworms, and it

decomposes to form humus. Humus is finally broken down by an oxidation process. The organic matter content of soil varies according to soil type, and usually increases with clay content. Peaty soils have a high organic matter content and some are totally made up of organic matter.

Organization for Economic Cooperation and Development *noun* an international intergovernmental association set up in 1961 to coordinate the economic policies of member nations. Abbr **OECD**

organochlorine /ˌɔːɡænəʊˈklɔːriːn/ *noun* a chemical compound containing chlorine, used as an insecticide

COMMENT: Organochlorine insecticides are very persistent, with a long half-life of up to 15 years. Chlorinated hydrocarbon insecticides can enter the food chain and kill small animals and birds which feed on insects.

organophosphate /ˌɔːɡənəʊˈfɒsfeɪt/ *noun* a synthetic insecticide that attacks the nervous system, e.g. chlorpyrifos

COMMENT: Organophosphates are not as persistent as organochlorines and do not enter the food chain. They are, however, very toxic and need to be handled carefully, as breathing in their vapour may be fatal.

organophosphorous insecticide /ˌɔːɡənəʊˌfɒsfərəs ɪnˈsektɪsaɪd/ *noun* same as **organophosphate**

organophosphorus compound *noun* an organic compound containing phosphorus

orphaned animal *noun* a young animal whose mother has died, and is therefore either fostered onto another animal or has to be hand-reared

Orpington /ˈɔːpɪŋtən/ ♦ **Buff Orpington**

Oryza /ˈɒrɪzə/ *noun* the Latin name for rice

OS *abbreviation* Ordnance Survey

osier /ˈəʊziə/ *noun* a species of willow, the shoots of which are used in making baskets

osmosis *noun* the movement of molecules of a solvent from a solution of one concentration to a solution of a higher concentration through a semi-permeable membrane until the two solutions balance in concentration

osmotic pressure /ɒzˌmɒtɪk ˈpreʃə/ *noun* the pressure required to prevent the flow of a solvent into a solution through a semi-permeable membrane

OSR *abbreviation* oilseed rape

osteo- /ɒstiəu/ *prefix* bone

osteomalacia /ˌɒstiəuməˈleɪʃə/ *noun* a condition where the bones become soft because of lack of calcium or phosphate

ostrich *noun* a large flightless bird (*Struthio camelus*) raised in farms for its meat

OTMS *noun* until 2005, a scheme under which all cattle slaughtered over the age of 30 months were incinerated or rendered for safe disposal. Full form **Over Thirty Month Scheme**

outbreeding /ˈautbriːdɪŋ/ *noun* **1.** breeding between individuals that are not related **2.** fertilisation between two or more separate plants, rather than within a flower or between flowers of the same plant ○ *Outbreeding occurs in broad beans.* ▶ compare **inbreeding**

outcrossing /ˈautkrɒsɪŋ/ *noun* the process of bringing some new genetic variation ('new blood') into a flock or herd, usually by introducing a new male

outfall *noun* a pipe from which sewage, either raw or treated, flows into a river, lake or the sea. Also called **outfall sewer**

outfields /ˈautfiːldz/ *noun* in hill farms, the fields furthest from the homestead, cropped only from time to time and allowed to lie fallow for long periods

outhouse /ˈauthaus/ *noun* a farm building which is not attached to the main farmhouse, and may be used for storage or for keeping poultry

outline planning permission *noun* permission in principle to build a property on a piece of land, but not the final approval because further details must be submitted

out-of-season *adjective, adverb* referring to a plant which is grown or sold at a time when it is not naturally available from outdoor cultivation ○ *Out-of-season strawberries are imported from Spain.* ○ *Glasshouses provide out-of-season tomatoes.*

outstation /ˈautsteɪʃən/ *noun* in New Zealand and Australia, a sheep station separate from the main station

outwinter /ˈautwɪntərɪŋ/ *verb* to keep cattle and sheep outdoors in fields during the winter months ○ *a herd of outwintered heifers*

'It was estimated that savings achieved by outwintering compared with inwintering would be in the region of pounds 30 per head in terms of feed costs plus savings in labour and time.' [*Farmers Guardian*]

ova /ˈəuvə/ plural of **ovum**

ovary *noun* **1.** one of two organs in a woman or female animal that produce ova or egg cells and secrete the female hormone oestrogen **2.** the part of a flower that contains the ovules, at the base of a carpel

oven-ready poultry *noun* poultry which has been slaughtered and dressed so that it can be cooked without any further preparation

overcropping /ˌəuvəˈkrɒpɪŋ/ *noun* the practice of growing too many crops on poor soil, which has the effect of greatly reducing soil fertility

overcultivated /ˌəuvəˈkʌltɪveɪtɪd/ *adjective* referring to land that has been too intensively cultivated and has reduced fertility

overexploit /ˌəuvərɪkˈsplɔɪt/ *verb* to cultivate soil too intensely

overexploitation /ˌəuvəreksplɔɪˈteɪʃ(ə)n/ *noun* the uncontrolled use of natural resources until there is very little left ○ *Overexploitation has reduced herring stocks by half.*

overfeed /ˌəuvəˈfiːd/ *verb* to give animals too much feed

overfertilisation /ˌəuvəfɜːtɪlaɪˈzeɪʃ(ə)n/, **overfertilization** *noun* the application of too much fertiliser to land (NOTE: Excess fertiliser draining from fields can cause pollution of the water in rivers and lakes.)

overgraze /ˌəuvəˈgreɪz/ *verb* to graze a pasture so much that it loses nutrients and is no longer able enough to provide food for livestock

overgrazing /ˌəuvəˈgreɪzɪŋ/ *noun* the practice of grazing a pasture so much that it loses nutrients and is no longer able to provide food for livestock ○ *Overgrazing has led to soil erosion and desertification.*

'Examples of the steps farmers will have to take include reducing the risk of soil erosion and avoiding the deterioration of habitats by preventing undergrazing as well as overgrazing.' [*Farmers Weekly*]

overgrown /ˌəuvəˈgrəun/ *adjective* referring to a seedbed or field which is covered with weeds or other unwanted vegetation

overlying /ˌəʊvəˈlaɪɪŋ/ *noun* the crushing of piglets by the sow which lies on top of them

overproduction /ˌəʊvəprəˈdʌkʃən/ *noun* the production of more of something than is wanted or needed

overshot wheel /ˈəʊvəʃɒt ˌwiːl/ *noun* a type of waterwheel where the water falls on the wheel from above. It is more efficient than an undershot wheel, where the water flows underneath the wheel.

overstorey /ˈəʊvəˌstɔːri/ *noun* the topmost vegetation layer in a forest, formed by the tallest trees. Also called **overwood**

Over Thirty Month Scheme *noun* full form of **OTMS**

overtopping /ˌəʊvəˈtɒpɪŋ/ *noun* cutting too much off the top of a plant when preparing it, e.g. when preparing sugar beet

overwinter /ˌəʊvəˈwɪntə/ *verb* **1.** to spend winter in a particular place ○ *The herds overwinter on the southern plains.* **2.** to remain alive though the winter ○ *Many plants will not overwinter in areas that have frost.*

overwood /ˈəʊvəwʊd/ *noun* same as **overstorey**

ovicide /ˈəʊvɪsaɪd/ *noun* a substance, especially an insecticide, that kills eggs

oviduct /ˈəʊvɪdʌkt/ *noun* a tube that transports eggs from the ovary to the uterus in mammals or in birds and reptiles secretes the eggshell and conveys the egg to the outside (NOTE: In mammals it is also called the fallopian tube.)

oviparous /əʊˈvɪpərəs/ *adjective* referring to an animal that carries and lays eggs. Compare **viviparous**

Ovis /ˈəʊvɪs/ *noun* the Latin name for the sheep genus

ovulate /ˈɒvjʊleɪt/ *verb* to release an ovum from the mature ovarian follicle into the fallopian tube

ovulation /ˌɒvjʊˈleɪʃ(ə)n/ *noun* the release of an ovum from the mature ovarian follicle into the fallopian tube

ovule /ˈɒvjuːl/ *noun* an immature egg or an unfertilised seed

ovum /ˈəʊvəm/ *noun* a female egg cell which, when fertilised by a spermatozoon, begins to develop into an embryo (NOTE:

The plural is **ova**. For other terms referring to ova, see words beginning with **oo-**.)

COMMENT: At regular intervals (in the human female, once a month) ova, or unfertilised eggs, leave the ovaries and move down the fallopian tubes to the uterus. Ovulation is regular in the mare, sow, ewe and cow.

ox *noun* a male or female beast from domestic cattle, and also the castrated male, especially when used as a draught animal (NOTE: The plural is **oxen**.)

ox-eye *noun* any flower with a round yellow centre, e.g. the ox-eye daisy

Oxford Down /ˈɒksfəd daʊn/ *noun* the largest of the down breeds of sheep, produced by crossing Southdown improved stock with the longwoolled Cotswold. It has a dark-brown face and legs and a conspicuous topknot.

oxidase /ˈɒksɪdeɪz/ *noun* an enzyme which encourages oxidation by removing hydrogen

oxidise /ˈɒksɪdaɪz/, **oxidize** *verb* to form an oxide by the reaction of oxygen with another chemical substance ○ *Over a period of time, the metal is oxidised by contact with air.*

oxygen *noun* a colourless, odourless gas, essential to human life, constituting 21% by volume of the Earth's atmosphere ○ *Our bodies obtain oxygen through the lungs in respiration.*

COMMENT: Oxygen is an important constituent of living matter, as well as water and air. It is formed by plants from carbon dioxide in the atmosphere during photosynthesis and released back into the air. Oxygen is absorbed from the air into the bloodstream through the lungs and is carried to the tissues along the arteries. It is essential to normal metabolism.

oxygenate /ˈɒksɪdʒəneɪt/ *verb* **1.** to treat blood with oxygen **2.** to become filled with oxygen

oxygenation /ˌɒksɪdʒəˈneɪʃ(ə)n/ *noun* the process of becoming filled with oxygen

oxytocin /ˌɒksɪˈtəʊsɪn/ *noun* a hormone which activates the release of milk in the udder and the contractions in the uterus during birth. It is also possibly important in contracting the uterus during mating. Its action is blocked by the release of adrenalin.

P

P *symbol* Poor (*in the EUROP carcass classification system*)

packhouse /'pækhaʊs/ *noun* a building used for grading, cleaning and packing produce on a farm, before it is sent to the customer

paddock *noun* a small enclosed field, usually near farm buildings

paddock grazing *noun* a rotational grazing system which uses paddocks of equal area for grazing, followed by a rest period

palatability /ˌpælətə'bɪlɪti/ *noun* the extent to which something is good to eat

palatable *adjective* good to eat ○ *Some types of grass are less palatable than others.* ○ *Big bales preserve the grass in an almost cut state which is very palatable.*

pale *noun* **1.** a pointed piece of wood used for fencing **2.** a husk on grass or cereal seeds

pale leaf spot *noun* white spots which form on leaves of clover plants due to potash deficiency

pale persicaria /ˌpeɪl ˌpɜːsɪ'keəriə/ *noun* a weed found in spring-sown crops

pale soft exudative muscle *noun* a condition where an animal's meat becomes pale and lacks firmness. Abbr **PSE**

palm *noun* **1.** a large tropical plant like a tree with branching divided leaves, that produces fruits which give oil and other foodstuffs **2.** the inner surface of the hand or the underside of a mammal's forefoot that is often in contact with the ground

palynology /ˌpælɪ'nɒlədʒi/ *noun* the scientific study of pollen, especially of pollen found in peat and coal deposits. Also called **pollen analysis**

pan *noun* **1.** a wide shallow pot for growing seeds **2.** a hard cemented layer of soil, impervious to drainage, lying below the surface. It is formed by the deposition of iron compounds or by ploughing at the same depth every year. Pan may be broken up by using a subsoiler.

pan- /pæn/ *prefix* affecting everything or everywhere

pandemic /pæn'demɪk/ *adjective, noun* referring to an epidemic disease which affects many parts of the world. ◊ **endemic, epidemic**

panemone /'pænɪməʊn/ *noun* a type of windmill in which flat surfaces spin round a vertical axis

panicle /'pænɪk(ə)l/ *noun* a flower head (**inflorescence**) with many branches that carry small flowers, e.g. the flower head of a rice plant

Panicum /'pænɪkəm/ *noun* the Latin name for millet

pannage /'pænɪdʒ/ *noun* **1.** pasturage for pigs in a wood or forest **2.** the corn and beech mast on which pigs feed

papain /pə'peɪɪn/ *noun* an enzyme found in the juice of the papaya, used as a meat tenderiser and in medicine to help wounds to heal

Papaver /pə'pɑːvə/ *noun* the Latin name for poppy

parameter *noun* **1.** a factor that defines the limits or actions of something **2.** a variable quantity or value for which a measurement is attempted, e.g. mean height

paraquat /'pærəkwɒt/ *noun* a herbicide that destroys a wide range of plants by killing their foliage and becomes inert on contact with the soil. It is poisonous to mammals, including humans.

parasite *noun* a plant or animal which lives on or inside another organism, the host, and derives its nourishment and other needs from it ○ *a water-borne parasite*

COMMENT: The commonest parasites affecting animals are lice on the skin and various types of worms in the intestines. Many diseases of humans such as malaria and amoebic dysentery are caused by infestation with parasites. Viruses are parasites on animals, plants and even on bacteria. Fungal diseases in plants, such as mildews and rusts, are caused by the action of parasitic fungi on their hosts.

parasitic /ˌpærəˈsɪtɪk/ *adjective* referring to animal or plant parasites ○ *a parasitic worm* ○ *Dodder is a parasitic plant.*

parasitic gastro-enteritis *noun* an infection of the stomach caused by roundworms, especially *Osteragia*. It can be cured by anthelmintics. Abbr **PGE**

parasiticide /ˌpærəˈsaɪtɪsaɪd/ *noun* a substance that kills parasites

parasitise /ˈpærəsɪtaɪz/, **parasitize** *verb* to live as a parasite on another organism ○ *Sheep are parasitised by flukes.*

parasitism /ˈpærəsaɪtɪz(ə)m/ *noun* a state in which one organism, the parasite, lives on or inside another organism, the host, and derives its nourishment and other needs from it

parasitoid /ˈpærəsaɪtɔɪd/ *noun* an organism that is a parasite only at one stage in its development

parasitology /ˌpærəsaɪˈtɒlədʒi/ *noun* the scientific study of parasites

parathion /ˌpærəˈθaɪən/ *noun* an organophosphorus insecticide no longer approved for use in the UK

paratyphoid /ˌpærəˈtaɪfɔɪd/ *noun* a disease of pigs caused by infection with salmonella bacteria. Young pigs run a high fever and may die within 24 hours.

parent *noun* a male or female that has produced offspring

parent material rock *noun* the unweathered base rock which breaks down to form a constituent part of the surface soil

parent plant *noun* a plant from which others are produced

parkland /ˈpɑːklænd/ *noun* grazed grassland or heathland with large individual trees or small groups of large trees, usually part of a designed and managed landscape

parlour systems *plural noun* the four basic designs of milking parlour: the herringbone parlour, the abreast parlour, the tandem parlour and the rotary parlour

'In normal milking parlour systems, the dairyman has to be present two or three times a day to carry out the milking.

Robotic milking does away with this structured, time-consuming job, but it relies on continuous operation of sophisticated machinery.' [*Farmers Weekly*]

parrot mouth *noun* a malformation of the upper jaw of horses, preventing proper mastication. The condition prevents the horse from grazing.

parsley *noun* a common herb (*Petroselinum crispum*) used for garnishing and flavouring

parsley piert /ˈpɑːsli ˌpiɜːt/ *noun* a common weed (*Aphanes arvensis*) affecting winter cereals. Also called **lamb's foot**

parsnip *noun* a plant (*Pastinaca sativa*) whose long white root is eaten as a vegetable

Parthenais /ˈpɑːtəneɪ/ *noun* a breed of cattle originating in France. It produces calves for a suckler herd, and is known for easy calving and high growth rate.

parthenocarpy /ˌpɑːθenəʊˈkɑːpi/ *noun* the production of seedless fruits without fertilisation having taken place

parthenogenesis /ˌpɑːθənəʊˈdʒenəsɪs/ *noun* a form of reproduction in which an unfertilised ovum develops into an individual

partial drought *noun* in the UK, period of at least 29 consecutive days when the mean rainfall does not exceed 2.54mm

particle *noun* a very small piece of a substance ○ *soil particles* ○ *Particles of volcanic ash were carried into the upper atmosphere.*

particle size distribution *noun* a way of measuring the composition of soil, which can be used in planning irrigation and crop arrangement

partition *noun* a moveable wall which divides a room, e.g. a partition in a stable

partly mixed ration *noun* a winter feed for livestock which combines total mixed ration and separate concentrate feeding, by mixing some concentrates with the roughage and keeping the rest aside as additional feed for higher-yielding animals. Abbr **PMR**

part-time farming *noun* a type of farming, where the farmer has a regular occupation other than farming and which is common throughout much of central and Eastern Europe. In the UK, part-time farmers are mainly wealthy people who farm as a hobby or as a second form of business.

parturition /ˌpɑːtjʊˈrɪʃ(ə)n/ *noun* the act of giving birth to offspring, when the foetus leaves the uterus, called by different names according to the animal. ◊ **calving, farrowing, foaling, lambing**

parvovirus /ˈpɑːvəʊˌvaɪrəs/ *noun* any of a group of viruses that have a single strand of DNA, especially those causing infertility in pigs

passive immunity *noun* immunity received in the womb from the mother. Antibodies produced by her immune system to resist diseases that she has experienced or been vaccinated against are transferred to the embryo.

'Colostrum management is becoming an increasing problem on many large US dairy farms with the result that up to 60% of calves do not have the required levels of passive immunity to protect them against disease.' [*Dairy Farmer*]

passport *noun* an official document issued to many types of animals or plants being moved from one country to another, certifying freedom from disease ■ *verb* to issue a plant or animal passport

passporting *noun* the provision of a plant or animal passport

pastern /ˈpæstɜːn/ *noun* the thin part of a horse's leg, between the fetlock and the hoof

pasteurellosis /ˌpɑːstʃərəˈləʊsɪs/ *noun* a clostridial disease mainly affecting young lambs, adult sheep and store lambs. It may be caused by contaminated food or water. Symptoms are high temperature and difficult breathing, and death may follow a few days after the symptoms become apparent.

pasteurisation /ˌpɑːstʃəraɪˈzeɪʃ(ə)n/, **pasteurization** *noun* the heating of food or food products for a specific period to destroy bacteria

COMMENT: Pasteurisation is carried out by heating food for a short time at a lower temperature than that used for sterilisation. The two methods used are heating to 72°C for fifteen seconds (the high-temperature-short-time method) or to 65°C for half an hour, and then cooling rapidly. This has the effect of killing tuberculosis bacteria.

pasteurise /ˈpɑːstʃəraɪz/, **pasteurize** *verb* to kill bacteria in food by heating it

pastoral *adjective* **1.** referring to agriculture based on grazing animals **2.** referring to land available for pasture

COMMENT: Pastoral farming ranges from large-scale highly scientific systems such as cattle- or sheep-ranching to primitive systems of nomadism. In relatively dry or inhospitable regions, the need to find pastures for grazing animals causes the farmers to lead a nomadic existence.

pastoralist /ˈpɑːstərəlɪst/ *noun* a farmer who keeps grazing animals on pasture ○ *The people most affected by the drought in the Sahara are nomadic pastoralists.*

pasturage /ˈpɑːstʃərɪdʒ/ *noun* a place where animals are pastured

pasture *noun* land covered with grass or other small plants, used by farmers as a feeding place for animals ○ *a mixture of pasture and woodland* ○ *Their cows are on summer pastures high in the mountains.* ■ *verb* to put animals onto land covered with grass or other small plants ○ *Their cows are pastured in fields high in the mountains.*

pastureland /ˈpɑːstʃələænd/ *noun* land covered with grass or other small plants, used by farmers as a feeding place for animals

pasture management *noun* the control of pasture by grazing, cutting, reseeding and similar techniques

'In terms of pasture management for early grazing, the recommendation is to ensure swards are grazed down fully this autumn, as leaving old grass over winter will increase the risks of winterkill and reduce productivity next spring.' [*Farmers Guardian*]

pasture topper *noun* a piece of machinery that is attached to the back of a tractor and used to keep grassy areas such as pastures free from weeds and coarse grasses

patch *noun* a small cultivated area with one type of plant growing in it, e.g. a pumpkin patch or onion patch

patent flour *noun* a very fine good-quality wheat flour

patho- /pæθəʊ/ *prefix* disease

pathogen /ˈpæθədʒən/ *noun* an agent, usually a microorganism, that causes a disease

pathogenesis /ˌpæθəˈdʒenəsɪs/ *noun* the origin, production or development of a disease

pathogenetic /ˌpæθədʒəˈnetɪk/ *adjective* referring to pathogenesis

pathogenic /ˌpæθəˈdʒenɪk/ *adjective* able to cause or produce a disease

pathogenicity /ˌpæθədʒəˈnɪsɪti/ *noun* the ability of a pathogen to cause a disease

pathogenic organism *noun* an organism responsible for causing a disease

pathology *noun* the study of diseases and the changes in structure and function which diseases can cause

pause *noun* a rest period in a bird's laying cycle

PCV2 *noun* a virus which is thought to be a key cause of PMWS in pigs. Full form **porcine coronavirus type 2**

PDA *abbreviation* Potash Development Association

PDNS *abbreviation* Porcine Dermatitis and Nephropathy Syndrome

pea *noun* an important grain legume

COMMENT: Peas are grown for pulses and for their immature seeds which are eaten fresh as a green vegetable and are also often frozen. The young pods are also occasionally eaten as mangetouts. Peas are also grown for forage and may be used for hay and silage. They are often grown following a cereal crop in rotation, and enrich the soil with nitrogen. Most peas are harvested, transported and processed on the same day and the majority are taken for freezing and canning. Vining peas are Britain's most important contract vegetable crop.

pea and bean weevil *noun* a pest (*Sitona* sp) affecting peas, beans and other legumes. The eggs are laid in soil near the plants, allowing the larvae to feed on the roots. The adult weevils feed on the leaves, making U-shaped notches in the edges of the leaves.

peach *noun* a small deciduous tree (*Prunus persica*) found particularly in Mediterranean areas, though it will grow as far north as southern England. The fruit are large and juicy, with a downy skin, but they cannot be kept for any length of time. (NOTE: Peaches are divided into two groups: the freestone (where the flesh is not attached to the stone), and the clingstone. The nectarine is a form of peach with a smooth skin.)

peach-leaf curl *noun* a fungal disease which affects peaches, where the leaves swell and become red

pear *noun* a pome fruit of the genus *Pyrus* used for dessert fruit, cooking or for fermenting to make perry. In the UK, William's Bon Chretien, Conference and Doyenne du Comice are popular dessert varieties, while William's is also commonly used for canning.

peat *noun* the accumulated partly decayed mosses and other plants which form the soil of a bog, often forming a deep layer

COMMENT: Acid peats are formed in waterlogged areas where marsh plants grow, and where decay of dead material is slow. Black fen soils found in East Anglia are very fertile. These soils contain silts and calcium carbonate in addition to the remains of vegetation. Peat can be cut and dried in blocks, which can then be used as fuel. It is also widely used in horticulture, after drying and sterilising. Peat was used as a fuel in some areas and was widely used in gardens to improve the texture of the soil or mixed with soil or other materials to grow plants in pots. These practices are now discouraged in order to prevent the overuse of peat bogs.

peat-free *adjective* referring to material such as compost that does not contain peat

peatland /ˈpiːtlænd/ *noun* an area of land covered with peat bog

peaty /ˈpiːti/ *adjective* containing peat ○ *peaty soil* ○ *peaty water*

pecan *noun* a North American tree (*Carya illinoensis*) which produces sweet nuts which are eaten as dessert nuts and used in many forms of confectionery

peck *noun* a measure of capacity of dry goods, equal to a quarter of a bushel or two gallons. Pecks are used as a measure of grain. ■ *verb* to pick up food with the beak

pecking order *noun* the order of social dominance in a group of birds, and also animals (NOTE: The equivalent in cattle is the 'bunt order'.)

pectin /ˈpektɪn/ *noun* a sticky mixture of various polysaccharides found in plant cell walls

ped /ped/ *noun* an aggregate of soil particles

pedigree *noun* the ancestral line of animals bred by breeders, or of cultivated plants ■ *adjective* descended from a line of animals whose pedigree has been recorded over several generations ○ *a pedigree dog*

pedigree market *noun* the market for animals sold for breeding rather than for slaughter. Compare **commercial market**

pedigree records *plural noun* records of pedigree stock kept by the breeder and by breed societies. Pedigree animals are registered at birth and given official numbers.

pedigree selection *noun* the selection of animals for breeding based on the records of their ancestors

pedologist /pə'dɒlədʒɪst/ *noun* a scientist who specialises in the study of the soil

pedology /pə'dɒlədʒi/ *noun* the study of the soil

peel *noun* 1. the outer layer of a fruit ○ *Oranges have a thick peel.* ○ *Lemon peel is used as flavouring.* 2. the skin of a potato ■ *verb* to remove the peel from a fruit or potato

PEG *abbreviation* production entitlement guarantee

Pekin /piː'kɪn/ *noun* a breed of table duck. It has buff coloured feathers and bright orange feet, legs and bill.

pellet *noun* a form of feedingstuff, usually mash, which has been moistened and pressed to form small grains

pelleted seed *noun* a seed coated with clay to produce pellets of uniform size and density. Pelleting is done to make the sowing of very fine seed easier.

pen *noun* a small enclosure for animals or poultry ■ *verb* to enclose animals such as sheep in a pen

penicillin *noun* an antibiotic, originally produced from a fungus, that controls bacterial and fungal infections (NOTE: Penicillin and the related family of drugs have names ending in **-cillin**: *amoxycillin.*)

COMMENT: Penicillin is effective against many microbial diseases, such as mastitis in cattle.

Penicillium /ˌpenɪ'sɪliəm/ *noun* the genus of fungus from which penicillin is derived

Penistone /'penɪstən/ *noun* same as **White-faced Woodland**

pen mating *noun* the practice of using one male animal to mate with a number of females

pepino mosaic virus /pəˌpiːnəʊ məʊ ˈzeɪɪk ˌvaɪrəs/ *noun* a highly contagious virus affecting tomato plants

pepper *noun* 1. a spice, either black or white, made from the berry-like fruit of the pepper vine 2. the fruit of the *Capsicum*, either red, yellow or green

pepper and salt *noun* same as **shepherd's purse**

peppermint *noun* an aromatic herb (*Mentha piperata*) which is cultivated to produce an oil used in confectionery, drinks and toothpaste

pepsin /'pepsɪn/ *noun* an enzyme in the stomach which breaks down the proteins in food

peptic /'peptɪk/ *adjective* referring to digestion or to the digestive system

peptone /'peptəʊn/ *noun* a substance produced by the action of pepsins on proteins in food

PER *abbreviation* protein efficiency ratio

Percheron /'pɜːʃərɒn/ *noun* a heavy breed of horse, developed in Normandy. It is grey in colour.

Perendale /'perəndeɪl/ *noun* a New Zealand breed of sheep

perennial *adjective* referring to plants which persist for more than two years ■ *noun* a plant that lives for a long time, flowering, often annually, without dying. ◊ **annual, biennial** (NOTE: In herbaceous perennials the parts above ground die back in winter, but the plant persists under the ground and produces new shoots in the spring. In woody perennials, permanent stems remain above the ground in the winter.)

perennial agriculture *noun* a system of agriculture in regions where there is no winter and several crops can be grown on the same land each year

perennial irrigation *noun* a system which allows the land to be irrigated at any time. This may be by primitive means such as shadufs, or by distributing water from barrages by canal and ditches.

perennial ryegrass *noun* a grass (*Lolium perenne*) which forms the basis of the majority of long leys in the UK. It is the most important grass in good permanent pasture and is often sown mixed with other grasses and clover (NOTE: Perennial ryegrass has a long growing season, is quick to become established and responds well to fertilisers. It is best suited to grazing and is highly palatable for animals.)

performance test *noun* a record of growth rate in an individual animal over a given period of time, when fed on a standard ration. Performance testing gives the breeder a better chance of identifying genetically superior animals.

pericarp /ˈperɪkɑːp/ *noun* the part of a fruit that encloses the seed or seeds

peri-urban /ˌperi ˈɜːb(ə)n/ *adjective* on the edge of a built-up area

permaculture /ˈpɜːməkʌltʃə/ *noun* a system of permanent agriculture, which involves carefully designing human habitats and food production systems

permanent grassland, permanent pasture *noun* land that remains as grassland for a long time and is not ploughed

permanent wilting point *noun* the soil water content below which plants wilt and are unable to recover

permeability /ˌpɜːmiəˈbɪlɪti/ *noun* **1.** the ability of a rock to allow water to pass through it **2.** the ability of a membrane to allow fluid or chemical substances to pass through it

perpetual-flowering *adjective* referring to a variety of plant which bears flowers more or less all year round

perry /ˈperi/ *noun* fermented pear juice

persimmon /pəˈsɪmən/ *noun* a native tree *(Diospyros kaki)* of Japan and China, which produces reddish-orange fruit, similar in appearance to tomatoes. The fruit are eaten either as dessert or may be cooked; they are very rich in vitamins.

persist *verb* **1.** to continue to exist ○ *Snow cover tends to persist on north-facing slopes of mountains.* **2.** *(of a chemical compound)* to remain active without breaking down in the environment for a period of time ○ *The chemical persists in the soil.* **3.** *(of a plant)* to grow for several seasons

persistence /pəˈsɪstəns/ *noun* **1.** the ability of a chemical to remain active without breaking down in the environment for a period of time **2.** *(of a plant)* the ability to grow for several seasons

persistency /pəˈsɪstənsi/ *noun* ability of a plant to survive for a long time, even when the soil is cultivated ○ *Ryegrasses are used for leys where persistency is not important.*

persistent *adjective* **1.** remaining active without breaking down for some time ○ *persistent chemicals* **2.** growing for several seasons ○ *persistent species*

pest *noun* **1.** an organism that carries disease or harms plants or animals ○ *a*

spray to remove insect pests **2.** the name given to some diseases, e.g. fowl pest

pest control *noun* the process of keeping down the number of pests by various methods

pesticide *noun* a chemical compound used to kill pests such as insects, other animals, fungi or weeds

COMMENT: There are four basic types of pesticide: 1. organochlorides, which have a high persistence in the environment of up to about 15 years (DDT, dieldrin and aldrin); 2. organophosphates, which have an intermediate persistence of several months (parathion, carbaryl and malathion); 3. carbamates, which have a low persistence of around two weeks (Tenik, Zectran and Zineb); 4. synthetic pyrethroids, which are non-persistent, contact and residual acting insecticides (cypermethrin, permethrin), suitable for a wide range of crops and target insects. Most pesticides are broad-spectrum, that is they kill all insects in a certain area and may kill other animals like birds and small mammals. Pesticide residue levels in food in the UK are generally low. Pesticide residues have been found in bran products, bread and baby foods, as well as in milk and meat. Where pesticides are found, the levels are low and rarely exceed international maximum residue levels.

pesticide residue *noun* the amount of a pesticide that remains in the environment after application

'Technical solutions and risk management tools have been developed for six water catchments to communicate best practice to reduce pesticide residues in water.' [*Arable Farming*]

Pesticide Safety Precaution Scheme *noun* an agreement between the agrochemical industry and the government, supported by Health and Safety regulations, which designates products as safe to use, provided recommended precautions are taken during their use. ◊ **FEPA**. Abbr **PSPS**

Pesticides Safety Directorate *noun* an executive agency of Defra which oversees the development, licensing and safe use of pesticides in the UK. Abbr **PSD**

Pesticides Trust *noun* a group that works to minimise and eventually eliminate the hazards of pesticides

pesticide tax *noun* a proposed tax to restrict the use of pesticides

Pesticide Usage Survey *noun* an annual survey of the range and amount of pesticides used on crops

petal *noun* a single part of the corolla of a flower ○ *A buttercup flower has yellow petals.*

petiole /'petiəʊl/ *noun* the stalk of a leaf

PGE *abbreviation* parasitic gastroenteritis

PGR *abbreviation* plant growth regulator

PGRO *abbreviation* Processors and Growers Research Organisation

pH *noun* a measure of the acidity of a solution, determined as the negative logarithm of the hydrogen ion concentration, on a scale from 0 to 14 ○ *soil pH*

COMMENT: A pH value of 7 is neutral, the same as that of pure water. Lower values indicate increasing acidity and higher values indicate increasing alkalinity: 0 is most acid and 14 is most alkaline. Acid rain has been known to have a pH of 2 or less, making it as acid as lemon juice. Plants vary in their tolerance of soil pH. Some grow well on alkaline soils, some on acid soils only, and some can tolerate a wide range of pH values. The soil pH value for rye and lupins is approximately 4.5, for oats and potatoes 5.0, for wheat, beans, peas, turnips and swedes 5.5, for clover, maize and oilseed rape 6.0, and for barley, sugar beet and lucerne 6.5.

phacelia /fə'siːliə/ *noun* a plant used as a ground cover crop. It was introduced into the UK from the USA.

-phage /feɪdʒ/ *suffix* eating

phago- /fægəʊ/ *prefix* eating

phagocyte /'fægəʊsaɪt/ *noun* a cell that can surround and destroy other cells such as bacteria, e.g. a white blood cell

phagocytic /ˌfægə'sɪtɪk/ *adjective* referring to phagocytes

pharming /'fɑːmɪŋ/ *noun* the production of proteins that have medicinal value in genetically modified livestock or crops

Phaseolus /ˌfæzɪ'əʊləs/ *noun* the Latin name for beans such as the French bean and butter bean

pheasant *noun* a game bird *(Phasianus colchicus)* with long tail feathers

phenolics /frɪ'nɒlɪks/ *noun* organic chemicals

phenotype /'fiːnətaɪp/ *noun* the physical characteristics of an organism, produced by its genes. Compare **genotype**

'For centuries, farmers have used phenotype to improve livestock. They selected the best looking and performing animals to produce their next generations.' [*Farmers Weekly*]

phenotypic /ˌfiːnəʊ'tɪpɪk/ *adjective* relating to a phenotype

phenylalanine /ˌfiːnaɪl'æləniːn/ *noun* an essential amino acid

pheromone /'ferəməʊn/ *noun* a chemical substance produced and released into the environment by an animal, influencing the behaviour of another individual of the same species ○ *Some insects produce pheromones to attract mates.*

phloem /'fləʊəm/ *noun* the vascular tissue in a plant that is formed of living cells and conducts organic substances from the leaves to the rest of the plant. ◊ **xylem**

phosphate /'fɒsfeɪt/ *noun* a salt of phosphoric acid which is formed naturally by weathering of rocks

COMMENT: Natural organic phosphates are provided by guano and fishmeal, otherwise phosphates are mined. Artificially produced phosphates are used in agriculture: the main types of phosphate fertiliser are ground rock phosphate, hyperphosphate, superphosphate, triple superphosphate and basic slags. Phosphate deficiency is one of the commonest deficiencies in livestock, and gives rise to osteomalacia (also known as creeping sickness). Phosphates escape into water from sewage, especially waste water containing detergents, and encourage the growth of algae by eutrophication.

phosphorus /'fɒsf(ə)rəs/ *noun* a chemical element that is essential to biological life

COMMENT: Phosphorus is an essential part of bones, nerve tissue, DNA and RNA and is important in many biochemical processes, although in its pure form it is highly toxic. When an organism dies the phosphorus contained in its tissues returns to the soil and is taken up by plants in the phosphorus cycle. Phosphorus deficiency in plants causes stunted growth, discoloration of leaves and small or misshapen fruit.

photo- *prefix* light

photoperiodicity /ˌfəʊtəʊpiːriə'dɪsɪti/ *noun* the degree to which plants and animals react to changes in the length of the period of daylight from summer to winter

photorespiration /ˌfəʊtəʊrespɪ'reɪʃ(ə)n/ *noun* a reaction that occurs in plants, alongside photosynthesis, in which the plant fixes oxygen from the air and loses carbon dioxide (NOTE: Photorespiration reduces the production of sugars by photosynthesis. Some crop plants have been bred to reduce their photorespiration rate.)

photosensitisation /ˌfəʊtəʊsensɪtaɪ'zeɪʃ(ə)n/, **photosensitization** *noun* a disease of livestock caused by activation of photodynamic agents in the skin by light. Skin becomes pink and inflamed and may develop deep cracks.

photosynthesis /ˌfəʊtəʊ'sɪnθəsɪs/ *noun* the process by which green plants convert carbon dioxide and water into sugar and oxygen using sunlight as energy

phototroph *noun* an organism that obtains its energy from sunlight. Compare **chemotroph**

phototrophic /ˌfəʊtəʊ'trɒfɪk/ *adjective* obtaining energy from sunlight. Compare **chemotrophic** (NOTE: Plants are phototrophic.)

PHSI *abbreviation* Plant Health and Seeds Inspectorate

phylloxera /fɪ'lɒksərə/ *noun* an aphid which attacks vines. It threatened to destroy the vineyards of Europe in the 19th century, but the vines were saved by grafting susceptible varieties onto resistant American rootstock.

physiology *noun* the scientific study of the functions of living organisms

phyto- /faɪtəʊ/ *prefix* plant

phytome /'faɪtəʊm/ *noun* a plant community

phytonutrient /'faɪtəʊˌnjuːtriənt/ *noun* a substance in plants that is beneficial to human health, e.g. a vitamin or antioxidant

phytophagous /faɪ'tɒfəgəs/ *adjective* referring to an animal that eats plants

phytotoxic /ˌfaɪtəʊ'tɒksɪk/ *adjective* poisonous to plants

phytotoxicant /ˌfaɪtəʊ'tɒksɪkənt/ *noun* a substance that is phytotoxic

phytotoxin /ˌfaɪtəʊ'tɒksɪn/ *noun* a poisonous substance produced by a plant

pick *verb* to take ripe fruit or vegetables from plants

picker /'pɪkə/ *noun* a person who picks fruit or vegetables ○ *a strawberry picker*

picking /'pɪkɪŋ/ *noun* the work of taking fruit or vegetables from plants ○ *Teams of workers are employed on raspberry picking in July.*

pickle *verb* to preserve food by keeping it in vinegar □ **pickling cabbage**, **pickling onions** cabbage *or* onions specially grown for pickling

pick-up attachment *noun* an attachment used on a combine to lift grass swath and feed it into the main elevator. It is fitted over the combine cutter bar.

pick-up baler *noun* a machine which picks up cut grass and makes small bales. A machine which makes big bales is called a 'big baler'.

pick-up reel *noun* a part of a combine harvester with spring tines, used to improve cutting efficiency in tangled crops

pick-your-own *noun* a system where customers are allowed onto the farm to pick what they need from the field. It is used with most types of soft-fruit and many vegetables. Abbr **PYO**

COMMENT: The existence of a large car-owning population helped in the growth of PYO farms, especially close to large urban populations.

PIDC *abbreviation* Potato Industry Development Council Order (1997)

piebald /'paɪbɔːld/ *adjective, noun* referring to an animal whose coat has two colours, especially black and white, in irregular shapes

piecework /'piːswɜːk/ *noun* work for which workers are paid for the products produced or for the piece of work done, and not at an hourly rate. Potato pickers and strawberry pickers are paid in this way.

Piedmontese /ˌpiːdmən'tiːz/ *noun* a breed of beef cattle from north-west Italy. The animals are light or dark grey, with black horns, ears and tail.

piert /'pɪɜːt/ ♦ **parsley piert**

Pietrain /'piːtreɪn/ *noun* a Belgian breed of pig imported into the UK for cross-breeding, which is very muscular and has prick ears and irregular dark spots over the whole body. It is very lean but carries the halothane gene.

pig *noun* an animal of the *Suidae* family kept exclusively for meat production (NOTE: The males are called **boars**, the females are **sows**, the young are **piglets**. Pigs reared for pork meat are called

porkers and those reared for bacon are **baconers**.)

COMMENT: There have been changes in pig types and breeds because of the needs of the meat industry. The main breeds used in the UK include the Large White, Landrace, Welsh, Hampshire, Duroc and Pietrain. In pig farming, the common practice is to use a white quality-type pig which may be crossed with other breeds to produce good-quality commercial pigs. A large number of crossbred sows are being used. Good-quality pigs are needed for fresh meat and for manufacturing, and they are reared to provide pork, bacon, hams, pies and sausages. Some pig production deals with the specialised business of pig breeding and some farmers only breed and sell the young pigs as weaners or stores. Most pig farming, however, consists of rearing pigs for pork and fresh meat, or for bacon, ham and other processed foods. Pigs mature quickly and are marketed at between four and six months of age. Bacon pigs are marketed at about 90kg liveweight and porkers from around 60kg. Most pigs are intensively housed in prefabricated buildings with mechanical ventilation and often automated feeding, although there is a movement towards 'free range' pig farming. Pig rearing is carried on in most parts of the UK, but is being increasingly concentrated in the Eastern part of the country. Rare breeds (non-commercial breeds) include the British Saddleback, Large Black, Gloucester Old Spot, Berkshire, Tamworth, Middle White and Chester White.

pigeon *noun* a bird which is regarded as a pest, and can cause severe damage to small plants. The wood pigeon eats cereal seed and causes damage to young and mature crops of peas and brassicas.

pig futures *plural noun* sales on the commodity market of pigmeat for future delivery

piggery /'pɪgəri/ *noun* a place where pigs are housed

COMMENT: Pigs can be kept successfully under all sorts of conditions, but housing systems must provide efficient feeding facilities. Individual feeding of sows is essential, and confining dry sows in stalls is common. Farrowing sows and their litters need careful housing; farrowing should take place in a special crate which prevents overlying of piglets (where the sow crushes the piglets by lying on them). Breeding pigs may be reared outdoors and housed in moveable arks. Fattening pigs can be housed in yards, usually covered with straw bedding and provided with a shelter, but more commonly they are kept in special buildings with controlled environments and ventilation, feeding, watering and efficient disposal of dung through special channels. There are many systems of pig housing, ranging from the traditional cottage piggery, with its low-roofed pens and open yard; the Danish type, a totally enclosed building with a controlled environment; the flat deck type, used for rearing young pigs, with special heating and ventilation controls; the Solari piggery, with fattening pens on each side of a central feeding passage, housed in an open-sided Dutch barn; and the mono-pitch type, incorporating artificially controlled natural ventilation.

piglet *noun* a young pig

piglet anaemia *noun* a metabolic disease caused by milk deficiency. It can cause death of piglets. Iron compounds are administered as treatment.

pigman /'pɪgmæn/ *noun* a term formerly used for a male farm worker who looks after pigs. It is now replaced, in advertisements, by 'pigperson'.

pigmeat /'pɪgmiːt/ *noun* a term used in the EU for meat from pigs

pigperson /'pɪgpɜːsən/ *noun* a farm worker who looks after pigs. This is a term used in job advertisements.

pig pox *noun* a disease of pigs caused by infection with one of two different viruses, most commonly the swinepox virus. Young pigs have red spots on belly, face and head which turn into blisters. Fever may follow.

pig production *noun* the commercial farming of pigs

COMMENT: Sows are mature enough to breed at around six months of age. They are usually mated with the boar every 140–150 days. After a gestation period of 115 days a litter is born and is usually allowed to suckle for 21–28 days. Four to seven days after separating the sow from her piglets, she comes into oestrus and allows the boar to mate. Each of the resulting litters will consist of about ten piglets, and over her lifetime the sow will mother six to eight litters. Sows may be kept in stalls, sometimes completely closed, and sometimes open-ended. Sows are also kept in covered yards on straw. New systems are being introduced, such as the use of a collar for each sow marked with its computer code number. The number is relayed to a computer which allocates a measured amount of feed to the sow. Herds of sows are also kept in open fields, using arks for shelter.

pigstock /'pɪgstɒk/ noun herds of pigs

pigstock person noun a farm worker who looks after pigs

pigsty noun same as **sty**

pig typhoid noun same as **swine fever**

pigweed /'pɪgwiːd/ noun same as **knot-grass**

pinch out verb to remove small shoots by pinching between finger and thumb. Pinching out is done when training young plants.

pine noun 1. an evergreen coniferous tree. Genus: *Pinus*. ○ *The north of the country is covered with forests of pine.* 2. same as **pining**

pine kernels plural noun seeds of the Mediterranean pine (*Pinus pinea*) eaten raw or roasted and salted, and also used in cooking and confectionery

pinewood /'paɪnwʊd/ noun a wooded area containing mainly pines

pining /'paɪnɪŋ/ noun a disease of sheep, caused by shortage of cobalt, which causes the animals to lose condition, and have poor fleece, dull eyes and general weakness

pint noun a non-metric measure of liquids and dry goods such as seeds. It is equal to 0.56 litres. (NOTE: usually written **pt** after figures: **4pts of water**)

pinworm /'pɪnwɜːm/ noun a thin parasitic worm *Enterobius,* which infests the large intestine. Also called **threadworm**

Pinzgauer /'pɪntsgaʊə/ noun a breed of cattle from Austria, used for both meat and draught. The animals are coloured reddish or dark brown with a distinctive white strip along the back.

pioneer crop noun a crop grown to improve general soil fertility, prior to the sowing of another more valuable crop. Pioneer crops are grazed by livestock with the result that their dung improves the soil's fertility.

piperazine /pɪ'perəziːn/ noun a drug used to treat worm infestation

pippin /'pɪpɪn/ noun a name given to several dessert apples

pistachio /pɪ'stæʃiəʊ/ noun a small tree (*Pistacia vera*) native of central Asia and now cultivated in Mediterranean regions. The nuts are eaten salted or in confectionery.

pistil /'pɪstɪl/ noun the female reproductive part of a flower, made up of the ovary, the style and the stigma

Pisum /'paɪsəm/ noun the Latin name for pea

pit noun the stone in certain fruit, e.g. in cherries, plums, peaches or dried fruit such as raisins and dates

pitch noun a dark sticky substance obtained from tar, used to make objects watertight

pitch pole noun a harrow with double-ended tines

placement drill noun a machine which drills seeds and fertiliser at the same time, and places the fertiliser close to the side of and below the rows of seeds

placenta /plə'sentə/ noun the tissue which grows inside the uterus in mammals during pregnancy and links the baby to the mother

placental /plə'sent(ə)l/ adjective referring to the placenta

placental mammal noun same as **eutherian**

plague noun 1. an infectious disease that occurs in epidemics which kill many organisms 2. a widespread infestation by a pest ○ *A plague of locusts has invaded the region and is destroying crops.*

plain plural noun **plains** a large area of flat country with few trees, especially in the middle of North America

plane tree noun a common temperate deciduous hardwood tree, frequently grown in towns because of its resistance to air pollution

planning authority noun a local authority which gives permission for development such as changes to existing buildings or new use of land

planning controls plural noun legislation used by a local authority to control building

planning department noun a section of a local authority which deals with requests for planning permission

planning inquiry noun a hearing before a government inspector relating to a decision of a local authority in planning matters

planning permission noun an official agreement allowing a person or company to plan new buildings on empty land or to alter existing buildings

plant *noun* an organism containing chlorophyll with which it carries out photosynthesis ■ *verb* to put plants in the ground ○ *to plant a crop of rice*

plantain /'plænteɪn/ *noun* **1.** a name given to various types of banana, used for cooking and brewing. It has a lower sugar content than dessert bananas. **2.** a common weed (*Plantago major*). Compare **ribwort**

plantation *noun* **1.** an estate, especially in the tropics, on which large-scale production of cash crops takes place (NOTE: Plantations specialise in the production of a single crop such as cocoa, coffee, cotton, tea or rubber.) **2.** an area of land planted with trees for commercial purposes. Also called **plantation forest**

plant breeder *noun* a person who produces new forms of ornamental or crop plants

plant breeding *noun* the practice of producing new forms of ornamental and crop plants by artificial selection

plant cover *noun* the percentage of an area occupied by plants ○ *Plant cover at these altitudes is sparse.*

plant ecology *noun* the study of the relationship between plants and their environment

planter /'plɑːntə/ *noun* **1.** a person who plants, especially a person who plants and looks after a plantation **2.** a device for planting. ◊ **potato planter**

plant food ratio *noun* the ratio of nitrogen to phosphate and potash in a fertiliser

plant genetic resources *plural noun* the gene pool of plants, especially of plants regarded as of value to humans for food or pharmaceuticals

plant growth regulator *noun* a chemical treatment which slows the growth of plants, used in low-maintenance areas such as roadside verges. Abbr **PGR**

plant health *noun* the areas related to the prevention of pests and diseases affecting plants and plant produce, including the control of imports and exports

Plant Health and Seeds Inspectorate *noun* a branch of Defra which deals with plant health in the UK. Abbr **PHSI**

plant hormones *plural noun* hormones such as auxin which particularly affect plant growth. They are more accurately called 'plant growth substances'.

plant nutrient *noun* a mineral whose presence is essential for the healthy growth of plants

plant passport *noun* ♦ **passport**

plant protection *noun* the activity of protecting plants from disease by biocontrol, cultivation practices and especially by the application of pesticides

plant protection product *noun* a general term for a chemical such as a pesticide or fungicide that is used to keep plants free from disease and pests

plant senescence *noun* the final stage in the life cycle of a plant, leading to the death of part or all of the plant (NOTE: Knowledge of plant senescence is important for farmers as it determines when they should harvest a crop in order to ensure it is of the highest possible quality.)

Plant Variety Rights Office *noun* the certifying authority for agricultural and horticultural seeds in England and Wales, based in Cambridge. Abbr **PVRO**

plastic *noun* a man-made material, used as a cover to protect young crops. Thin films of polythene may be used to cover and warm soil, while black plastic sheeting is used as a form of mulch, and also to cover clamps and bales. ■ *adjective* describing the state of a soil when it is too wet. The soil deforms and does not recover.

plate and flicker *noun* a type of machine used for distributing fertiliser

plateau *noun* an area of high flat land (NOTE: The plural is **plateaux**.)

plate mill *noun* a type of mill used for grinding grain. The machine is made of two circular plates, one of which is fixed, while the other rotates against it.

plot *noun* a small area of cultivated land, which has been clearly defined

plough *noun* an agricultural implement used to turn over the surface of the soil in order to cultivate crops ■ *verb* to turn over the soil with a plough (NOTE: [all senses] The US spelling is **plow**.)

COMMENT: The modern plough is usually fully mounted on a tractor's hydraulic system, though some are semi-mounted, with the rear supported by one or more wheels, and some may be trailed. The principal parts of a plough are the beam or frame, made of steel, to which are attached a number of parts which engage the soil, such as the disc coulter, the share and the mouldboard which turns the furrow slice. There are three main types of plough: conventional, with right-handed mouldboards, reversible, with left- and right-handed mouldboards, and disc ploughs, which are used for deep rapid cultivation of hard dry soils, and are not common in the UK. There are many designs of plough body, the main ones being the general purpose plough, the semi-digger, the digger and the Bar Point. The three main methods of ploughing are: **systematic,** where the field is divided into lands by shallow furrows; **round and round ploughing,** in which fields are ploughed from the centre to the outside or from the edge to the centre (see also **square ploughing**); **reversible ploughing,** where the field is ploughed up and down the same furrow, giving a very level surface.

plough body *noun* the main part of the plough, consisting of the frog, mouldboard, share and landside

plough in *verb* to cover a crop, stubble or weeds with soil, by turning over the surface with a plough

ploughland /ˈplaʊlænd/ *noun* arable or cultivated land

ploughman /ˈplaʊmən/ *noun* a man who ploughs

plough pan *noun* a hard layer in the soil caused by ploughing at the same depth every year

ploughshare /ˈplaʊʃeə/ *noun* a heavy metal blade of a plough, which cuts the bottom of the furrow. It can be in one piece or three sections, the point, wing and shin. Also called **share**

plough to plate *noun* same as **farm to fork**

plough up *verb* to plough a pasture, usually in order to use it for growing crops

pluck *verb* **1.** to remove the feathers from a bird's carcass. This is usually done by machine, but still also done by hand. **2.** to remove the internal organs from an animal carcass after slaughter **3.** to remove the leaves from a plant such as the tea plant

plum *noun* a stone fruit (*Prunus domestica*)

COMMENT: There are many varieties of cultivated plums, and they vary in colour, shape, size and flavour. Pond's Seedling, Monarch and Pershore are cooking varieties, while the rich-flavoured dessert varieties include Victoria, Laxton and Kirke's Blue. American varieties include Chickasaw and the Oregon Plum. In the UK, plums are grown mainly in the south and in the west Midlands; in the USA, they are common in the Pacific states.

plumage *noun* the feathers of a bird

plum pox *noun* a viral disease affecting plums, damsons and peaches. The fruit has dark blotches and ripens prematurely, and it is often sour.

plumule /ˈpluːmjuːl/ *noun* the tiny structure in a plant embryo from which a shoot will develop

Plymouth rock /ˌplɪməθ ˈrɒk/ *noun* a large heavy hardy dual-purpose breed of poultry, originally coming from the USA. The feathers are rich lemon-buff.

PMG *abbreviation* processing and marketing grant

PMR *abbreviation* partly mixed ration

PMWS *noun* a disease which causes wasting, paleness and diarrhoea in pigs between 6 and 14 weeks old and is often fatal. Full form **post-weaning multisystemic wasting syndrome**

'The Danish sector is still suffering from the pig wasting disease PMWS with the number of infected herds now at 270, but Lundholt is optimistic that farmers can control the condition and manage their way clear.' [*The Grocer*]

pneumatic distributor *noun* a machine which conveys fertiliser from a hopper to nozzles for spreading by a stream of air. Both trailed and mounted models are made.

pneumatic grain drill, pneumatic drill *noun* a machine which sows grain, the seed being moved from a hopper down the drill pipe by compressed air

pneumonia *noun* the inflammation of a lung, where the tiny alveoli of the lung become filled with fluid

Poaceae /pəʊˈæsiaɪ/ *noun* the grasses, which is a very large family of plants including bamboo and cereals such as wheat and maize. Former name **Gramineae**

poach *verb* **1.** to catch animals, birds or fish illegally on someone else's property **2.** to trample the ground in wet weather. Heavy soils such as clays are particularly susceptible to poaching.

poacher *noun* a person who catches animals, birds or fish illegally on someone else's land

pocket *noun* a large sack of dry hops

pod *noun* a container for several seeds, e.g. a pea pod or bean pod

podsol /'pɒdsɒl/, **podzol** /'pɒdzɒl/ *noun* a type of acid soil where organic matter and mineral elements have been leached from the light-coloured top layer into a darker lower layer through which water does not flow and which contains little organic matter. Compare **chernozem**

COMMENT: On the whole podsols make poor agricultural soils, owing to their low nutrient status and the frequent presence of an iron pan. Large areas of the coniferous forest regions of Canada and Russia are covered with podsols.

podsolic /pɒd'sɒlɪk/, **podzolic** /pɒd'zɒlɪk/ *adjective* referring to podsol

podsolisation /ˌpɒdsɒlaɪ'zeɪʃ(ə)n/, **podsolization** *noun* the process by which a podsol forms

pod up *verb* to begin to develop pods

point *noun* the forward end of a ploughshare

point of lay *noun* a term referring to pullets that are approaching the time when they will lay their first eggs

'The new building was erected last year and the first batch of hens, 16 weeks old and at point of lay, were brought in.' [*Farmers Weekly*]

polder /'pəʊldə/ *noun* a piece of low-lying land which has been reclaimed from the sea and is surrounded by earth banks, especially in the Netherlands

Policy Commission on Farming and Food *noun* ♦ Curry Report

poll *noun* the top of an animal's head ■ *verb* to dehorn an animal

pollard /'pɒləd/ *noun* a tree of which the branches have been cut back to a height of about 2 m above the ground ■ *verb* to cut back the branches on a tree every year or every few years to a height of about 2 m above the ground. Compare **coppice**

COMMENT: Pollarding allows new shoots to grow, but high enough above the ground to prevent them from being eaten by animals. Willow trees are often pollarded.

Poll Dorset *noun* an Australian breed of sheep similar to the Dorset Horn, but with no horns

polled stock *noun* **1.** animals which are naturally hornless **2.** animals which have had their horns removed

pollen *noun* the mass of small grains in the anthers of flowers which contain the male gametes

COMMENT: Grains of pollen are released by trees in spring and by flowers and grasses during the summer. The pollen is released by the stamens of a flower and floats in the air until it finds a female flower. Pollen in the air is a major cause of hay fever. It enters the eyes and nose and releases chemicals which force histamines to be released by the sufferer, causing the symptoms of hay fever to appear.

pollen analysis *noun* same as **palynology**

pollen beetle *noun* a pest of Brassica, which makes buds wither. The beetle feeds on buds and flower parts.

pollinate /'pɒlɪneɪt/ *verb* to transfer pollen from the anther to the stigma in a flower

pollination /ˌpɒlɪ'neɪʃ(ə)n/ *noun* the action of pollinating a flower (NOTE: There is no English noun 'pollinisation'.)

pollinator /'pɒlɪneɪtə/ *noun* **1.** an organism which helps pollinate a plant, e.g. a bee or bird ○ *Birds are pollinators for many types of tropical plant.* **2.** a plant from which pollen is transferred by bees to pollinate another plant, especially a fruit tree, that is not self-fertile ○ *Some apple and pear trees need to be planted with pollinators.*

pollinosis /ˌpɒlɪ'nəʊsɪs/ *noun* inflammation of the nose and eyes caused by an allergic reaction to pollen, fungal spores or dust in the atmosphere. Also called **hay fever**

pollutant *noun* **1.** a substance that causes pollution **2.** noise, smell or another unwanted occurrence that affects a person's surroundings unfavourably

pollute *verb* to discharge harmful substances in unusually high concentrations into the environment ○ *Polluting gases react with the sun's rays.* ○ *Polluted soil must be removed and buried.*

polluter /pə'luːtə/ *noun* a person or company that causes pollution

pollution *noun* **1.** the presence of unusually high concentrations of harmful substances or radioactivity in the environment, as a result of human activity or a natural process such as a volcanic eruption ○ *In terms of pollution, gas is by far the cleanest fuel.* ○ *Pollution of the atmosphere has increased over the last 50 years.* ○ *Soil pollution round mines poses a problem for land reclamation.* **2.** the unwanted presence of something such as noise or artificial light

Polwarth /ˈpɒlwɔːθ/ *noun* an Australian breed of sheep (from Lincoln and Merino) which gives a fine wool

poly- /pɒli/ *prefix* **1.** many **2.** made of polythene **3.** touching many organs

polyculture /ˈpɒlikʌltʃə/ *noun* the rearing or growing of more than one species of plant or animal on the same area of land at the same time

polyethylene /ˌpɒliˈeθiliːn/ *noun* same as **polythene**

polyphagous /pəˈlɪfəgəs/ *adjective* referring to an organism that eats more than one type of food. Compare **monophagous**

polysaprobic /ˌpɒlisæˈprəʊbɪk/ *adjective* referring to organisms that can survive in heavily polluted water

polythene *noun* a type of plastic used to make artificial fibres, packaging, boxes and other articles. Also called **polyethylene**

polytunnel /ˈpɒlitʌnəl/ *noun* a cover for growing plants, like a large greenhouse, made of a rounded plastic roof attached to semi-circular supports

'The family are also looking to install more polytunnels, which will increase the duration of the picking season and allow pickers to work in poor weather conditions. The confidence for these moves has been brought about by the investment of around pounds 90,000 in a new packing house at the farm.' [*Farmers Guardian*]

pome /pəʊm/ *noun* a fruit with a core containing the seeds enclosed in a fleshy part that develops from the receptacle of a flower and not from the ovary (NOTE: The fruit of apples and pears are pomes.)

pomegranate *noun* a semi-tropical tree (*Punica granatum*) native to Asia, but now cultivated widely. The fruit are round and yellow, with masses of seeds surrounded by sweet red flesh.

pomology /pəˈmɒlədʒi/ *noun* the study of fruit cultivation

pond *noun* a small area of still water formed artificially or naturally. ◊ **dew pond**

pony *noun* a small breed of horse, often ridden by children, but also living wild in some parts of the world

pony-trekking *noun* a recreational activity where people hire ponies to ride along country paths, now sometimes organised from farms as a form of diversification

porcine /ˈpɔːsaɪn/ *adjective* referring to pigs

porcine coronavirus type 2 *noun* full form of **PCV2**

Porcine Dermatitis and Nephropathy Syndrome *noun* a disease of pigs causing blotches under the skin. The cause of this disease is not yet known. Abbr **PDNS**

porcine reproductive respiratory syndrome *noun* a viral disease of pigs which leads to fertility and breathing problems, as well as to high mortality rates in piglets. Abbr **PRRS**

porcine somatotropin *noun* a hormone administered to feeder pigs, which has been shown to increase feed efficiency, the ratio of lean meat to carcass weight, and market weight. Abbr **PST**

porcine spongiform encephalopathy *noun* a brain disease which has been induced in pigs experimentally. Abbr **PSE**

porcine stress syndrome *noun* a group of conditions which are associated with the halothane gene and which cause rapid respiration, twitching and sudden death in affected pigs. The condition is usually triggered by stress and can be screened for by exposing the animals to the anaesthetic halothane. Abbr **PSS**

pore *noun* **1.** a tiny hole in the skin through which sweat passes **2.** a tiny space in a rock formation or in the soil **3.** same as **stoma**

pore space *noun* the space in the soil not filled by soil particles, but which may be filled with water or air

pork *noun* fresh meat from pigs, as opposed to cured meat, which is bacon or ham. ◊ **pigmeat**

pork belly *noun* the part of a pig which is processed to produce bacon

porker /ˈpɔːkə/ *noun* a pig specially reared for fresh meat, as opposed to bacon or other processed meats

Portland /'pɔːtlənd/ *noun* a rare breed of sheep. Both sexes are horned, with brown or tan faces and legs.

post *noun* a solid wooden or concrete pole, placed in a hole in the ground, and used to support a fence or a gate

post hole digger *noun* an implement driven by a tractor, shaped like a very large screw which bores holes in the ground in which posts are placed

post-weaning multisystemic wasting syndrome *noun* full form of **PMWS**

pot *verb* to put a plant into a pot

potash /'pɒtæʃ/ *noun* any potassium salt (NOTE: Potash salts are crude minerals and contain much sodium chloride.)

COMMENT: The main types of potash used are muriate of potash (potassium chloride), sulphate of potash (potassium sulphate), kainite, nitrate of potash (potassium nitrate) or saltpetre, and potash salts. The quality of a potash fertiliser is shown as a percentage of potassium oxide K_2O equivalent. Potash deposits are mined in Russia, the USA, Germany and France.

Potash Development Association *noun* an organisation which provides industry information for those working in fertiliser sales. Abbr **PDA**

potash fertiliser *noun* a fertiliser based on potassium, e.g. potassium sulphate

potassium *noun* a soft metallic element, essential to biological life

COMMENT: Potassium is one of the three major soil nutrients needed by growing plants. The others are nitrogen and phosphorus. potassium also plays an important part in animal physiology and occurs in a variety of plant and animal foodstuffs. Potassium deficiency causes spotted leaves, weak growth and small fruit.

potassium chloride *noun* a colourless crystalline salt used as a fertiliser and in photography and medicine. Formula: KCl.

potassium nitrate *noun* a white crystalline salt, used as a fertiliser and meat preservative and in fireworks, explosives and matches. Formula: KNO_3.

potassium sulphate *noun* a fertiliser made from the muriate of potash. It contains about 50% potash and is used by potato growers and market gardeners. Formula: K_2SO_4. Also called **sulphate of potash**

potato *noun* a tuber of *Solanum tuberosum*, one of the most important starchy root crops. Apart from being grown for human consumption, potatoes also are used to produce alcohol and starch, and are used as stock feed.

COMMENT: The potato originated in South America. Its introduction into almost every country of the world has proved of great importance on account of its extreme productiveness, its easy cultivation and its remarkable powers of acclimatisation. Varieties of potato can be cultivated from the tropics to the furthest limits of agriculture, even further than the polar limit of barley. Huge quantities are grown in Russia, Poland, Germany, and the USA. Potatoes vary in shape and may be round, oval, kidney-shaped or just irregular. They also vary in colour, from cream to brown or red. In the UK, varieties are classified as the first and second earlies (harvested from May to August), and early and late maincrop (harvested in September/October). About 20% of the potatoes grown in the UK are earlies. Seed potatoes are grown under a certification scheme. In the UK, potatoes are grown mainly for human consumption, either in tubers, or for processing into chips, crisps, for canning or dehydration.

potato blight *noun* a fungus disease *(Phytophthora infestans)* which kills potato foliage and rots the tubers

potato clamp *noun* a heap of stored potatoes covered with straw and earth

potato cyst nematode *noun* a pest found in most soils that have grown potatoes. The eggs hatch in the spring, and the larvae invade the roots. The leaves of plant eventually yellow and are stunted.

potato elevator digger *noun* a machine which lifts potatoes

COMMENT: A wide flat share runs under the potato ridge and lifts the soil onto a rod-link conveyor. Most of the soil is returned to the ground and the potatoes move on to a second elevator, which returns the potatoes to the ground for hand picking.

potato harvester *noun* a machine which lifts the crop onto a sorting platform, where up to six pickers sort the potatoes from soil and stones. The potatoes are then raised onto a trailer.

Potato Industry Development Council Order (1997) *noun* the statute which allowed the British Potato Council to be set up. Abbr **PIDC**

Potato Marketing Board *noun* former name for **British Potato Council**

potato planter *noun* a machine for planting potatoes

COMMENT: A coulter makes a furrow and the seed potatoes are carried forward by a conveyor mechanism from the hopper and placed at regular intervals in the row. Potatoes are carried in self-filling cups from the hopper to the ground. Many potato planters have a fertiliser attachment. Automatic planters are best for handling unchitted seed.

potato spinner *noun* a machine for lifting the potato crop

COMMENT: A spider wheel rotates and the loosened potato ridge is pushed sideways. The potatoes hit a net at the side of the spinner and fall to the ground, where they are left for the hand pickers.

pot-bound plant *noun* a plant which is in a pot which is too small and which its roots fill

potential transpiration *noun* the calculated amount of water taken up from the soil and transpired through the leaves of plants. The amount varies according to the climate and weather conditions.

pot on *verb* to take a plant from one pot and repot it in a larger one so that it can develop

potting compost *noun* a mixture usually of soil and fibrous matter used to fill containers in which plants are grown (NOTE: Peat has often been used in such composts, but alternatives such as coir fibre from coconut husks are being introduced in an attempt to conserve peat bogs.)

poult /pəʊlt/ *noun* a young turkey, less than 8 weeks old

poultry *noun* a general term for domestic birds kept for meat and egg production. Chickens are the most common. Turkeys and geese are mainly kept as meat birds, while ducks and guinea fowl also produce significant quantities of meat and eggs. (NOTE: The word **poultry** does not have a plural form.)

poultryman /'pəʊltrɪmən/ *noun* a farm worker who looks after and raises poultry

pound *noun* a measure of weight, equal to 16 ounces or 453.592 grams. Symbol **lb**

poussin /'puːsæn/ *noun* a young chicken killed for the table

powder *noun* a substance made of ground or otherwise finely dispersed solid particles ○ *Dry chemical fire-extinguishers contain a non-toxic powder.*

powdered sulphur *noun* sulphur which is used to dust on plants to prevent mildew

powdery mildew /'paʊd(ə)ri ˌmɪldjuː/ *noun* a fungal disease *(Erysiphe graminis)* affecting cereals and grasses. Another form also affects sugar beet and brassicas.

'Trials have shown it gives excellent control of botrytis and blackspot in strawberries, along with additional protection against powdery mildew.' [*Farmers Guardian*]

power *noun* the energy, especially electricity, which makes a machine or device operate

power harrow *noun* a harrow with a rotary system of bars driven from the power take-off. Power harrows are used to prepare seedbeds of fine tilth.

power take-off *noun* a mechanism providing power to drive field machines from a tractor. Abbr **p.t.o.**

pox /pɒks/ *noun* ♦ **fowl pox**, **sheep pox**, **cowpox**

prairie *noun* a large area of grass-covered plains in North America, mainly without trees (NOTE: The prairies of the United States and Canada are responsible for most of North America's wheat production. In Europe and Asia, the equivalent term is **steppe**.)

Préalpes-du-Sud /ˌpreɪælp dʊ 'suːd/ *noun* a hardy breed of sheep found in Alpes-Provençales region of France

precision chop forage harvester *noun* a type of harvester which cuts the crop with flails, chops it into precise lengths and blows it into a trailer. It may be self-propelled, off-set trailed or mounted. It is used for harvesting green material for making silage.

precision drill *noun* a seed drill which sows the seed separately at set intervals in the soil

predation /prɪ'deɪʃ(ə)n/ *noun* the killing and eating of other animals

pre-emergent *adjective* before a plant's leaves appear from the seed in the soil

pre-emergent herbicide *noun* a herbicide such as paraquat which is used to clear weeds before the crop leaves have emerged

pregnancy toxaemia *noun* a metabolic disorder affecting ewes and does during late pregnancy. Animals wobble and fall, breathing is difficult, and death may follow. It is associated with lack of feed in late pregnancy. Also called **twin lamb disease**

premium *noun* a special extra payment

preservation *noun* the process of protecting something from damage or decay ○ *the preservation of herbarium specimens* ○ *Food preservation allows some types of perishable food to be eaten during the winter when fresh food is not available.*

preservative *noun* a substance added to food to preserve it by slowing natural decay caused by microorganisms (NOTE: In the EU, preservatives are given E numbers E200 – E297.)

preserve *verb* **1.** to protect something from damage or decay **2.** to stop food from changing or rotting

press *verb* to crush fruit or seeds to extract juice or oil

prick ears *plural noun* ears of an animal which stand up straight. Compare **lop ears**

prick out *verb* to transplant seedlings from trays or pans into pots or flowerbeds

prill /prɪl/, **prilled fertiliser** /ˌprɪld ˈfɜːtɪlaɪzə/ *noun* a form of fertiliser sold in small round granules

primaries *plural noun* the main feathers on a bird's wing. Also called **flight feathers**

primary *adjective* **1.** first, basic or most important **2.** being first or before something else. ◊ **secondary, tertiary**

primary commodity *noun* a basic raw material or food

primary industry *noun* an industry dealing with raw materials such as coal, food, farm produce or wood

primary producer *noun* a farmer who produces basic raw materials, e.g. wood, milk or fish

primary product *noun* a product which is a basic raw material, e.g. wood, milk or fish

primary productivity *noun* the amount of organic matter produced in a specific area over a specific period, e.g. the yield of a crop during a growing season

primed seed *noun* seed which has been moistened to start the germination process before sowing

primitive *adjective* referring to an early stage in an organism's development

primitive breeds *plural noun* old breeds of livestock which have not been bred commercially, but which are the descendants of wild livestock

prion /ˈpriːɒn/ *noun* a variant form of a protein found in the brains of mammals and causing diseases such as scrapie in sheep, BSE in cattle and variant CJD in humans

process *verb* to treat produce in a way that will make it keep longer or become more palatable

processed meats *plural noun* meat products such as bacon, sausages, etc.

'The firm, which owns several farms, a hatchery and feed mill, produces about 400,000 chickens a week as well as sausages and other processed meats.' [*Farmers Weekly*]

processing and marketing grant *noun* a sum of money given to farmers under the ERDP to help with packaging and marketing their products

Processors and Growers Research Organisation *noun* a crop research and development centre funded by levies. Abbr **PGRO**

prod *noun* a spiked metal rod, used to make cattle move forward

produce *verb* to make something using materials contained within itself or taken from the outside world ○ *a factory producing agricultural machinery* ○ *a drug which increases the amount of milk produced by cows* ■ *noun* foodstuffs such as vegetables, fruit and eggs which are produced commercially

producer *noun* **1.** a person or company that produces something. Compare **consumer** **2.** an organism that takes energy from outside an ecosystem and channels it into the system, e.g. green plants (**primary producers**) and herbivores (**secondary producers**) (NOTE: Producers are the first level in the food chain.)

producer-retailer *noun* a person who produces a commodity for sale directly to the public, as through a farm shop or by milk delivery

'Over the past three years – when the UK wholesale pool fell below quota – producer-retailers used temporary transfers to move quota away from wholesale to cover their direct sales. That reduced the wholesale pool by an average of 41m litres.' [*Farmers Weekly*]

product *noun* **1.** something that is produced by manufacture or in a chemical reaction **2.** the result or effect of a process

production *noun* the act of manufacturing or producing something

production diseases *plural noun* metabolic disorders of animals which are caused by high levels of production

production entitlement guarantee *noun* a proposed alternative to current agricultural subsidy schemes, in which each farmer's subsidy payment is limited to a fixed proportion of their historical output, with market forces determining any payment on top of this. Abbr **PEG**

production ration *noun* the quantity of food needed to make a farm animal produce meat, milk or eggs, which is always more than the basic maintenance ration

'The production ration for the winter is based on grass and whole crop fermented silage, pressed pulp, brewers grains, and rolled wheat plus a balancer.' [*Farmers Guardian*]

productive *adjective* producing a lot of something that can be used or sold ○ *making productive use of waste ground* ○ *highly efficient and productive farms*

productive agriculture *noun* same as **intensive agriculture**

productive soil *noun* soil which is very fertile and produces large crops

productivity *noun* the rate at which something is produced ○ *With new strains of rice, productivity per hectare can be increased.*

profusion *noun* a very large number or quantity of something ○ *a profusion of small white flowers*

progeny /ˈprɒdʒəni/ *noun* the young or children produced by any living thing

progeny test *noun* the evaluation of the breeding value of an animal or plant variety by examining the performance of its progeny

progesterone /prəʊˈdʒestərəʊn/ *noun* a female sex hormone produced by the corpus luteum of the ovary to prepare the lining of the womb for a fertilised ovum. Formula: $C_{21}H_{30}O_2$.

prolactin /prəʊˈlæktɪn/ *noun* a hormone produced by the pituitary gland that stimulates milk production

prolific *adjective* referring to an animal or plant which produces a large number of offspring or fruit

prolonged *adjective* lasting for a long time ○ *prolonged drought*

promote *verb* to encourage or enable something to take place ○ *Growth-promoting hormones are used to increase the weight of beef cattle.*

promotion *noun* the activity of encouraging or enabling something to take place ○ *the promotion of recycling schemes*

prong *noun* one of the long pointed pieces of metal which make up the end of a fork

propagate *verb* to produce new plants by a technique such as taking cuttings, grafting, budding or layering ○ *fuschias propagated from cuttings*

propagation /ˌprɒpəˈɡeɪʃ(ə)n/ *noun* the production of new plants ○ *propagation by runners*

COMMENT: The most common of the various forms of plant propagation are **seminal propagation**, growing from seed or from tubers as with potatoes, and **vegetative propagation**, by taking cuttings, grafting, dividing, budding roses, layering, etc.

propagator /ˈprɒpəˌɡeɪtə/ *noun* a closed but transparent container in which seed can be sown or cuttings grown in a moist, warm atmosphere

prostaglandin /ˌprɒstəˈɡlændɪn/ *noun* a hormone that is used to make oestrus happen in many animals at the same time and to start the birth process or abortion

protease /ˈprəʊtieɪz/ *noun* a digestive enzyme which breaks down proteins

protectant fungicide /prəˈtektənt ˌfʌŋɡɪsaɪd/ *noun* a fungicide applied to the leaves of plants. It can be washed off by rain, so removing the protection.

protected cropping, protected cultivation *noun* the growing of crops under some form of protection, e.g. in greenhouses or under polythene sheeting

'Meanwhile New Zealand is seeing a resurgence in its blueberry industry with the introduction of new varieties which, with protected cropping, have extended the season by as much as six weeks, according to its HortResearch Institute. It is also looking at the potential for boysenberries and cranberries.' [*The Grocer*]

Protection of Animals Act 1911 noun legislation which makes it an offence to mistreat a domestic or captive animal or to cause it unnecessary suffering

protein noun a nitrogen compound formed by the condensation of amino acids that is present in and is an essential part of living cells

COMMENT: Proteins are necessary for the growth and repair of the body's tissue. They are mainly formed of carbon, nitrogen and oxygen in various combinations as amino acids. Foods such as beans, meat, eggs, fish and milk are rich in protein. Although ruminant animals can make use of cellulose and non-protein nitrogen (which human beings cannot metabolise), animal feeds are often supplemented with grains, pulses and other foods that contain protein.

protein efficiency ratio noun a measure of the nutritional value of proteins carried out on young growing animals. The protein efficiency ratio is defined as the gain in weight per gram of protein eaten. Abbr **PER**

protein equivalent noun a measure of the digestible nitrogen of an animal feedingstuff in terms of protein

protein quality noun a measure of the usefulness of a protein food for various purposes, including growth, maintenance, repair of tissue, formation of new tissue and production of eggs, wool and milk

proteolysis /ˌprəʊtiˈɒləsɪs/ noun the breaking down of proteins in food by digestive enzymes

proteolytic /ˌprəʊtiəʊˈlɪtɪk/ adjective referring to proteolysis ○ a proteolytic enzyme

protoplasm /ˈprəʊtəʊˌplæz(ə)m/ noun a substance like a jelly which makes up the largest part of each cell

protoplasmic /ˌprəʊtəʊˈplæzmɪk/ adjective referring to protoplasm

protoplast /ˈprəʊtəʊplɑːst/ noun a basic cell unit in a plant formed of a nucleus and protoplasm

proud adjective referring to excessive growth or development in crops or livestock

provender /ˈprɒvəndə/ noun a general terms used to describe dry feeds for livestock

proven sire noun a bull, boar or ram which has been shown to sire progeny that produce milk, meat or wool of high quality

proventriculus /ˌprəʊvenˈtrɪkjələs/ noun the gizzard of birds, or the thick-walled stomach of insects and crustaceans

proximate analysis noun a method of chemical analysis used on animal feeding-stuffs, which measures the amounts of ash, crude fibre, crude protein, ether extract, moisture content and nitrogen-free extract

PRRS abbreviation porcine reproductive respiratory syndrome

prune noun a black-skinned dried plum ■ verb to remove pieces of a plant, in order to keep it in shape, or to reduce its vigour

pruners /ˈpruːnəz/ noun a type of secateurs, used for pruning fruit trees

pruning knife noun a special knife with a curved blade, used for pruning fruit trees

Prunus /ˈpruːnəs/ noun the Latin name for the family of trees including the plum, peach, almond, cherry, damson, apricot

PSD abbreviation Pesticides Safety Directorate

PSE abbreviation 1. pale soft exudative muscle 2. porcine spongiform encephalopathy

PSPS abbreviation Pesticide Safety Precaution Scheme

PSS abbreviation porcine stress syndrome

PST abbreviation porcine somatotropin

p.t.o. abbreviation power take-off ○ The p.t.o. shaft from a tractor drives a baler or rotavator.

public elevator noun an elevator which is used for storage by several farmers, and does not belong to one farmer alone

pullet /ˈpʊlɪt/ noun a young female fowl, from hatching until a year old

pulling peas plural noun peas harvested by removing the pods when fresh and sold as young peas in pods

pulp noun 1. the soft inside of a fruit or vegetable 2. a thick soft substance made by crushing ○ wood pulp

pulpy kidney disease /ˌpʌlpi ˈkɪdni dɪˌziːz/ noun a disease caused by a strain of the same bacteria which cause lamb dysentery. It occurs in older lambs and can be fatal.

pulsator /pʌlˈseɪtə/ noun the part of a milking machine which causes the suction action and release of the milk from the udder

pulse noun a general term for a certain type of seed that grows in pods, and which is often applied to an edible seed of a leguminous plant used for human or animal consumption, e.g. a lentil, bean or pea

pumpkin *noun* a large round yellow vegetable, eaten both as a vegetable and in pies as a dessert

pungent *adjective* with a sharp taste or smell, like, e.g., mustard

pupa /ˈpjuːpə/ *noun* a stage in the development of some insects such as butterflies when the larva becomes encased in a hard shell (NOTE: The plural is **pupae**.)

pupal /ˈpjuːpəl/ *adjective* referring to a pupa

pupate /pjuːˈpeɪt/ *verb* (*of an insect*) to move from the larval to the pupal stage

purebred /ˈpjʊəbred/ *adjective* referring to an animal which is the offspring of parents which are themselves the offspring of parents of the same breed

pure breeding *noun* the mating of pure-bred animals of the same breed

put to *verb* to bring a female animal to be impregnated by a male ○ *Ewes are put to the ram.*

PVRO *abbreviation* Plant Variety Rights Office

pygmy beetle *noun* a beetle pest affecting sugar beet (*Atomaria linearis*)

PYO *abbreviation* pick-your-own

pyrethrum /paɪˈriːθrəm/ *noun* 1. an organic pesticide, developed from a form of chrysanthemum, which is not very toxic and is not persistent 2. an annual herb, grown for its flowers which are used in the preparation of pyrethrum

pyridoxine /ˌpɪrɪˈdɒksɪn/ *noun* vitamin B_6

Pyrus /ˈpaɪrəs/ *noun* the Latin name for the pear

Q

QMS *abbreviation* Quality Meat Scotland

quadrat /'kwɒdrət/ *noun* an area of land measuring one square metre, chosen as a sample for research on plant populations ○ *The vegetation of the area was sampled using quadrats.*

quail /kweɪl/ *noun* a small game bird (*Coturnix coturnix*), now reared to produce oven-ready birds and also for their eggs

quality assurance *noun* the system of procedures used in checking that the quality of a product is good

'British producers have been forced to achieve higher standards in terms of quality assurance, traceability and environmental protection, yet they are getting a price that fails to cover costs.' [*Farmers Weekly*]

quality grain *noun* the application of quality standards when selling grain. Good quality is indicated by a high specific weight.

Quality Meat Scotland *noun* the red meat marketing board for Scotland. Abbr **QMS**

quarantine *noun* the period when an animal, person, plant or ship just arrived in a country is kept separate in case it carries a serious disease, to allow the disease time to develop and so be detected ■ *verb* to put a person, animal or ship in quarantine

COMMENT: Some animals coming into Britain are quarantined for six months because of the danger of rabies, but pets such as cats and dogs can now have 'pet passports' to prove that they are free from infection. This only applies to pets coming from European countries, and does not apply to pets coming from North America. People suspected of having an infectious disease can be kept in quarantine for a period which varies according to the incubation period of the disease. The main diseases concerned are cholera, yellow fever and typhus.

quart *noun* a measure of liquids and dry goods such as grain equal to two pints, or 1.14 litres

Quercus /'kwɜːkəs/ *noun* the Latin name for the oak tree

quick-freeze *verb* to preserve food products by freezing them rapidly

quicklime /'kwɪklaɪm/ *noun* a calcium compound made from burnt limestone (NOTE: It is used in the composition of cement and in many industrial processes.)

quince /kwɪns/ *noun* a small tree (*Cydonia vulgaris*) native of western Asia, the hard pear-shaped sour fruit of which are rich in pectin and used to make jellies and other preserves

quintal /'kwɪntəl/ *noun* a unit of weight sometimes used to measure bulk agricultural commodities, equal to 100kg

quota *noun* a fixed amount of something which is allowed ○ *A quota has been imposed on the fishing of herring.*

COMMENT: Quotas were introduced in the European Community in 1984, and were based on each state's 1981 production, plus 1%. A further 1% was allowed in the first year. A supplementary levy, or superlevy, was introduced to penalise milk production over the quota level. In the UK, milk quotas can be bought and sold, either together with or separate from farmland, and are a valuable asset.

quota system *noun* a system where imports or supplies are regulated by fixing maximum or minimum amounts

R

R *symbol* Average (*in the EUROP carcass classification system*)

RAAW *noun* an independent section of the Transport and General Workers Union representing the interests of farmworkers in negotiating terms and conditions of their employment. Full form **Rural, Agricultural and Allied Workers'**

rabbit *noun* a common furry herbivorous rodent (*Oryctolagus cuniculus*)

COMMENT: Rabbits are raised for meat and for their fur. Wild rabbits are a major pest in some parts of the world, especially in Australia, where myxomatosis was introduced to attempt to eradicate the wild rabbit population.

rabi /'rɑːbi/ *noun* ◆ **kohlrabi**

rabies *noun* a frequently fatal notifiable viral disease transmitted to humans by infected animals

COMMENT: Rabies affects the mental balance of a person or animal, and the symptoms include difficulty in breathing or swallowing and an intense fear of water (hydrophobia) to the point of causing convulsions at the sight of water. Rabies is not present in Britain.

race *noun* **1.** a group of individuals within a species that are distinct, especially physiologically or ecologically, from other members of the species. ◊ **landrace 2.** an improvised wooden way along which animals are made to walk, such as when being loaded into a vehicle. Also called **raceway**

raceme /'ræsiːm/ *noun* an inflorescence in which flowers are borne on individual stalks on a main flower stem with the youngest flowers at the top of the main stalk

raceway /'reɪsweɪ/ *noun* same as **race**

rack *noun* a frame of wooden or metal bars which holds fodder, and from which animals can eat

raddle /'ræd(ə)l/ *noun* a flexible length of wood used for making hurdles or fences

radicle /'rædɪk(ə)l/ *noun* the tiny structure in a plant embryo from which the root will develop

radish *noun* a small plant with red or white roots used mainly in salads

Radnor /'rædnə/ *noun* a breed of small hill sheep similar to the Welsh Mountain

rafter *verb* to plough land, leaving a space between the furrows

ragwort /'rægwɜːt/ *noun* a weed (*Senecio jacobea*) found in grassland. It can cause poisoning of cattle, horses and sheep, and must therefore be controlled.

'Ragwort is one of five injurious weeds specified in the Weeds Act 1959, which empowers the Secretary of State to take action to prevent the spread of common ragwort, creeping or field thistle, spear thistle, curled dock and broad-leaved dock.' [*Farmers Guardian*]

rain *noun* water that falls from clouds as small drops ■ *plural noun* **rains** in some countries, repeated heavy falls of rain during a season of the year

COMMENT: Rain is normally slightly acid (about pH 5.6) but becomes more acid when pollutants from burning fossil fuels are released into the atmosphere.

rainfall *noun* the amount of water that falls as rain on an area over a period of time ◊ *an area of high/low rainfall*

rain gun *noun* a spraying device used for applying irrigation water, which it shoots out in a powerful jet

rainmaking /'reɪnmeɪkɪŋ/ *noun* the attempt to create rain by releasing crystals of salt, carbon dioxide and other substances into clouds

rainwater *noun* the water which falls as rain from clouds

raise *verb* 1. to make plants germinate and nurture them as seedlings ○ *The plants are raised from seed.* 2. to breed and keep livestock

rake *noun* an implement with a handle, a crossbar with several prongs, used for pulling hay together, or for smoothing loose soil to form a seedbed ■ *verb* 1. to pull hay or dead grass with a rake 2. to smooth loose soil to form an even seedbed 3. to move a flock of sheep from one pasture to another

ram *noun* a male sheep or goat, that has not been castrated

ranch *noun* 1. a large farm, specialising in raising cattle, sheep or horses 2. a large farm, specialising in raising any type of animal or growing any type of crop

rancher /'rɑːntʃə/ *noun* 1. the owner of a ranch 2. somebody who works on a cattle ranch

ranching /'rɑːntʃɪŋ/ *noun* 1. an agricultural system based on commercial grazing on ranches 2. the raising of cattle on large grassland farms

range *noun* 1. a large area of grass-covered farmland used for raising cattle or sheep 2. open space, particularly for poultry. ◊ **free-range eggs** (NOTE: Eggs produced on a range are called 'free-range eggs'.)

rape *noun* same as **oilseed rape**

rapeseed /'reɪpsiːd/ *noun* the seed of the rape

rare breed *noun* a breed of farm animal which is protected because its numbers are falling and it is in danger of becoming extinct

'In the months before coming to Hardwick Hall Mr Aldis had started to source his livestock, with a view to introducing rare breeds. Hardwick Hall has a long association with Longhorn cattle, having had a famous showing herd many years ago.' [*Farmers Guardian*]

Rare Breeds Survival Trust *noun* a trust established in 1973 to foster interest in breeds which have historical importance and may prove useful in the future

RAS *abbreviation* Royal Agricultural Societies

RASE *abbreviation* Royal Agricultural Society of England

raspberry *noun* a cane (*Rubus idaeus*) which provides a most important soft fruit, sold fresh, sent for freezing and also used for processing into jams

raspberry beetle *noun* a serious pest (*Byturus tomentosus*) whose larvae feed on young raspberry fruit

rat *noun* a rodent *Genus Rattus* with a long tail, similar to but larger than a mouse, which can be very destructive of growing and stored crops and also carry disease to cattle and pigs. ◊ **brown rat**

ration *noun* an amount of food given to an animal or person

ration formulation *noun* the process of putting together different types of feedstuff in order to provide the amount of nutrients required by a particular animal or type of animal

ratoon /ræ'tuːn/, **ratoon crop** *noun* the second and later crops taken from the regrowth of a crop after it has been harvested once. Sugar cane plants, e.g., can be harvested many times.

ray fungus *noun* a bacterium which affects grasses and cereals, and can cause actinomycosis in cattle

RCGM *abbreviation* rectified concentrated grape must

RCVS *abbreviation* Royal College of Veterinary Surgeons

RDA *abbreviation* 1. Recommended Daily Amount 2. Regional Development Agency

REACH *abbreviation* Registration, Evaluation, Authorisation and Restrictions of Chemicals

reafforestation /riːæˌfɒrɪs'teɪʃ(ə)n/ *noun* the planting of trees in an area which was formerly covered by forest

reap *verb* to cut a grain crop

reaping hook *noun* a short-handled semicircular implement with a sharp blade, formerly used for cutting corn by hand

rear *verb* to look after young animals such as they are old enough to look after themselves

rearer /'rɪərə/ *noun* a person who rears livestock

receptacle /rɪ'septək(ə)l/ *noun* the top part of a flower stalk that supports the flower (NOTE: In some plants such as strawberry it develops into the fruit.)

recessive /rɪ'sesɪv/ *adjective* (*of a gene or genetically controlled characteristic*) suppressed by the presence of a corresponding dominant gene. Compare **dominant**

'Breed society secretary Michael Woodhouse believes the breed will compete as a terminal sire. "Blue Texels resemble both the larger type of Texel common to the UK and the smaller well-muscled Dutch type. The blue gene is recessive, meaning crossbred lambs will normally be white. However, the blue gene will breed true in pedigree flocks".' [*Farmers Weekly*]

COMMENT: Since each physical characteristic is governed by two genes, if one gene is dominant and the other recessive, the resulting trait will be that of the dominant gene. Traits governed by recessive genes will appear if genes from both parents are recessive.

recessiveness /rɪˈsesɪvnəs/ *noun* the characteristic of a gene that leads to its not being expressed in the individual carrying it when a corresponding dominant gene is also present. Compare **dominance**

reclaim *verb* to make land usable for agricultural or commercial purposes, usually marshy land, a waste site, land which has previously been built on or used for industry, or land which has never been cultivated ○ *to reclaim land from the sea*

reclamation /ˌrekləˈmeɪʃ(ə)n/ *noun* **1.** the act of reclaiming land ○ *land reclamation schemes in urban centres* **2.** land which has been reclaimed

COMMENT: Reclamation includes the drainage of marshes and lakes, and the improvement of heathland and moorland.

Recommended Daily Amount *noun* the amount of a substance, e.g. a vitamin or mineral, that should be consumed each day for a person or animal to be healthy. Abbr **RDA**

record keeping *noun* the act of making records such as a livestock register, which are open to examination by the Government and welfare authorities

rectal palpation /ˌrekt(ə)l pælˈpeɪʃ(ə)n/ *noun* a technique used to diagnose pregnancy in cows

rectified concentrated grape must /ˌrektɪfaɪd ˌkɒnsəntreɪtɪd ˈgreɪp ˌmʌst/ *noun* a form of grape sugar produced by distillation from surplus wine, used to add to new wine during chaptalisation

rectum *noun* the last part of the large intestine, where waste material accumulates before leaving the body through the anus

recumbent /rɪˈkʌmbənt/ *adjective* referring to animals which are lying down, as in the case of cows after illness or injury ○ *Reduced phosphorus levels may also play a part in keeping affected animals recumbent.*

Red *noun* the English name for the Rouge de l'Ouest breed of sheep

red clover *noun* a short-lived deep-rooting species of clover (*Trifolium pratense*)

red corpuscle *noun* a red blood cell which contains haemoglobin and carries oxygen to the tissues

red currant *noun* a soft fruit, growing on bushes, and used mainly for making jams

Red Data Book *noun* a catalogue formerly published by the IUCN, listing species which are rare or in danger of becoming extinct. The information is now available in a searchable database. ◊ **Red list**

red deadnettle *noun* a weed (*Lamium purpureum*) which is common in gardens, and now affects cereals and oilseed rape. Also called **French nettle**

red fescue *noun* a species of grass (*Festuca rubra*), used on hill and marginal land and in fine-leaved lawns

red grouse *noun* ♦ **grouse**

redistribution of land *noun* the practice of taking land from large landowners and splitting it into smaller plots for many people to own

redlegs /ˈredlegz/ *noun* a common weed (*Polygonum persicaria*) which affects spring crops, and causes problems when harvesting. Also called **redshanks**

Red list *noun* **1.** a searchable database maintained by IUCN that records the conservation status of different organisms throughout the world. Full form **IUCN Red List of Threatened Species 2.** a list recording the conservation status of a particular type of organism in a specific geographical area ○ *the Red list of the epiphytic lichens of Switzerland*

Red Poll *noun* a dual-purpose breed of cattle, which originated in East Anglia. It is deep red in colour, with a white swish at the end of the tail.

redshanks /'redʃæŋks/ *noun* same as **redlegs**

redwater /'redwɔːtə/ *noun* a parasitic disease of cattle transmitted by the common tick. The affected animal becomes very dull, feverish, salivates freely and often staggers and falls. The acute form of the disease is often fatal.

reed /riːd/ *noun* an aquatic plant growing near the shores of lakes, used to make thatched roofs

reedbed /'riːdbed/ *noun* a mass of reeds growing together

reedbed filter *noun* a reedbed used as part of a system of cleaning sewage or dirty water

reel *noun* part of the mechanism of a combine harvester, which holds the crop against the cutter bar for cutting (NOTE: The reel directs the crop after it has been cut onto the cutter bar table or platform. Most combines have a pick-up reel which can be adjusted to deal with inlaid or tangled crops.)

refection /rɪ'fekʃən/ *noun* consumption by an animal of its own faeces

reference price *noun* the minimum price at which certain fruit and vegetables can be imported into the EU

refine *verb* to process something to remove impurities ○ *a by-product of refining oil*

refined *adjective* having had impurities removed ○ *refined oil*

refrigerate *verb* to cool produce and keep it at a cool temperature

refrigerated lorry *noun* a special lorry which carries produce under refrigeration

refrigerated processed foods of extended durability *noun* prepared and chilled food such as ready meals, which can be kept for longer than fresh food. Abbr **REPFEDs**

refrigeration /rɪˌfrɪdʒə'reɪʃ(ə)n/ *noun* a method of prolonging the life of various foods by storing them at very low temperatures

COMMENT: Low temperature retards the rate at which food spoils, because all the causes of deterioration proceed more slowly. In freeze-drying, the food has to be quick-frozen and then dried by vacuum, so removing the moisture. Pre-cooked foods should be cooled rapidly down to −3°C and eaten within five days of production. Certain high-risk chilled foods should be kept below 5°C. These foods include soft cheese and various pre-cooked products. Eggs in shells can be chilled for short-term storage (i.e. up to one month) at temperatures between −10°C and −16°C. Bakery products, including bread, have storage temperatures between −18°C and −40°C; bread goes stale quickly at chill temperatures which are above these. Potatoes in the form of pre-cooked chips can be stored at −18°C or colder, but ordinary potatoes must not be chilled at all. Apples and pears can be kept in air-cooled boxes at between −1°C and +4°C (this is known as 'controlled temperature storage'). Lettuces and strawberries (which normally must not be chilled) can be kept fresh by vacuum cooling, while celery and carrots can be chilled by hydrocooling.

refrigerator *noun* a device for cooling produce and keeping it cool

refrigerator ship *noun* a ship which carries produce under refrigerated conditions

regenerate *verb* to grow again, or grow something again ○ *A forest takes about ten years to regenerate after a fire.* ○ *Salamanders can regenerate limbs.*

regeneration /rɪˌgenə'reɪʃ(ə)n/ *noun* the process of vegetation growing back on land which has been cleared or burnt ○ *Grazing by herbivores prevents forest regeneration.*

regenerative /rɪ'dʒenərətɪv/ *adjective* allowing new growth to replace damaged tissue

Regional Development Agency *noun* an organisation which promotes the social and economic benefits of living in a region and undertakes projects to bring new industries and jobs to the region. Abbr **RDA**

Registration, Evaluation, Authorisation and Restrictions of Chemicals *noun* a proposed review of the EU's chemicals policy, in which comprehensive health and safety research will be carried out for all substances. Abbr **REACH**

regrowth /rɪ'grəυθ/ *noun* the growth that occurs after a cut or harvest, or after accidental damage or fire

regulate *verb* **1.** to change or maintain something by law ○ *Development is regulated by local authorities.* **2.** to control the growth of a plant

regulation *noun* **1.** a rule made by a government or official body **2.** a rule made by the Council of Ministers or the

Commission of the EU, which has legal force in all member countries. European regulations require, e.g., that animals being transported should be rested, fed and watered every 24 hours.

regulator /'regjʊleɪtə/ *noun* something or someone controlling a process or activity

rein *noun* a long narrow strap used to control a horse, each end of which is attached to the bit in the horse's mouth

RELU *abbreviation* rural economy and land use

rendzina /ren'ziːnə/ *noun* soil developed on chalk and limestone rocks characterised by its shallowness and lack of true subsoil

renewable *adjective* referring to something that can be replaced or can renew itself by regrowing, reforming or breeding ○ *Herring stocks are a renewable resource if the numbers being caught are controlled.* □ **renewable sources of energy** energy from the sun, wind, waves, tides or from geothermal deposits or from burning waste, none of which uses up fossil fuel reserves

renewable energy *noun* energy from the Sun, wind, waves, tides, from geothermal deposits or from burning waste

'In a recent report to the DTI, it called for government to ensure that more renewable energy is sourced straight from UK farms. The NFU also urged ministers to take advantage of crops such as wheat and sugar beet, to produce renewable biofuels.' [*Farmers Weekly*]

renewable resource *noun* a natural resource that replaces itself unless overused, e.g. animal or plant life, fresh water or wind energy

rennet /'renət/ *noun* an extract from the stomach of a calf; it contains the enzyme rennin, which clots milk. It is used in the production of certain milk products such as cheese.

rennin /'renɪn/ *noun* an enzyme which makes milk coagulate in the stomach, so as to slow down the passage of the milk through the digestive system

rent *noun* money paid to use a farm or land for a period of time ■ *verb* to pay money to hire a farm *or* land for a period of time

COMMENT: Since 1950, there has been a decline in the area and number of farm

holdings which are rented in Great Britain. In 1950, rented agricultural land in England, Wales and Scotland accounted for 60% of the holdings. By 2000 the figure was 37%.

REPFEDs /'repfedz/ *abbreviation* refrigerated processed foods of extended durability

replacement milk *noun* milk which is used to feed young animals which cannot be fed by their mothers, e.g. 'lamb replacement milk'

replacement rate *noun* the rate of introduction of heifers into a dairy herd to replace ageing cows or cows with low milk yields

replant /riː'plɑːnt/ *verb* **1.** to grow plants in an area again ○ *After the trees were felled, the land was cleared and replanted with mixed conifers and broadleaved species.* **2.** to put a plant in the ground again

replant disease *noun* a condition affecting apple trees planted in an orchard which has been grubbed out

repot /riː'pɒt/ *verb* to take a plant out of its pot and plant it in another, changing or adding to the soil at the same time

reproduce *verb* **1.** to produce offspring **2.** (*of bacteria*) to produce new cells

reproduction *noun* the production of offspring

COMMENT: Service by the male is only allowed by the females of most animals during the heat period or oestrus. This acts as a natural check on the breeding rate of animals. The length of the oestrus varies with the animal.

reproductive /ˌriːprə'dʌktɪv/ *adjective* referring to the production of offspring ○ *Pollination is a reproductive process.*

reproductive organs *plural noun* parts of the bodies of animals which are involved in the conception and development of a foetus

RES *abbreviation* Rural Enterprise Scheme

reseed /riː'siːd/ *verb* to reestablish a ley by sowing seed again

COMMENT: Reseeding is carried out to improve permanent pasture. This is done by direct reseeding which involves sowing again without a cover crop, or by undersowing, where the seed mixtures are sown with another crop, usually a cereal.

reserve *noun* an area of land maintained for the benefit of plant or animal life where no commercial exploitation is allowed. ◊ **nature reserve, wildlife reserve**

reservoir *noun* an artificial or natural area of water, used for storing water for domestic or industrial use ○ *The town's water supply comes from reservoirs in the mountains.* ○ *After two months of drought the reservoirs were beginning to run dry.*

residual *adjective* referring to the amount of something that is left behind

residual herbicide *noun* a herbicide applied to the surface of the soil which acts through the roots of existing plants and also new plants as they germinate

residue *noun* the material left after a process has taken place or after a material has been used

resin *noun* a sticky oil secreted by some conifers or other trees, especially when they are cut

resinous /'rezɪnəs/ *adjective* like resin, or producing resin

resist *verb* to fight off or not be subject to the effects of something

resistance *noun* the ability of an organism not to be affected by something such as a disease, stress factor, process or treatment ○ *Increasing insect resistance to chemical pesticides is a major problem.* ○ *Crop plants have been bred for resistance to disease.*

resistant *adjective* referring to something which is unaffected by a disease, stress factor, process or treatment ○ *Some alloys are less resistant to corrosion than others.* ○ *The plants were not resistant to mildew.*

COMMENT: Resistant strains develop quite rapidly after application of the treatment. Some strains of insect have developed which are resistant to DDT. The resistance develops as non-resistant strains die off, leaving only individuals which possess a slightly different and resistant chemical makeup. Hence a pesticide will select out only resistant individuals. This can be avoided by using pesticides in combination or by not using the same chemical (or chemicals with a similar mode of action) repeatedly.

-resistant *suffix* not adversely affected by something ○ *a DDT-resistant strain of insects* ○ *disease-resistant genetic material* ○ *a new strain of virus-resistant rice*

respiration /ˌrespə'reɪʃ(ə)n/ *noun* the action of breathing

COMMENT: Respiration includes two stages: breathing in (inhalation) and breathing out (exhalation). Air is taken into the respiratory system through the nose or mouth and goes down into the lungs through the pharynx, larynx and windpipe. In the lungs, the bronchi take the air to the alveoli (air sacs) where oxygen in the air is passed to the bloodstream in exchange for waste carbon dioxide which is then breathed out.

respiratory *adjective* referring to respiration

respiratory quotient *noun* the ratio of the amount of carbon dioxide passed from the blood into the lungs to the amount of oxygen absorbed into the blood from the air. Abbr **RQ**

respiratory system *noun* a series of organs and passages that take air into the lungs and exchange oxygen for carbon dioxide

response *noun* the beneficial reaction of a growing crop to the application of fertiliser

response curve *noun* a graph showing the yield (or some associated factor) against fertiliser input, level of feed, antibiotics, etc.

responsible care *noun* an initiative of the chemical industry which requires member firms to follow codes of conduct on such matters as toxic materials, waste reduction, chemical-accident minimisation, worker safety and community consultation

rest *verb* □ **to rest land** to let land lie fallow, without growing any crops

restore *verb* to give something back, or put something back to a previous state or position ○ *By letting the land lie fallow for a couple of years, farmers hope to restore some of the natural nutrients which have been removed from the soil.*

retail *noun* the sale of small quantities of goods to the general public

retained placenta /rɪˌteɪnd plə'sentə/, **retained afterbirth** *noun* a disease of cattle caused by interference at calving, premature calving or milk fever. The placenta should be removed by a veterinary surgeon.

retard /rɪ'tɑːd/ *verb* to make something happen later ○ *The injections retard the effect of the anaesthetic.*

reticulum /rɪˈtɪkjʊləm/ *noun* the second stomach compartment of ruminants such as cows and sheep

retinol /ˈretɪnɒl/ *noun* a vitamin that is soluble in fat and can be formed in the body but is mainly found in food such as liver, vegetables, eggs and cod liver oil. Also called **vitamin A**

retting /ˈretɪŋ/ *noun* a process used in the preparation of flax, where flax is soaked in water and allowed to rot, so freeing the fibres from the plant stems

reversible plough *noun* a plough with left- and right-handed mouldboards, which make it possible to plough up and down the same furrow

rhizoctonia root rot /ˌraɪzɒktəʊniə ˈruːt ˌrɒt/ *noun* a common soil fungus (*Rhizoctonia solani*) which attacks the roots of seedlings and retards growth

rhizomania /ˌraɪzəʊˈmeɪniə/ *noun* a notifiable virus disease affecting sugar beet, in which hairs grow on the roots and the leaves turn yellow. The disease is endemic in the Netherlands, and some cases have been reported in the UK.

rhizome /ˈraɪzəʊm/ *noun* a plant stem that lies on or under the ground and has leaf buds, adventitious roots and sometimes branches

'Recent announcement of grant aid for planting miscanthus can help to offset establishment costs, and Bical is seeking to reduce the cost of rhizomes by sourcing them from the West Indies where they can be multiplied more rapidly. Import clearance depends on the authorities being convinced that no pests and diseases will be imported with the rhizomes.' [*Arable Farming*]

rhizosphere /ˈraɪzəʊsfiə/ *noun* the soil surrounding the roots of a plant

Rhode Island Red /ˌrəʊd ˌaɪlənd ˈred/ *noun* a heavy breed of fowl, with red feathers on the body, and black tail and wing feathers. It produces large brown eggs.

RHS *abbreviation* Royal Horticultural Society

rhubarb *noun* a perennial plant (*Rheum rhaponticum*), of which the leaf stalks are cooked and eaten as dessert. It has a high oxalate content and the leaves are toxic.

Ribes /ˈraɪbiːs/ *noun* the Latin name for blackcurrant

rib grass *noun* a palatable deep-rooting herb with a high mineral content, which may benefit pasture

riboflavin /ˌraɪbəʊˈfleɪvɪn/, **riboflavine** *noun* a vitamin found in eggs, liver, green vegetables, milk and yeast and also used as an additive (E101) in processed food. Also called **vitamin B$_2$** (NOTE: Lack of riboflavin will affect a child's growth and can cause anaemia and inflammation of the mouth and tongue.)

ribwort /ˈrɪbwɜːt/ *noun* same as **plantain**

rice *noun* a plant that is the most important cereal crop and the staple food of half the population of the world. Latin name: *Oryza sativa*.

COMMENT: Wet rice is by far the commonest method of cultivation: the paddies are enclosed by low banks and are kept flooded during the growing season. They are allowed to dry out before the crop is harvested. Dry land rice is cultivated in a similar way to wheat or barley. Rice is classified according to the length of the grains: long-grain rice is grown in tropical climates such as India, while short-grain rice is grown in colder climates such as Japan. There are over 120,000 varieties of rice grown world-wide, with more than 40,000 varieties being cultivated in India alone. Rice is an important crop in most countries of Asia, and is becoming increasingly important in Africa, South America, the USA and Australia. In Europe, Italy, France and Hungary grow considerable amounts of rice. The world's leading rice exporters are the USA and Thailand.

rich *adjective* (*of soil*) having many nutrients that are useful for plant growth

rick /rɪk/ *noun* a stack, usually of hay, with a sloping roof

rickets /ˈrɪkɪts/ *noun* a disease of young animals due to deficiency of Vitamin D. Bones fail to ossify and joints become swollen.

riddle *noun* a coarse sieve for sieving soil ■ *verb* to grade and sort produce according to size, using a sieve ○ *Potatoes are riddled to separate the best potatoes, called 'wares' from the small potatoes, called 'chats'.*

ridge *noun* **1.** a long raised section of ground, occurring as part of a mountain range, in a field, on a beach or on the ocean floor **2.** a long narrow band of high pressure leading away from the centre of an anticyclone ○ *A ridge of high pressure is lying across the country.* **3.** a long raised section of earth, made by ploughing up and down on either side of the furrow. In systematic ploughing, ridges first mark out land in a field before the plough is reset for normal work and the field is ploughed.

ridger /'rɪdʒə/ *noun* a type of plough used to form ridges for earthing up crops such as potatoes

rig *noun* a male animal in which one or both testicles have not descended into the scrotum at the usual time

right of access *noun* **1.** the right of someone to be able to get to land by passing over someone else's property **2.** the right of the public to walk in areas of the countryside, providing they do not harm crops or farm animals

right of way *noun* a legal right to go across someone else's property

rill /rɪl/ *noun* **1.** a very narrow stream **2.** a small channel eroded in soil by rainwater. It can be removed during ordinary cultivation.

ring *noun* a metal circle which goes through the nose of an animal ■ *verb* **1.** to attach a numbered ring to the leg of a bird so that its movements can be recorded **2.** to attach a ring to an animal, such as to the nose of a bull

COMMENT: Some animals can be ringed to allow them to be led, while others are ringed to prevent excessive grubbing in the ground.

ring-barking *noun* the cutting of a strip of bark from a tree as a means of making the tree more productive. It restricts growth and encourages fruiting.

ring bone *noun* a growth of bony tissue in the joints of a horse's foot

ring rot *noun* a disease affecting potatoes

ringworm /'rɪŋwɜːm/ *noun* any of various infections of the skin by a fungus, in which the infection spreads out in a circle from a central point (NOTE: Ringworm is very contagious and very difficult to get rid of. In animals, it is most common in young store cattle, but it also affects humans.)

riparian /rɪ'peəriən/ *adjective* referring to the bank of a river ○ *riparian fauna*

ripe *adjective* referring to fruit or grain that is ready for eating ○ *When the corn is ripe the harvest can start.* ○ *The early varieties of apple are ripe in August.* ○ *Ripe peaches cannot be kept very long.* ○ *Bananas should be picked before they are* ripe, and allowed to ripen during transport and storage. (NOTE: The opposite is **unripe**.)

ripen *verb* to become ready for eating, or to make something, especially a fruit, ready for eating ○ *Unripe bananas are shipped in special containers and will ripen in storage.* ○ *Tomatoes can be picked when still pink and allowed to ripen off the plant.*

ripper /'rɪpə/ *noun* a heavy cultivator consisting of a strong frame with long tines attached to it. It is used to break up compacted soil to allow free passage of air and water. Also called **subsoiler**

risk *noun* **1.** a combination of the likelihood of injury, damage or loss being caused by a potentially dangerous substance, technology or activity, or by a failure to do something, and the seriousness of the possible consequences **2.** something that is regarded as likely to cause injury, damage or loss ○ *a fire risk* ○ *a health risk* Compare **hazard**

COMMENT: A substance or practice may have the potential to cause harm, i.e. may be a hazard, but risk only arises if there is a likelihood that something will be harmed by it in a specific set of circumstances. A highly dangerous thing may in fact present only a small risk. Risk assessment is used to decide what the degree and nature of the risk, if any, may be so that measures to reduce or avoid it can be taken.

river *noun* a large flow of water, running from a natural source in mountains or hills down to the sea

riverine /'rɪvəraɪn/ *adjective* referring to a river ○ *The dam has destroyed the riverine fauna and flora for hundreds of kilometres.*

RIW *abbreviation* Rural Inspectorate Wales

RLR *abbreviation* Rural Land Register

roan /rəʊn/ *noun* the coat of an animal in which the main colour is mixed with another, as e.g. red and white, or black and white

robot milker *noun* a system used in a completely automated milking parlour. Lasers, mirrors and cameras are used to put all four caps on the teats simultaneously.

rock *noun* a solid mineral substance which forms the outside crust of the Earth

rock phosphate *noun* a natural rock ground to a fine powder, used as a fertiliser

rod *noun* an old measurement of land. When used as a measurement of length a rod equals 5 metres, and when used as a measurement of area it equals 25 square metres.

rodent *noun* a mammal that has sharp teeth for gnawing, e.g. a rat or mouse

roe deer /ˈrəʊ dɪə/ *noun* one of the breeds of deer which are found wild in the UK

rogue *noun* a plant of a different variety found growing in a crop ■ *verb* to remove unwanted plants from a crop, usually by hand

roguing glove *noun* a glove impregnated with herbicide, used to destroy wild oats

roll *noun* a tractor-drawn implement used for breaking clods, firming the soil, pushing stones into the soil and providing a smooth firm surface for drilling (NOTE: The two main types are the Cambridge roll, with a number of cast iron rings on an axle which leave a corrugated surface, and a flat roll which leaves a smooth surface.)

rolled grain *noun* a grain which has been through a roller mill before it is fed to livestock. Rolled grain, usually barley, is more easily digested.

roller *noun* same as **roll** ○ *The most common bearings used in gas turbine engine are the ball or roller type.*

roller crusher *noun* a machine used to condition freshly-cut grass. The swath of cut grass is picked up by the rolls and the stems are flattened as the grass is passed between them. With the sap removed from the stems, the drying process is much faster.

roller mill *noun* a piece of equipment used in the preparation of flour and animal feed. It has two smooth steel rollers which crush the grain.

roller table *noun* a machine, consisting of a horizontal line of rotating rollers, used for removing stones and clods from a crop such as potatoes

Romagnola /ˌrɒməˈnjəʊlə/ *noun* a large docile hardy breed of beef cattle from north-east Italy. The animals are grey with a black muzzle and hooves.

Roman *noun* a breed of white goose, now quite rare

Romney /ˈrɒmni/ *noun* a hardy breed of sheep found in large numbers on Romney Marsh, which has heavy fine-quality long wool fleece. The Romney half-breed has been developed by crossing Romney ewes with North Country Cheviot rams, and has been widely exported. Also called **Kent**

rook *noun* a crow-like bird which causes much damage to crops

rookery /ˈrʊkəri/ *noun* breeding place for a colony of rooks

roost /ruːst/ *noun* a place where birds rest at night ■ *verb* to sleep on a perch at night

rooster /ˈruːstə/ *noun* a cock, a male domestic fowl (*especially US*)

root *noun* a part of a plant which is usually under the ground and absorbs water and nutrients from the surrounding soil ■ *verb* (*of a plant*) to produce roots ○ *The cuttings root easily in moist sand.*

root crop *noun* a plant that stores edible material in a root, corm or tuber and is grown as food (NOTE: Root crops include carrots, parsnips, swedes and turnips. Starchy root crops include potatoes, cassavas and yams.)

'Some 40 acres of root crops are grown for sheep feed over the winter and the cereals from 40 acres of combinable crops are also used on-farm.' [*Farmers Guardian*]

root harvester *noun* a machine for lifting root crops out of the ground, e.g. a sugar beet harvester

rooting compound *noun* a powder containing plant hormones (**auxins**) into which cuttings can be dipped to encourage the formation of roots

rooting depth *noun* the depth of soil from which plant roots take up water, or the depth of soil to which roots reach

rootlet /ˈruːtlət/ *noun* a little root which grows from a main root

rootstock /ˈruːtstɒk/ *noun* **1.** same as **rhizome 2.** a plant with roots onto which a piece of another plant is grafted. ◊ **scion**

root system *noun* all the roots of a plant

rosemary *noun* an aromatic herb (*Rosemarinus officinalis*) used for flavouring and also as a source of oil used in soaps and cosmetics

rot *verb* (*of organic tissue*) to decay or become putrefied because of bacterial or fungal action

rotary cultivator *noun* a mounted or trailed machine with a shaft bearing a number of L-shaped blades. Rotary cultivators are used for stubble-clearing, seedbed work and general land reclamation and cleaning.

rotary mower *noun* a machine used for cutting grass and other upright crops. Rotary mowers have two or four rotors each with three or four swinging blades. The rotors rotate in opposite directions and leave a single swath of cut grass.

rotary parlour *noun* the most expensive and complex of the four milking systems, where the cows stand on a rotating platform with the milker in the middle. The operator may work on the inner or outer side of the circle. ◊ **abreast parlour, herringbone parlour**

rotary sprinkler *noun* a machine used for irrigation purposes. Sprinklers can be fitted with fine spray nozzles for protection of fruit crops and potatoes against frost damage.

rotate *verb* to grow different crops from year to year in a field (NOTE: The advantages of rotating crops include: different crops utilising soil nutrients differently, pests specific to one crop being discouraged from spreading, and some crops such as legumes increasing the nitrogen content of the soil if their roots are left in the soil after harvesting.)

rotating flails *plural noun* parts used on manure spreaders to distribute materials and on machines for cutting crops or grass verges. Used also in mixing machines such as composters.

rotating tines *plural noun* spikes used on machines such as rotavators and power harrows for cultivation purposes. They are also used on machine pick-ups.

rotational grazing *noun* the movement of livestock around a number of fields or paddocks in an ordered sequence. Also called **on-off grazing**

'While set stocking has been an extremely popular and successful grazing system for many years, there is now considerable evidence to suggest that rotational grazing can give greater grass growth throughout the season.' [*Farmers Guardian*]

rotation design *noun* a method of conserving soil nutrients in organic farming by planting different crops in different years

rotation of crops *noun* same as **crop rotation**

rotavator /ˈrəʊtəveɪtə/ a trademark for a type of rotary cultivator

rotavirus /ˈrəʊtəvaɪrəs/ *noun* a wheel-shaped RNA virus that causes diarrhoea in piglets, calves and foals

rotenone /ˈrəʊtənəʊn/ *noun* the active ingredient of the insecticide derris

Rothamsted /ˈrɒθəmsted/ the site of the Agricultural Experimental station, established in 1843 by John Bennett Lawes. The station specialised in research into plant nutrition, and demonstrated the importance of nitrogen, phosphorus and potassium to plants. Today it is important for its research into biotechnology and is to a large extent sponsored by the BBRSC.

Rouen /ˈruːɒ̃/ *noun* a breed of table duck. The drake has a green head and neck, rich claret-coloured breast and grey-black body. The female is mostly brown.

Rouge de l'Ouest /ˌruːʒ də ˈlwest/ *noun* a breed of sheep originating in France. Also called **Red**

roughage /ˈrʌfɪdʒ/ *noun* **1.** fibrous matter in food, which cannot be digested. Also called **dietary fibre 2.** animal feedingstuffs with high fibre content, e.g. hay or straw

Rough Fell *noun* a hardy moorland breed of horned sheep, closely related to the Swaledale. It has a dark-coloured face with irregular patterns. The wool is of coarse quality.

rough grazing *noun* unimproved grazing, found in mountain, heath and moorland areas

rough stalked meadow grass *noun* a type of grass, highly palatable but low in production compared to ryegrass; common in lowland pastures on rich moist soils. When found in cereal crops it is treated as a weed.

rough terrain vehicle *noun* a vehicle specially designed to travel over difficult ground. Abbr **RTV**

round and round ploughing *noun* a system of ploughing in which fields are ploughed from the centre to the outside or from the edge to the centre

round baler *noun* a tractor-drawn machine which straddles the swath with a pickup cylinder. The crop is passed over a system of belts to form a round bale; when the bale is complete, twine is wrapped round it and it is thrown out of the machine.

rounds *plural noun* circular walls built to protect sheep from snow drifts

roundworm /ˈraʊndwɜːm/ *noun* a type of worm with a round body, some of which are parasites of animals, others of roots of plants

Roussin /ˈruːsæn/ *noun* a breed of sheep imported into the UK from France

row crop *noun* a crop planted in rows wide enough to allow cultivators between the rows. Most farm crops are drilled in rows, in preference to broadcasting.

row crop tractor *noun* a lightweight tractor with a narrow turning circle and adjustable wheel track widths, used by market gardeners and farmers who grow row crops

Royal Agricultural Societies *plural noun* an alliance between the Royal Agricultural Society of England, the Royal Highland and Agricultural Society of Scotland, the Royal Welsh Agricultural Society and the Royal Ulster Agricultural Society. Abbr **RAS**

Royal Agricultural Society of England *noun* an organisation whose main task is running the annual Royal Show held at The National Agricultural Centre, Stoneleigh, Kenilworth, Warwickshire. Abbr **RASE**

Royal College of Veterinary Surgeons *noun* a body which organises the examinations for veterinary surgeons and represents them. Abbr **RCVS**

Royal Horticultural Society *noun* a national society which organises the Chelsea Flower Show and has permanent gardens at Wisley in Surrey. Abbr **RHS**

Royal Society for the Prevention of Cruelty to Animals *noun* a UK charity that runs centres to take care of animals in distress and lobbies the government on animal welfare legislation. Abbr **RSPCA**

Royal Society for the Protection of Birds *noun* a UK charity that works to ensure a good environment for birds and wildlife. Abbr **RSPB**

royalty *noun* a payment made to plant breeders for the use of seed of registered plant varieties

RPA *abbreviation* Rural Payments Agency

RQ *abbreviation* respiratory quotient

RRA *abbreviation* Rothamsted Research Association

RSPB *abbreviation* Royal Society for the Protection of Birds

RSPCA *abbreviation* Royal Society for the Prevention of Cruelty to Animals

RTV *abbreviation* rough terrain vehicle

rubbed seed *noun* same as **graded seed**

rubber *noun* **1.** a material which can be stretched and compressed, and is made from a thick white fluid (**latex**) from a tropical tree **2.** the rubber tree, a tropical tree grown for its latex. In commercial practice, trees are grafted onto suitable rootstock.

Rubus /ˈruːbəs/ *noun* a genus of plants including cane fruits such as raspberries and blackberries

ruddle /ˈrʌd(ə)l/ *noun* a red colouring material on a harness worn by rams so that ewes which have been mated will be marked and identified

Rules of Good Husbandry *plural noun* an unwritten set of 'rules' which, if they are deemed to have been broken by a tenant, can give a landlord the excuse to evict him

rumen /ˈruːmən/ *noun* the first stomach of ruminating animals such as cows, sheep or goats, all of which have four stomachs. It is used for storage of food after it has been partly digested and before it passes to the second stomach. ◊ **abomasum, omasum, reticulum**

rumen liquor *noun* a concentrated liquid found in the rumen of an animal, used to test the digestibility of feed or the nutrient balance of an animal's diet

ruminant /ˈruːmɪnənt/ *noun* an animal that has a stomach with several chambers, e.g. a cow

COMMENT: Ruminants have stomachs with four sections, the rumen, the reticulum, the omasum and the abomasum. They take in foodstuffs into the upper chamber where it is acted upon by bacteria. The food is then regurgitated into their mouths where they chew it again before passing it to the last two sections where normal digestion takes place.

rumination /ˌruːmɪˈneɪʃ(ə)n/ *noun* the process by which food taken to the stomach of a ruminant is returned to the mouth, chewed again and then swallowed

run *noun* **1.** an enclosure for animals, e.g. a chicken run **2.** an extensive area of land used for sheep grazing ■ *verb* to keep animals

runch /rʌnʃ/ *noun* a common weed (*Raphanus raphanistrum*). Also called **wild radish**

runholder /ˈrʌnhəʊldə/ *noun* a farmer who owns a sheep run

runner *noun* a long shoot that grows sideways from a plant such as a strawberry, ending in a tuft of leaves which will take root

runner bean *noun* a garden bean (*Phaseolus coccineus*) grown exclusively for the fresh trade

runoff /ˈrʌnɒf/ *noun* **1.** the flow of rainwater or melted snow from the surface of land into streams and rivers **2.** the flow of excess fertiliser or pesticide from farmland into rivers ○ *Nitrate runoff causes pollution of lakes and rivers.* ○ *Fish are extremely susceptible to runoff of organophosphates.* **3.** the portion of rainfall which finally reaches a stream

'A land drain gravity-feeds the dirty water (together with runoff from the yards and field drains) into a settling lagoon situated away from the farm to avoid the smell. The water then flows into the adjacent dirty water reservoir where it is stored until required.' [*Dairy Farmer*]

runoff rate *noun* the amount of excess fertiliser or pesticide from farmland that flows into rivers in a specific period

runt /rʌnt/ *noun* **1.** a small individual animal, one that it is smaller than average for its kind **2.** the smallest animal in a litter

Rural, Agricultural and Allied Workers *noun* full form of **RAAW**

rural affairs *plural noun* the activities and concerns of rural communities

rural area *noun* an area in the countryside where the main activities are farming or forestry and where relatively few people live

rural depopulation /ˌrʊərəl ˌdiːpɒpjʊˈleɪʃ(ə)n/ *noun* the loss of population from the countryside due to various causes, including decline in agriculture and increased mechanisation

rural development *noun* a programme of activities undertaken to ensure that rural areas remain economically and socially sustainable

rural development agency *noun* an official body set up to develop policies and oversee rural development

rural development policy *noun* a set of aims and guidelines issued by an authority, used when planning rural development

rural economy *noun* farming and other businesses in rural areas

Rural Enterprise Scheme *noun* a system of government support for the adaptation and development of the rural economy, community, heritage and environment. It is part of the England Rural Development Programme.

rural environment *noun* the countryside

Rural Inspectorate Wales *noun* the organisation that is responsible for administering CAP schemes in Wales. Abbr **RIW**

Rural Land Register *noun* a digitised map of all registered land parcels in the UK, kept by the Rural Payments Agency. Abbr **RLR**

Rural Payments Agency *noun* the organisation that is responsible for all CAP schemes in England and for some schemes in the rest of the UK. Abbr **RPA**

rural planning *noun* same as **country planning**

rural recreation *noun* same as **countryside recreation**

Rural Stewardship Scheme *noun* in Scotland, a scheme of payments to encourage farmers to be involved in the protection and enhancement of the environment, to support sustainable rural development and to maintain the prosperity of rural communities

rural sustainability *noun* the act of trying to make sure that rural development does not use too many natural resources or cause other damage

rural tourism *noun* holiday and leisure activities carried out in the countryside

rurban /ˈrɜːb(ə)n/ *adjective* referring to areas that combine the characteristics of agricultural activities found in rural zones with those of suburban living areas and industrialised zones

COMMENT: Rapid extension of urbanisation the world over has created areas that have both urban and rural characteristics. Land planning specialists talk about 'rurban areas' and the 'rurbanisation' process.

rush *noun* a common weed (*Juncus*) growing near water, in moors and marshes, of little nutritional value

russet /ˈrʌsɪt/ noun a type of dessert apple with a rough brown skin

russeting /ˈrʌsɪtɪŋ/ noun the formation of brown patches on the skin of an apple ○ *Cox's Orange often have some russeting on them, the amount depending on the weather conditions*

rust noun a fungal disease that gives plants a reddish powdery covering

rustle verb to steal livestock, especially cows and horses

rustler /ˈrʌs(ə)lə/ noun a person who steals livestock ○ *a cattle rustler*

rustling /ˈrʌs(ə)lɪŋ/ noun the crime of stealing cattle or horses

rut noun a period of intense sexual activity that occurs in males of various mammals such as cattle, sheep and particularly deer

rye noun a hardy cereal crop grown in temperate areas. Latin name: *Secale cereale*.

ryegrass /ˈraɪɡrɑːs/ noun a term for a most important group of grasses

COMMENT: Many varieties of hybrid ryegrass are now used. They are crosses between perennial and Italian ryegrasses, and often also tetraploids.

Ryeland /ˈraɪlənd/ noun a rare breed of sheep. It is a medium-sized animal, white faced and without horns. The sheep has a very symmetrical shape and a thick growth of wool.

S

S *symbol* sulphur

Saanen /ˈsɑːnən/ ♦ **British Saanen**

sack *noun* a measure of capacity, particularly for cereals, equal to four bushels. Also called **coomb**

saddle *noun* **1.** a leather seat for a rider placed on the back of a horse **2.** a coloured patch on the back of an animal such as a pig which looks a little as if the animal is carrying a saddle **3.** the part of a shafthorse's harness that bears the shafts

saddleback /ˈsæd(ə)lbæk/ *noun* **1.** a breed of pig now known as the British Saddleback **2.** any pig with a white saddle, such as the American-bred Hampshire breed

saddle bow *noun* a high part of a saddle in front of the rider

safety cab *noun* a protective cab fitted to a tractor to prevent injury to the driver if the tractor turns over

safflower /ˈsæflaʊə/ *noun* an oilseed crop *(Carthamus tinctorius)* grown mainly in India. The oil is used in the manufacture of margarine, and the residual oilseed cake has a limited use as a livestock feed.

saffron *noun* a spice obtained from the dried flowers of the crocus plant *Crocus sativus*

sage *noun* an aromatic herb *(Salvia officinalis)*, the leaves of which are dried and used for flavouring

sainfoin /ˈsænfɔɪn/ *noun* a forage legume very similar to lucerne, grown mainly in areas with calcareous soil

Saler /ˈseɪlə/ *noun* a hardy breed of French cattle, found in the Cantal department of central France. The animals are reddish in colour and are reared both for meat and for milk production. The Saler is one of the best French suckler cows.

salination /ˌsælɪˈneɪʃ(ə)n/ *noun* a process by which the salt concentration of soil or water increases, especially as a result of irrigation in hot climates. Also called **salinisation**

saline *adjective* referring to salt

salinisation /ˌsælɪnaɪˈzeɪʃ(ə)n/, **salinization** *noun* same as **salination**

salinised /ˈsælɪnaɪzd/, **salinized** *adjective* referring to soil where evaporation leaves salts as a crust on the dry surface

salinity /səˈlɪnɪti/ *noun* the concentration of salt in an amount of water or soil

saliva *noun* a clear fluid secreted by the salivary glands into the mouth, and containing water, mucus and enzymes to lubricate food and break down starch into sugars

COMMENT: Saliva is a mixture of a large quantity of water and a small amount of mucus, secreted by the salivary glands. Saliva acts to keep the mouth and throat moist, allowing food to be swallowed easily. It also contains the enzyme ptyalin, which begins the digestive process of converting starch into sugar while food is still in the mouth.

salivary /səˈlaɪv(ə)ri/ *adjective* referring to saliva

salivary digestion *noun* the first part of the digestive process, which is activated by the saliva in an animal's mouth

salivary gland *noun* a gland which secretes saliva

salivate *verb* to produce saliva

salivation /ˌsælɪˈveɪʃ(ə)n/ *noun* the production of saliva

Salmonella /ˌsælməˈnelə/ *noun* a genus of bacteria found in the intestines, which are acquired by eating contaminated food (NOTE: Different species cause food poisoning and typhoid fever.)

salmonellosis /ˌsælmənelˈləʊsɪs/ *noun* a disease caused by Salmonella bacteria

salsify /ˈsælsəfi/ *noun* a plant with a long white root which is used as a vegetable

salt *noun* sodium chloride as part of the diet ○ *a salt-restricted diet* ○ *He should reduce his intake of salt.* ■ *verb* to preserve food by keeping it in salt or in salt water

COMMENT: Salt forms a necessary part of diet, as it replaces salt lost in sweating and helps to control the water balance in the body. It also improves the working of the muscles and nerves. An adequate intake of sodium chloride is necessary to all animals. Grass is relatively poor in sodium and its high potassium content induces secretion of sodium in urine. This loss causes a craving for salt, which may be supplied by salt licks.

salt poisoning *noun* a disease of pigs usually caused by inadequate provision of water, but which may also be caused by increased salt in the ration. Pigs become constipated before twitching, fits and death.

salty *adjective* **1.** containing salt ○ *Excess minerals in fertilisers combined with naturally saline ground to make the land so salty that it can no longer produce crops.* **2.** tasting of salt

sand *noun* fine grains of weathered rock, usually round grains of quartz, found especially on beaches and in the desert

sandstone *noun* a sedimentary rock formed of round particles of quartz

sandy soil *noun* soil containing a high proportion, approximately 50%, of sand particles (NOTE: Sandy soil feels gritty. These soils drain easily and are naturally low in plant nutrients through leaching. They are often called 'light' soils, as they are easy to work and also 'hungry' soils since they need fertiliser. Market gardening is particularly well-suited to sandy soils.)

SAOS *abbreviation* Scottish Agricultural Organisation Society

sap *noun* a liquid carrying nutrients which flows inside a plant

sapling *noun* a young tree

sappy /ˈsæprəʊ/ *adjective* referring to tree trunks or branches, or wood, that are full of sap

sapro- *prefix* decay or rotting

saprophagous /səˈprɒfəgəs/ *adjective* referring to organisms that feed on decaying organic matter

saprophyte /ˈsæprəfaɪt/ *noun* an organism that lives and feeds on dead or decaying organic matter, e.g. a fungus

saprophytic /ˌsæprəˈfɪtɪk/ *adjective* referring to organisms that live and feed on dead or decaying organic matter

sapwood /ˈsæpwʊd/ *noun* an outer layer of wood on the trunk of a tree, which is younger than the heartwood inside and carries the sap. ◊ **heartwood**

Sardinian /sɑːˈdɪniən/ *noun* an important breed of Italian sheep which provides milk for the manufacture of cheese

SAS *abbreviation* set-aside scheme

satellite *noun* a man-made device that orbits the Earth, receiving, processing and transmitting signals and generating images such as weather pictures

sativus /sæˈtiːvəs/, **sativa, sativum** *adjective* a Latin word meaning 'sown' or 'planted', used in the generic names of many plants

saturate *verb* to fill something with the maximum amount of a liquid that can be absorbed ○ *Nitrates leached from forest soils, showing that the soils are saturated with nitrogen.*

COMMENT: When continuous rainfall occurs for long periods, the soil becomes saturated with water; no more water can enter the soil and it ponds on the surface or runs off down slopes.

saturation /ˌsætʃəˈreɪʃ(ə)n/ *noun* the point at which air contains 100% humidity ○ *The various types of fog are classified by the manner in which saturation is reached.*

savoy /səˈvɔɪ/ *noun* a type of winter cabbage with crinkly leaves

sawdust *noun* powder produced when sawing wood. Sawdust is used both as a mulch for plants and as bedding for animals.

sawfly /ˈsɔːflaɪ/ *noun* a family of insects, the larvae or caterpillars of which cause serious damage to fruit and crops. They include the apple sawfly, gooseberry sawfly, pear and cherry sawfly and rose sawfly.

saw-toothed beetle *noun* a dark brown beetle which lives in stored grain. The eggs are laid in the grain and the larvae feed on it, causing mould.

scab *noun* **1.** a disease of which the scab is a symptom. It affects the skin of animals. **2.** a fungal disease of fruit and vegetables, including potato scab and apple and pear scab

scabies /'skeɪbiːz/ *noun* a very irritating infection of the skin caused by a mite which lives under the skin

scald *noun* **1.** a defect in stored apples, where brown patches appear on the skin and the tissue underneath becomes soft **2.** a bacterial disease of sheep. It causes lameness in lambs.

scale *noun* a thin membranous leaf structure

scaly leg /'skeɪli leg/ *noun* a disease affecting the legs of poultry, caused by a mite which burrows under the leg scales causing considerable itching. Large hard scales develop on the unfeathered parts of the legs.

scar *noun* a mark left on the skin surface after a wound has healed

scarecrow /'skeəkrəʊ/ *noun* a figure shaped like a man, wearing old clothes, put in fields to keep birds way from growing crops

scarifier /'skærɪfaɪə/ *noun* a machine with tines which stirs the soil surface without turning it over

scarify /'skærɪfaɪ/ *verb* **1.** to stir the surface of the soil with an implement with tines, e.g. a wire rake, but without turning the soil over. Lawns can be scarified to remove moss and matted grass. **2.** to slit the outer coat of seed in order to speed up germination

SCF *abbreviation* **1.** Scottish Crofting Foundation **2.** separate concentrate feeding

Schistosoma /ˌʃɪstə'səʊmə/ *noun* a fluke which enters the patient's bloodstream and causes schistosomiasis

schistosomiasis /ˌʃɪstəsəʊ'maɪəsɪs/ *noun* a tropical disease caused by flukes taken in from water affecting the intestine or bladder. Also called **bilharziasis**

Schleswig-Holstein system /ˌʃlesvɪg 'hɒlstaɪn ˌsɪstəm/ *noun* a system of cereal cultivation practised in North Germany, giving high average yields

COMMENT: The system involves careful management of the crop and includes high seed rates and high amounts of fertiliser. Crops are carefully monitored and visited each day. Disease is controlled by spraying.

Schwarzkopf /'ʃvɑːtskɒpf/ *noun* a breed of German sheep found mainly in Hesse and Westphalia

sciarid fly /'saɪərɪd flaɪ/ *noun* a pest (*Bradysia*) affecting greenhouse pot plants. The larvae feed on fine roots causing plants to wilt.

scion /'saɪən/ *noun* a piece of a plant which is grafted onto a rootstock

sclerotic *adjective* referring to the hardening and thickening process in plant cell walls that makes stems woody

sclerotinia /ˌsklerə'tɪniə/ *noun* a soilborne disease affecting many crops, including potatoes, oilseed rape and peas

Scots pine *noun* a common commercially grown European conifer. Latin name: *Pinus sylvestris*.

Scottish Agricultural Organisation Society *noun* a consultancy agency which promotes and advises on joint ventures between Scottish farmers. Abbr **SAOS**

Scottish Blackface *noun* a very hardy breed of small mountain sheep. The fleece gives a long coarse springy wool, valued for making carpets. Older ewes are crossed with Border Leicester rams to give Greyface hybrids.

Scottish Crop Research Institute *noun* a company which researches fertilisers and plant growth regulators, pests, pesticides and the genetic modification of crops. Abbr **SCRI**

Scottish Enterprise *noun* the main economic development agency for Scotland, dealing with education, communications and the expansion of businesses. Abbr **SE**

Scottish Environment Protection Agency *noun* a public body with responsibility for the protection of Scotland's natural landscape and resources. Abbr **SEPA**

Scottish Executive Environment and Rural Affairs Department *noun* the department of the devolved Scottish Executive government which deals with farming, the environment, animal welfare and rural development in Scotland. Abbr **SEERAD**

Scottish halfbreed *noun* a crossbred type of sheep obtained by using a Border Leicester ram on a Cheviot ewe. They are used widely in lowland Britain.

Scottish Natural Heritage *noun* an official body responsible for the conservation of fauna and flora in Scotland

Scottish Rural Property and Business Association *noun* a group which represents the interests of Scotland's rural businesses. Abbr **SRPBA**

scour *verb* to wash wool to remove grease and contaminants

scouring *noun* diarrhoea in livestock. It may be a symptom of other diseases such as Johne's disease, dysentery or coccidiosis, or it may simply be due to a chill or to poor diet.

SCP *abbreviation* sustainable consumption and production

SCPS *abbreviation* Suckler Cow Premium Scheme

scraper /'skreɪpə/ *noun* a steel-framed attachment for a tractor. It has a rubber scraping edge, and is used for heavy duty work, clearing slurry from farmyards.

scrapie /'skreɪpi/ *noun* a brain disease of sheep and goats. Affected animals twitch, then suffer intense itching and thirst. They become extremely thin, and death follows. It is a notifiable disease.

screen *noun* a hedge or row of trees grown to shelter other plants, to protect something from the wind or to prevent something from being seen ■ *verb* **1.** to pass grain through a sieve to grade it **2.** to protect plants from wind, e.g. by planting windbreaks

screenings /'skriːnɪŋz/ *plural noun* grains which are small and pass through the sieve when grain is screened

screwworm /'skruːwɜːm/ *noun* a fly similar to the bluebottle, but dark green in colour, common in Central and South America. It devastated cattle in the USA in the 1950s, but has now been eradicated there.

SCRI *abbreviation* Scottish Crop Research Institute

scrub *noun* **1.** small trees and bushes **2.** an area of land covered with small trees and bushes

scrubland /'skrʌblænd/ *noun* land covered with small trees and bushes

scurs /skɜːz/ *plural noun* small horns that are not part of the animal's skull but are attached to the skin

scutch /skʌtʃ/ *noun* same as **couch grass**

scythe *noun* a hand implement with a long slightly curved blade attached to a handle with two short projecting hand grips. Scythes are now used for cutting grass and were formerly used for reaping.

SDA *abbreviation* Severely Disadvantaged Area

Se *symbol* selenium

SE *abbreviation* Scottish Enterprise

seakale /'siːkeɪl/ *noun* a plant of the cabbage family whose leaves are used as vegetable

season *noun* **1.** one of the four parts into which a year is divided, i.e. spring, summer, autumn and winter **2.** the time of year when something happens ○ *the mating season* **3.** the oestrus period of a female animal □ **in season** ready for mating

seasonal *adjective* referring to or occurring at a season ○ *seasonal changes in temperature* ○ *Plants grow according to a seasonal pattern.*

seaweed *noun* any of the large algae that grow in the sea and are usually attached to a surface

Secale /sɪ'keɪli/ *noun* rye

secateurs /ˌsekə'tɜːz/ *noun* a cutting tool, like small shears with sharp curved blades, used for pruning

second *adjective* coming after the first

secondary *adjective* **1.** less important than something else ○ *a secondary reason* **2.** coming after something else ○ *secondary thickening of stems in plants* ◊ **primary**, **tertiary**

secondary substances *plural noun* chemical substances found in plant leaves, believed to be a form of defence against herbivores

second cut *noun* grass which has been cut a second time in the season for hay or silage

second early potatoes *plural noun* the crop of potatoes that follows the first early crop

seconds *plural noun* grain of medium size

secrete *verb* (*of a gland*) to produce a substance such as a hormone, oil or enzyme

secretion *noun* a liquid which is secreted

sedentary agriculture *noun* subsistence agriculture practised in the same place by a settled farmer

sedge /sedʒ/ *noun* one of a number of grass or rushlike herbs of the family *Cyperaceae*, common in marshlands and poorly drained areas. They have minimal nutritional value.

sedimentary rocks *plural noun* rocks which were formed by deposition of loose material such as sand and gravel, mainly in water

seed *noun* a fertilised ovule that forms a new plant on germination ■ *verb* **1.** (*of a plant*) to produce offspring by dropping seed which germinates and grows into plants in following seasons ○ *The poppies seeded all over the garden.* ○ *The tree was left standing to allow it to seed the cleared area around it.* **2.** to sow seeds in an area ○ *The area of woodland was cut and then seeded with pines.* □ **to seed itself** to sow seeds naturally and grow the following year **3.** to drop crystals of salt, carbon dioxide and other substances onto clouds from an aeroplane in order to encourage rain to fall

COMMENT: EU regulations require all seed sold to farmers to be tested and to be guaranteed to meet certain standards of purity and to be free from pests and diseases.

seedbed /'siːdbed/ *noun* an area of land tilled to produce a fine tilth, firm and level, into which seeds will be sown. Some crops such as potatoes do not need a fine tilth and a rough damp bed is preferable.

'While delaying drilling into April will lose some yield potential, seedbeds should not be forced or losses will be even greater. Patience is key where soils need to dry out.' [*Farmers Guardian*]

seedbed wheels *plural noun* a set of wheels bolted onto the front of a tractor which will give even compaction and a uniform sowing depth

seed-borne *adjective* carried by seeds

seed-borne disease *noun* a disease which is carried in the seed of a plant

seedbox /'siːdbɒks/ *noun* **1.** a box in which seeds can be planted for cultivation in a greenhouse **2.** the part of the plant head which contains the seeds

seedcase /'siːdkeɪs/ *noun* a hard outside cover that protects the seeds of some plants

seed certification *noun* the testing, sealing and labelling of seed sold to farmers. This ensures that the seed is free from disease and from weeds.

seed coat *noun* same as **testa**

'Subsequently the grain is passed through a crimping machine which cracks the seed coat and applies an additive. The grain can then be ensiled in layers and compacted in a silo as for grass silage.' [*Farmers Guardian*]

seed dormancy *noun* a period when a seed is not active

seed dressing *noun* the treatment of seeds with a fungicide and/or an insecticide to prevent certain soil and seed-borne diseases

seed drill *noun* a machine consisting of a hopper carried on wheels with a feed mechanism which delivers grain to seed tubes

COMMENT: Most seed drills are designed to plant seed in prepared seed beds. Drills for cereals and grasses sow the seed at random, while precision drills, used mainly for sugar beet and vegetable crops, place seed at preset intervals in the rows. Precision drills are also called 'seeder units'.

seeder /'siːdə/ *noun* a machine for sowing seeds

seeder unit *noun* a seed drill which sows the seed separately at set intervals in the soil

seedhead /'siːdhed/ *noun* the top of a stalk with seeds, either in a seedcase or separately attached to the stem

seeding year *noun* the calendar year in which the seed is sown

seedless /'siːdləs/ *adjective* with no seeds

seedless hay *noun* hay obtained from a grass crop after threshing out the seedheads

seedling *noun* a young plant that has recently grown from a seed

seed mixture *noun* seeds of different plants supplied by seed merchants to farmers to produce a new ley. It will include grasses and legumes.

seed plot *noun* same as **nursery plot**

seed potato *noun* a potato tuber which is sown to produce new plants. In the UK, these are grown mainly in Scotland, and produced under a certification scheme (the Seed Potato Classification Scheme).

seed rate *noun* the amount of seed sown per hectare shown as kilos per hectare (kg/ha)

seed ripeness *noun* the stage at which the seed can be harvested successfully

seed royalties *plural noun* money paid by seed growers to breeders of seeds

seed tree *noun* a tree left standing when others are cut down, to allow it to drop seeds on the cleared land around it

seed trials *plural noun* tests of new seeds to see if they germinate correctly

seed weevil *noun* a pest affecting brassica seed crops. Seeds are destroyed in their pods by the larvae.

seep *verb* (*of a liquid*) to flow slowly through a substance ○ *Water seeped through the rock.* ○ *Chemicals seeped out of the container.*

seepage /'siːpɪdʒ/ *noun* slow oozing out of ground water from the soil surface

SEERAD /'siːræd/ *abbreviation* Scottish Executive Environment and Rural Affairs Department

select *verb* to identify plants or animals with desirable characteristics such as high yield or disease resistance as part of the activity of breeding new varieties

selection *noun* 1. the process of identifying plants or animals with desirable characteristics such as high yield or disease resistance as part of the activity of breeding new varieties 2. an individual chosen from a group in a breeding programme on the basis of distinctive characteristics

selective herbicide, selective weedkiller *noun* a weedkiller which is designed to kill only plants with specific characteristics and not others

selective pesticide *noun* a pesticide which takes toxic action against specific pests without affecting the growing crop

selenium /sə'liːniəm/ *noun* a trace element, an essential part of the diet for all animals. White muscle disease is the symptom of selenium deficiency.

'Northumberland-based suckler producer Charles Armstrong agrees that breeding herd health status is vital for maximising calf returns. His 400-cow suckler herd are blood sampled every year for copper and selenium status, as well as for appropriate vaccination regimes.' [*Farmers Weekly*]

self-blanching celery *noun* a variety of celery where the stalks are naturally white, and do not need to be earthed up

self-contained herd *noun* a dairy herd which breeds its own replacements, the calves being kept and reared

self-feed *verb* to take a controlled amount of feed from a large container as required

self-feed silage *noun* a feeding system where stock feed from silage, the amount of silage available being centrally controlled

self-fertile *adjective* referring to a plant that fertilises itself with pollen from its own flowers

self-fertilisation /ˌself ˌfɜːtəlaɪ'zeɪʃ(ə)n/, **self-fertilization** *noun* the fertilisation of a plant or invertebrate animal with its own pollen or sperm

selfing /'selfɪŋ/ *noun* same as **self-fertilisation**

self-pollination *noun* the pollination of a plant by pollen from its own flowers. Compare **cross-pollination**

self-purification *noun* the ability of water to clean itself of polluting substances

self-raising flour *noun* a type of flour with baking powder added to it

self-regulating *adjective* (*of an ecosystem*) controlling itself without outside intervention ○ *Most tropical rainforests are self-regulating environments.*

self-seeded *adjective* referring to a plant that grows from seed that has fallen to the ground naturally rather than being sown intentionally ○ *Several self-seeded poppies have come up in the vegetable garden.*

self-sterile *adjective* referring to a plant that cannot fertilise itself from its own flowers

self-sufficiency *noun* a simple traditional way of farming with little use of modern technology that provides only enough food and other necessary materials for a family

self-sufficient *adjective* 1. able to provide enough food and other necessary materials for a family, often by means of a simple traditional way of farming with little use of modern technology ○ *We're self-sufficient in salad crops from the garden in the summer time.* 2. able to provide the required quantity of a product locally or for yourself, without needing to purchase or import it ○ *The country is self-sufficient in barley.*

sell-by date *noun* a date on the label of a food product which is the last date on which the product should be sold and can be guaranteed as of good quality

semen *noun* in mammals, a thick pale fluid containing spermatozoa, produced by the testes and ejaculated from the penis

semi- *prefix* **1.** half **2.** partly

semi-digger *noun* a type of mouldboard on a plough

semi-mounted *adjective* referring to an implement which is supported by a tractor but also has its own wheels

seminal /ˈsemɪn(ə)l/ *adjective* referring to semen or to seed

seminal propagation *noun* the process of growing new plants from seed or from tubers such as potatoes

seminal roots *plural noun* the secondary roots of a plant which support the primary root. This root system is then replaced by adventitious roots.

semiochemical /ˌsemiəʊˈkemɪk(ə)l/ *noun* a chemical released by animals, especially insects, as a means of communication, e.g. a pheromone

semolina /ˌseməˈliːnə/ *noun* a coarse flour made from wheat after the fine flour has been ground. It is used to make puddings.

senescence /sɪˈnesəns/ *noun* the process of growing older. ◊ **plant senescence**

sentient /ˈsentiənt/ *adjective* capable of feeling and perception (NOTE: Since 1997 EU law has recognised that animals are sentient, and this concept lies behind the animal welfare codes that set out guidelines for the treatment of farm animals in the United Kingdom.)

SEPA *abbreviation* Scottish Environment Protection Agency

sepal /ˈsep(ə)l/ *noun* a part of the calyx of a flower, usually green and sometimes hairy

separate concentrate feeding *noun* a winter feeding system for livestock in which the animals are allowed free feeding of roughage and concentrates are fed separately in restricted quantities. Abbr **SCF**

separated milk *noun* milk from which the cream has been removed. Also called **skimmed milk**

septic tank *noun* an underground tank for household sewage that is not connected to the main drainage system and in which human waste is decomposed by the action of anaerobic bacteria

septoria /sepˈtɔːriə/ *noun* a fungal disease which affects the leaves of wheat crops

sericulture /ˈserɪkʌltʃə/ *noun* raising silkworms for the production of silk

serve *verb* of a male animal, to mate with a female

service *verb* of a male animal, to mate with a female

sessile /ˈsesaɪl/ *adjective* attached directly to a branch or stem without a stalk ○ *The acorns of a sessile oak tree have no stalks or very short stalks.*

set *noun* **1.** a seed potato **2.** a seed onion **3.** a badger's burrow ■ *verb* **1.** to harden ○ *The resin sets in a couple of hours.* **2.** to form fruit or seed **3.** to plant something

set aside *verb* to use a piece of formerly arable land for something other than growing food crops

set-aside *noun* a piece of formerly arable land used for something other than growing food crops

set on *verb* to foster an orphaned animal to another female, as a lamb onto a ewe

set stocking *noun* a grazing system associated with extensive grazing. Livestock graze an area where they remain for an indefinite period. This is the traditional practice in Britain.

set to *noun* an orphan lamb given to a foster mother

Severely Disadvantaged Area *noun* land which is extremely poor and difficult to farm. Abbr **SDA**. ◊ **disadvantaged area**

sewage *noun* waste water and other material such as faeces, carried away in sewers. Also called **sewage waste**

sewage sludge *noun* the solid or semi-solid part of sewage

sex *noun* one of the two groups, male and female, into which animals and plants can be divided

COMMENT: In mammals, females have a pair of identical XX chromosomes and males or have one X and one Y chromosome. Out of the twenty-three pairs of chromosomes in each human cell, only two are sex chromosomes. The sex of a baby is determined by the father's sperm. While the mother's ovum only carries X chromosomes, the father's sperm can carry either an X or a Y chromosome. If the ovum is fertilised by a sperm carrying an X chromosome, the embryo will contain the XX pair and so be female.

sex linkage *noun* an existence of characteristics which are transmitted through the X chromosomes

sex-linked *adjective* referring to a genetically inherited characteristic that appears in only one sex

sex organs *plural noun* organs which are associated with reproduction and sexual intercourse

SFA *abbreviation* Small Farms Association

SFPA *abbreviation* Scottish Fisheries Protection Agency

SFPS *abbreviation* Single Farm Payment Scheme

SGM *abbreviation* Standard Gross Margin

shade *noun* a place sheltered from direct sunlight

shade plants *plural noun* plants which prefer to grow in the shade

shading /'ʃeɪdɪŋ/ *noun* the action of cutting off the light of the sun ○ *Parts of the field near tall trees suffer from shading.*

'In Scotland, shading of weed growth by late-drilled wheats was more important in the spring and early summer than in southern Europe where growers placed more importance on preventing weed growth in the winter and early spring.' [*Farmers Guardian*]

shaggy *adjective* referring to an animal such as Highland cattle which has long hair

shank /ʃæŋk/ *noun* the lower part of a horse's leg between the knee and the foot

share *noun* same as **ploughshare**

sharecropper /'ʃeəkrɒpə/ *noun* a tenant farmer who pays a part of his crop to the landlord as a form of rent

sharecropping /'ʃeəkrɒpɪŋ/ *noun* a system of land tenure, whereby tenants pay an agreed share of the crop to the landlord as a form of rent

sharefarming /'ʃeəfɑːmɪŋ/ *noun* a joint enterprise between a party with an interest in the land and another party involved in farming operations. Usually one party provides the capital and the other the farm management inputs such as labour and equipment.

sharp eyespot *noun* a soil-borne fungus (*Rhizoctonia solani*) affecting cereals, which can cause lodging and shrivelled grain

shavings *plural noun* thin curled pieces of wood removed when planing, used as litter for animals

sheaf *noun* a bundle of corn stalks tied together after reaping (NOTE: The plural is **sheaves.**)

shear *verb* to clip the fleece from a sheep

COMMENT: Shearing nowadays is done with electric shears. Using these, a skilled shearer can fleece a ewe in under two minutes. Shearers often work in gangs. The fleece is clipped off in one piece. That of a longwool breed may weigh over 10kg.

shearer /'ʃɪərə/ *noun* a person who clips the fleece off a sheep

shearling /'ʃɪəlɪŋ/ *noun* a young sheep which has been sheared for the first time

shears *plural noun* large clippers used to shear a sheep

shed *verb* **1.** to separate one or more animals from a flock or herd **2.** to let leaves or grain fall

sheen *noun* the bright shiny appearance of a surface, used of fruit, animals' coats, meat, etc.

sheep *noun* a ruminant of the genus Ovis, family Bovidae. It is one of many domesticated varieties, farmed for their wool, meat and milk.

COMMENT: Most sheep in the UK are kept for meat, and milk production is relatively unimportant. Wool is an important by-product of sheep farming in the UK, but is the main product of sheep in some other countries, such as Australia. In 2005 the total sheep population of the UK was approximately 40 million, accounting for 30% of all sheep in the EU member states. Sheep are kept under a wide range of environmental and management conditions, from coastal lowland areas such as Romney Marsh to the upland areas of Wales, Scotland and the North of England. Lambs from the upland areas are moved to lowland farms for fattening. In the UK, a great many breeds of sheep have survived and there are some 50 recognised breeds as well as a variety of local types and many crossbreds. More recently, the introduction of continental breeds has increased the variety. A broad classification into three main categories may be made: the **long-woolled breeds** which include the Romney, Lincoln and Leicester; the **short-woolled breeds** including the Southdown, Dorset Down and Suffolk, and the **mountain, moorland and hill breeds** which include the Cheviot, Radnor, Scottish Blackface, Swaledale and Welsh Mountain.

Sheep Annual Premium Scheme *noun* until 2005, a subsidy for breeding ewes (NOTE: Now superseded by the **Single Payment Scheme**.)

sheep-dip *noun* a chemical preparation used in a dipping bath to disinfect sheep to control diseases such as sheep scab

COMMENT: All sheep in Britain are dipped for scale once a year, following the ministerial decision to have a single national dip. Dipping ceased to be compulsory in the UK in 1992. It is illegal to buy organophosphorous sheep-dip without a certificate of competence.

sheepdog *noun* a breed of dog trained and used by shepherds in controlling sheep

sheep ked /ˈʃiːp ked/ *noun* same as **sheep tick**

sheep maggot fly *noun* a type of fly that lays its eggs on the wool of sheep. The eggs hatch into maggots that burrow into the flesh causing a condition known as 'strike'.

sheepman /ˈʃiːpmən/ *noun* a shepherd, a farm worker who looks after sheep

sheepmeat /ˈʃiːpmiːt/ *noun* a term used in the EU for meat from a sheep or lamb

sheep pox *noun* a highly contagious viral disease. Symptoms include fever, loss of appetite, difficulty in breathing and in the final stages scabs and ulcers appear. It is a notifiable disease.

sheep run *noun* an extensive area used for sheep grazing, especially in New Zealand and Australia

sheep scab *noun* a serious disease of sheep, caused by a parasitic mite, which results in intense irritation, skin ulcers, loss of wool and emaciation. It is a notifiable disease.

sheep's fescue *noun* a species of grass useful under hill and marginal conditions

sheep's sorrel *noun* a common weed (*Rumex acetosella*)

sheep tick *noun* a small wingless dipterous insect, parasitic on sheep. Also called **sheep ked**

sheep walk *noun* an area of land on which sheep are pastured

sheet erosion *noun* erosion that takes place evenly over the whole area of a slope, caused by the runoff from saturated soil after heavy rainfall

shelf-life *noun* the number of days or weeks for which a product can stay on the shelf of a shop and still be good to use

shelter *noun* a structure or feature providing protection from wind, sun, rain or other weather conditions ■ *verb* to protect something from weather conditions

shelter belt *noun* a row of trees planted to give protection from wind

'Over the years the family has carried out extensive improvements, putting in shelter belts, new buildings, land drainage and farm road layouts.' [*Farmers Guardian*]

shelterwood /ˈʃeltəwʊd/ *noun* a large area of trees left standing when others are cut, to act as shelter for seedling trees

shepherd *noun* a person who looks after sheep

shepherd's purse *noun* a common weed (*Capsella bursa-pastoris*) in gardens and market gardens, found particularly among vegetables and root crops. Also called **pepper and salt**

Shetland /ˈʃetlənd/ *noun* 1. a rare breed of cattle, native to the Shetland Isles. It is medium-sized, black and white, with short legs, short horns and a bulky body. 2. a breed of sheep, native to the Shetland Isles. The colour varies from white, through grey and black to light brown; the ewes are polled and the rams horned; it produces fine soft wool of high quality, used in the Shetland wool industry. A small Shetland ewe yields a fleece 1.5–2 kilos in weight. 3. a breed of pony, used as a riding horse for children

shifting cultivation *noun* 1. an agricultural practice using the rotation of fields rather than of crops. Short cropping periods are followed by long fallows and fertility is maintained by the regeneration of vegetation. ◊ **fallow** 2. a form of cultivation practised in some tropical countries, where land is cultivated until it is exhausted and then left as the farmers move on to another area

COMMENT: In shifting cultivation, the practice of clearing vegetation by burning is widespread. One of the simplest forms involves burning off thick and dry secondary vegetation. Immediately after burning, a crop like maize is planted and matures before the secondary vegetation has recovered. Where fire clearance methods are used, the ash acts as a fertiliser.

shigella /ʃɪˈgelə/ *noun* a bacillus which causes dysentery

shin *noun* 1. the lower part of the foreleg of cattle 2. the upper part of a ploughshare

shire horse /'ʃaɪə hɔːs/ *noun* a tall heavy breed of draught horse. The coat may be of various colours, but there is always a mass of feather at the feet.

shivering /'ʃɪvərɪŋ/ *noun* an affliction of the nervous system with involuntary muscular contractions, usually of the hind legs. It is a progressive condition found in horses.

shock *noun* same as **stook**

shoddy *noun* a waste product of the wool industry. It contains up to 15% nitrogen and is used as a fertiliser, particularly in market gardens.

shoe *verb* to make, fit and fix horseshoes to the feet of a horse

shoot *noun* 1. a new growth from the stem of a plant 2. part of a young seed plant, the stem and first leaves which show above the surface of the soil ■ *verb* to kill something with a gun

short-day plant *noun* a plant that flowers as the days get shorter in the autumn, e.g. a chrysanthemum. Compare **long-day plant**

short duration ley *noun* a ley which is kept only for a short time

short duration ryegrass *noun* a class of grasses which are important to the farmer, including Westerwolds, Italian and Hybrid. These grasses are quick to establish and give early grazing. They are used where persistency is not important.

short-grain rice *noun* varieties of rice with short grains, grown in cooler climates such as Japan. See Comment at **rice**

shorthorn /'ʃɔːthɔːn/ *noun* a breed of cattle, with short horns

COMMENT: In the 18th century, Charles Colling used many of the breeding principles established by Robert Bakewell to develop the shorthorn breed, which became the most common in Britain and remained so for over a hundred years. It has later developed into three different strains: the Beef Shorthorn, the Dairy Shorthorn and the Lincoln Shorthorn.

short rotation coppice *noun* varieties of willow or poplar which yield a large amount of fuel and are grown as an energy crop

'Though there are plenty of bio-feedstocks around, for example short rotation coppice, few can be cost- and carbon-effectively turned into transport fuel.' [*Arable Farming*]

short ton *noun US* same as **ton**

shred *verb* to tear something into tiny pieces ○ *Farmyard manure is shredded before being spread on fields.*

shredder *noun* a machine for shredding waste vegetable matter before composting

shrivel *verb* to become dry and wrinkled ○ *The leaves shrivelled in the prolonged drought.*

Shropshire /'ʃrɒpʃə/ *noun* a medium-sized breed of sheep with a black face and heavy fleece, now rare

shrub *noun* a perennial plant with several woody stems

shrubby /'ʃrʌbi/ *adjective* growing like a shrub

sickle /'sɪk(ə)l/ *noun* a curved knife-edged metal tool with a wooden handle, used for harvesting cereals

sideland /'saɪdlænd/ *noun* a strip of land left at the side of a field during ploughing. It may be ploughed up with the headlands.

side rake *noun* a machine which picks up two swaths and combines them into one before baling

sidewalk farmer *noun US* a farmer who cultivates land some way away from his or her house in a town

sieve *noun* a garden implement with a base made of mesh or with perforations through which fine particles can pass while coarse material is retained. Compare **riddle** ■ *verb* to pass soil, etc. through a sieve to produce a fine tilth, or to remove the soil from root crops such as potatoes

silage /'saɪlɪdʒ/ *noun* food for cattle formed of grass and other green plants, cut and stored in silos

COMMENT: Silage is made by fermenting a crop with a high moisture content under anaerobic conditions. It may be made from a variety of crops, the most common being grass and maize, although grass and clover mixtures, green cereals, kale, root tops, sugar beet pulp and potatoes can also be used. Trials indicate that very high-quality grass silage can be fed to adult pigs.

silage additive *noun* a substance containing bacteria and/or chemicals, used to speed up or improve the fermentation process in silage or to increase the amount of nutrients in it

silage effluent *noun* an acidic liquid produced by the silage process which can be a serious pollutant, especially if it drains into a watercourse

silage liquor *noun* a liquid which forms in silage and drains away from the silo

silage tower *noun* a container used for making and storing silage

silk *noun* a thread produced by the larvae of a moth to make its cocoon (NOTE: It is used to make a smooth light fabric.)

silks *plural noun* the long thin styles of the female flowers of the maize plant

silkworm /'sɪlkwɜːm/ *noun* a moth larva which produces silk thread

silo *noun* a large container for storing grain or silage

COMMENT: There are many different types of silo. Some are pits dug into the ground, others are forms of surface clamp, while built silos are towers which may be either top- or bottom-loaded and are built of wood, concrete or steel.

silopress /'saɪləʊpres/ *noun* a polythene 'sack' into which silage is forced. As the sack fills up, it gradually grows longer and when completely full is sealed. A 'sack' may contain up to 80 tonnes of silage.

silt *noun* 1. soft mud which settles at the bottom of water 2. particles of fine quartz with a diameter of 0.002 – 0.06mm

silty soil /'sɪlti sɔɪl/ *noun* soil containing a high proportion of silt. Such soils are difficult to work and drainage is a problem.

silver-laced Wyandotte *noun* a dual-purpose breed of poultry. The feathers are silvery, with black edges, especially on the tail.

silvi- /sɪlvi/ *prefix* trees or woods

silvicide /'sɪlvɪsaɪd/ *noun* a substance which kills trees

silviculture /'sɪlvɪkʌltʃə/ *noun* the cultivation of trees as part of forestry

Simmental /'sɪməntɑːl/ *noun* a breed of cattle originating in Switzerland, the colour of which is yellowish-brown or red. It is a dual-purpose breed, with a high growth rate potential and good carcass quality.

Single Farm Payment Scheme *noun* ♦ Single Payment Scheme

single flower *noun* a flower with only one series of petals, as opposed to a double flower

Single Payment Scheme *noun* an initiative under the CAP which calculates farmers' subsidies with reference to the amount of land used in production, as well as the total eligible livestock or crop output. It replaces individual subsidy schemes. Abbr **SPS**

single-suckling *verb* a natural method of rearing beef cattle, where calves are permitted to suckle their own mothers

singleton /'sɪŋg(ə)ltən/ *noun* a single offspring

singling /'sɪŋg(ə)lɪŋ/ *noun* 1. the process of reducing the number of plants in a row 2. the process of reducing the number of plants from a multigerm seed to a single plant

sire /saɪə/ *noun* 1. the male parent of an animal 2. a male animal selected for breeding

sisal /'saɪs(ə)l/ *noun* a tropical plant (*Agave rigida*) which yields a hard fibre used for making binder twine and mats

site *noun* a geographically defined area whose extent is clearly marked

Site of Special Scientific Interest *noun* in England, Wales and Scotland, an area of land which is officially protected to maintain its fauna, flora or geology. Abbr **SSSI**

Sitka spruce /'sɪtkə spruːs/ *noun* a temperate softwood coniferous tree, that is fast-growing. It is used for making paper. Latin name: *Picea sitchensis*.

six-tooth sheep *noun* a sheep between two and three years old

skim coulter *noun* the part of a plough which turns a small slice off the corner of the furrow about to be turned and throws it into the bottom of the one before. It is attached to the beam behind the disc coulters.

skimmed milk, skim milk *noun* milk which has had both fat and fat-soluble vitamins removed. It is used as a milk substitute for calves and lambs.

skin *noun* the outer layer on an animal, fruit or vegetable

skin spot *noun* a potato disease causing pimple-like dark brown spots which can harm the buds in the eyes of seed tubers

slaked lime *noun* same as **hydrated lime**

slapmark /'slæpmɑːk/ *noun* the herdmark allocated by Defra, put on both shoulders of a pig. There are no specifications on the size of the slapmark, but it must be legible.

slash and burn agriculture *noun* a form of agriculture in which forest is cut down and burnt to create open space for growing crops. Also called **swidden farming** (NOTE: The space is abandoned after several crops have been grown and then more forest is cut down.)

slatted mouldboard noun a type of mouldboard which breaks up the soil as it is being ploughed

slaughter noun the killing of animals for food (NOTE: Animal welfare codes lay down rules for how animals should be slaughtered to ensure that they are not caused any avoidable and unnecessary pain or distress.) ▪ verb to kill animals for food

slaughterhouse noun same as **abattoir**

Slaughter Premium Scheme noun until 2005, a subsidy that provided direct support to all producers of domestic cattle (NOTE: Now superseded by the **Single Payment Scheme.**)

slender foxtail noun same as **blackgrass**

sling noun a type of harness that is used to support the weight of an animal that is suffering from some kind of disability

slink calf noun a calf born early, before the normal period of gestation is complete

slip noun a small piece of plant stem, used to root as a cutting, or in budding ▪ verb (of an animal) to miscarry

sloe /sləʊ/ noun the wild plum *Prunus spinosa*. Also called **blackthorn**

sludge noun 1. a thick wet substance, especially wet mud or snow 2. the solid or semi-solid part of sewage

sludge composting noun the decomposition of sewage for use as a fertiliser or mulch

slug noun an invertebrate animal without a shell. It causes damage to plants by eating leaves or underground parts, especially in wet conditions.

slug pellet noun a small hard piece of a mixture containing a substance such as metaldehyde which kills slugs. Slug pellets are usually coloured blue-green.

sluice noun a channel for water, especially through a dam or other barrier

slurry /ˈslʌri/ noun liquid or semi-liquid waste from animals, stored in tanks or lagoons and treated to be used as fertiliser

'All the slurry from the cows and pigs is spread on the grass and maize ground. Regular soil testing is used to check the nutrient content, but there is seldom any need for artificial fertiliser.' [*Farmers Weekly*]

COMMENT: Slurry can be spread on the land using pumps, a pipeline system and slurry guns. More often, slurry spreaders are used, which can load, transport and spread the material. New regulations to control the pollution from slurry were introduced in 1989. These require the base and walls of silage clamps to be impervious. An artificial embankment must be constructed around slurry tanks. Silage clamps in the middle of fields with covers held down with rubber tyres will be banned.

slurry gun noun a powerful spraying device that spreads slurry. Compare **rain gun**

slurry injector noun a tractor-hauled machine which injects slurry into the soil

slurry spreader noun a machine which spreads slurry

small and medium-sized enterprises plural noun organisations that have between 10 and 500 employees and are usually in the start-up or growth stage of development

Small Farms Association noun an organisation which represents the interests of small farmers on a national scale. Abbr **SFA**

smallholder /ˈsmɔːlhəʊldə/ noun a person who farms a smallholding

smallholding /ˈsmɔːlhəʊldɪŋ/ noun a small agricultural unit under 20 hectares in area

small nettle noun a weed (*Urtica urens*) which is common on rich friable soils. It affects vegetables and other row crops. Also called **annual nettle**, **burning nettle**

SMD abbreviation 1. soil moisture deficit 2. standard man day

SMEs abbreviation small and medium-sized enterprises

smoke verb to preserve food by hanging it in the smoke from a fire (NOTE: Smoking is used mainly for fish, but also for some bacon and cheese.)

smooth-stalked meadowgrass noun a species of grass which can withstand quite dry conditions. It is a perennial grass with smooth greyish-green leaves and green purplish flowers.

SMR abbreviation Statutory Management Requirement

smudging /ˈsmʌdʒɪŋ/ noun the process of burning oil to produce smoke to prevent loss of heat from the ground and so to minimise or prevent frost damage to crops and orchards

smut /smʌt/ *noun* a disease of cereal plants, caused by a fungus, that affects the development of the grain and makes it look black

snail *noun* a mollusc of the class Gastropoda. It is a pest which can be controlled by molluscicides.

snap beans *noun US* beans which are eaten in the pod, e.g. green beans or French beans, or of which the seed is eaten after drying, e.g. haricot beans. As opposed to broad beans or Lima beans, the seeds of which are eaten fresh.

SNF percentage *abbreviation* solid-not-fat percentage

snout *noun* the nose and mouth of some animals, including the pig

snow mould *noun* a fungal pre-emergent blight and root rot of cereals (*Micronectriella nivalis*)

snow rot *noun* a white mould growth affecting wheat, causing leaves to turn brown and shrivel (*Typhula incarnata*)

soakaway /'səʊkəweɪ/ *noun* a channel in the ground filled with gravel, which takes rainwater from a downpipe or liquid sewage from a septic tank and allows it to be absorbed into the surrounding soil

'It is essential to investigate the history of the site to avoid the risk of flash flooding, and to ensure that the old clay land drain system is fully operational and accessible. If it discharges into soakaways it should be diverted if possible into open ditches.' [*Farmers Weekly*]

Soay /'səʊeɪ/ *noun* a rare horned breed of sheep which sheds its fleece naturally, thought to be the link between wild and domesticated breeds. The short hairy fleece is tan or dark brown. (NOTE: The breed originally came from the island of Soay in the Outer Hebrides.)

sodium *noun* a chemical element which is a constituent of common salt and essential to animal life

sodium chloride *noun* common salt

soft fruit *noun* a general term for all fruits and berries that grow on bushes and canes, have a relatively soft flesh, and so cannot be kept, except in some cases by freezing. Typical soft fruit are raspberries, strawberries, blueberries and blackberries, and the various currants.

soft rot *noun* a bacterium, *Erwinia carotovora*, which affects stored potatoes and carrots. The cell walls dissolve causing the vegetables to become mushy, slimy and foul-smelling.

soft wheat *noun* wheat containing grains which, when milled, break down in a random manner. Soft wheats have less protein than hard wheats and have poor milling qualities.

softwood /'sɒftwʊd/ *noun* **1.** the open-grained wood produced by pine trees and other conifers **2.** a pine tree or other conifer that produces such wood. Compare **hardwood**

soil *noun* the earth in which plants grow. ◊ **chernozem**, **loess**, **podsol**, **subsoil**, **topsoil**

COMMENT: Soil is a mixture of mineral particles, decayed organic matter and water. Topsoil contains chemical substances which are leached through into the subsoil where they are retained. Without care, soils easily degrade, losing the few nutrients they possess and become increasingly acid or sour.

soilage /'sɔɪlɪdʒ/ *noun* green forage crops that are cut and carried to feed animals grazing on unproductive pastures, in order to supplement their diets. Crops commonly used for soilage are clovers and lucerne.

soil air *noun* the air content of the soil. It contains the same gases as the atmosphere, but in different amounts, because it is modified by the constituent parts of the soil. Also called **soil atmosphere**

soil association *noun* a group of soils associated with one area and which occur in a predictable pattern

Soil Association *noun* a UK organisation that certifies organically grown food

soil atmosphere *noun* same as **soil air**

soil ball *noun* the rooting system of a plant, complete with the soil attached to it, as when a plant is lifted from a pot or seedbed

soil-borne fungus *noun* a fungus whose spores are carried in the soil

soil capping *noun* a hard crust on the surface of the soil which can be caused by heavy rain drops or the passage of heavy farm machinery

soil classification *noun* in soil surveys, the classification of soils into groups with broadly similar characteristics

COMMENT: Soils are classified according to the areas of the world in which they are found, according to the types of minerals they contain or according to the stage of development they have reached. All forms of soil classification are artificial, however, as soils vary in three dimensions and in time. Therefore no clear boundaries exist between soil types.

soil compaction *noun* the process in which soil is pressed down, e.g. by heavy loads, and becomes very firm with little space between its particles

'The healthy world market price for soybeans has led to the development of continuous production of this crop and consequently some problems have developed, particularly soil compaction, weed species development, and reduction of soil fertility.' [*Arable Farming*]

soil conservation *noun* the use of a range of methods to prevent soil from being eroded or overcultivated, by irrigation, mulching, etc. Also called **conservation of soil**

soil contamination *noun* the presence of chemical or biological elements which affect the soil's natural function

soil drainage *noun* the flow of water from soil, either naturally or through pipes and drainage channels inserted into the ground

soil erosion *noun* the removal of soil by the effects of rain, wind, sea or cultivation practices

soil fertility *noun* the potential capacity of soil to support plant growth based on its content of nitrogen and other nutrients

soil horizon *noun* a layer of soil that is of a different colour or texture from other layers (NOTE: There are four soil horizons: the A horizon or topsoil containing humus; the B horizon or subsoil containing minerals leached from the topsoil and little organic matter; the C horizon or weathered rock; and the D horizon or bedrock.)

soil improvement *noun* the practice of making the soil more fertile by methods such as draining and manuring

soilless gardening /ˌsɔɪlləs ˈɡɑːd(ə)nɪŋ/ *noun* same as **hydroponics**

soil loosener *noun* a trailed implement which loosens the surface of the soil

soil management *noun* the study of soil's physical properties and how to maintain a healthy and functional soil system

soil mapping *noun* the process of making maps showing different types of soil in an area

soil moisture deficit *noun* the difference between the amount of water that is in a soil and the amount needed for crops to grow successfully. Abbr **SMD**

soil nutrition *noun* **1.** the condition of soil in terms of the plant nutrients it contains **2.** the action of putting nutrients into soil through the application of fertilisers ○ *Use muck and some seaweed for soil nutrition.*

soil organic matter *noun* decayed or decaying vegetation that forms part of soil. Abbr **SOM**

soil pan *noun* a hard layer in the soil

'Greater resistance in digging soils of similar texture and moisture content indicates poor structure. Concentration of roots indicates hard to penetrate layers or blocks of soil. A soil pan may be present if roots grow horizontally or do not penetrate to any depth.' [*Farmers Guardian*]

soil parent material *noun* material from which soil is formed

soil salinity *noun* the quantity of mineral salts found in a soil (NOTE: High soil salinity is detrimental to most agricultural crops, although some plants are adapted to such conditions.)

soil series *noun* the classification of soils based on their similarities, used in soil mapping (NOTE: Soil series are defined using a combination of three main properties: the parent material; the texture of the soil material and the presence or absence of material with a distinctive mineralogy; and the presence or absence of distinctive horizons.)

soil sterilant *noun* something used to remove microorganisms from soil, e.g. a chemical or steam

soil sterilisation *noun* the treatment of glasshouse, greenhouse and other horticultural soils in order to kill weed seeds, plant disease organisms and pests

soil survey *noun* the mapping of soil types using a soil classification system

Solanum /səˈleɪnəm/ *noun* the Latin name for the potato family

solar dryer *noun* a device for drying crops using the heat of the sun

Solari piggery /səʊˈlɑːri ˌpɪɡəri/ *noun* a type of housing for pigs, with fattening pens on each side of a central feeding passage, housed in an open-sided Dutch barn

solarisation /ˌsəʊləraɪˈzeɪʃ(ə)n/, **solarization** *noun* exposure to the rays of the sun, especially for the purpose of killing pests in the soil, by covering the soil with plastic sheets and letting it warm up in the sunshine

sole crop *noun* a crop grown in a stand with no other crops

sole furrow *noun* the last slice cut during ploughing

solid-not-fat percentage *noun* a measure of milk quality, showing the percentages of all substances other than fat in the milk. Abbr **SNF percentage**

solum /ˈsəʊləm/ *noun* soil, including both topsoil and subsoil

SOM *abbreviation* soil organic matter

soot *noun* a black deposit of fine particles of carbon which rise in the smoke produced by the burning of material such as coal, wood or oil

sooty mould *noun* a fungal disease of wheat (*Cladosporium*)

sorghum /ˈsɔːgəm/ *noun* a drought-resistant cereal plant grown in semi-arid tropical regions such as Mexico, Nigeria and Sudan. Latin name: *Sorghum vulgare*.

sorrel /ˈsɒrəl/ *noun* a plant with a sour juice sometimes eaten as a salad. Varieties include the sheep sorrel and the wood sorrel.

source *verb* to get materials or products from a particular place or supplier

'It was time people started buying food more ethically, said Mr Mitchell, to give them a sense of pride in their region. The extra work needed to source from multiple small suppliers was well worth it.' [*Farmers Weekly*]

sour soil *noun* soil which is excessively acid and hence needs liming to restore the correct balance between acidity and alkalinity

South Devon *noun* the heaviest breed of British cattle, with a light brownish-red colour. It was originally a dual-purpose breed, but now is mainly raised for beef.

Southdown /ˈsaʊθdaʊn/ *noun* the smallest of the Down breeds of sheep. It has a compact body and a dense fleece of high-quality short wool. The Southdown is an early maturing breed and produces meat of high quality.

sow *verb* to put seeds into soil so that they will germinate and grow ■ *noun* a female pig. ◊ **pig, pig production**

soya /ˈsɔɪə/ *noun* a plant that produces edible beans which have a high protein and fat content and very little starch. Latin name: *Glycine max*. Also called **soya bean, soybean**

soya bean, soybean *noun* **1.** a bean from a soya plant **2.** same as **soya**

COMMENT: Soya beans are very rich in protein and apart from direct human consumption are used for their oil and as livestock feed. After the oil has been extracted, the residue is used as a high protein feedingstuff. Other by-products are soya bean milk and soy sauce, both widely used in China and Japan. Soya beans are widely grown in China, where they are the most important food legume, in Brazil and in the USA.

soyoil /ˈsɔɪɔɪl/ *noun* an oil extracted from the soya bean

sp. *abbreviation* species (NOTE: The plural, for several species, is **spp.**)

space allowance *noun* the amount of space a farmed animal should have in which to move around, feed and rest (NOTE: Guidelines on minimum space allowances are set out in the animal welfare codes.)

spaced plant *noun* a plant grown in a row so that its canopy does not touch or overlap that of any other plant

spacing *noun* the process of making places between things, e.g. between plants in a row

spacing drill *noun* a precision seed drill

spade *noun* a common garden tool, with a wide square blade at the end of a strong handle. it is used for making holes or digging by hand.

spading machine *noun* a machine which uses rotating digger blades to cultivate compacted topsoil and dig out pans created by other cultivators

spay /speɪ/ *verb* to remove the ovaries of a female animal

SPCS *abbreviation* Seed Potato Classification Scheme

spear *noun* a shoot of a green plant such as asparagus or broccoli

speciation /ˌspiːsiˈeɪʃ(ə)n/ *noun* the process of developing new species

species *noun* a group of organisms that can interbreed. A species is a division of a genus. Abbr **sp.** (NOTE: The plural is **species.**)

specific *adjective* **1.** clearly defined and definite ○ *The airframe has to be built to very specific requirements.* **2.** characteristic of something **3.** referring to species. ◊ **interspecific, intraspecific**

specificity /ˌspesɪˈfɪsəti/ *noun* the characteristic of having a specific range or use ○ *Parasites show specificity in that they live on only a limited number of hosts.*

'In his view the sensitivity and specificity of the current testing method needed to be improved in order to increase its accuracy and research work was concentrating on the problem.' [*Farmers Guardian*]

specific weight *noun* the bulk density of a grain sample measured in hectolitres or bushels

speckle /ˈspek(ə)l/ *noun* a small spot. ◊ **Beulah speckle face**

speckled yellowing *noun* a disease of sugar beet caused by a deficiency of manganese

speedwell /ˈspiːdwel/ *noun* a widespread weed (*Veronica persica*) found in cereal crops and oilseed rape. Because it spreads rapidly it is a hazard in row crops.

spermatozoon /ˌspɜːmətəˈzəʊɒn/ *noun* a mature male sex cell that is capable of fertilising an ovum (NOTE: The plural is **spermatozoa**.)

COMMENT: A spermatozoon is very small and comprises a head, neck and very long tail. It can swim by moving its tail from side to side.

spermicide /ˈspɜːmɪsaɪd/ *noun* a substance that kills spermatozoa

sphagnum /ˈsfægnəm/ *noun* a type of moss that grows in acid conditions

spice *noun* a substance used as a flavouring in cooking, made from the pungent or aromatic parts of plants. Spices are obtained from seeds, fruit, flowers, roots, bark or buds of plants. The commonest are pepper, mustard, ginger, cloves and nutmeg.

spider *noun* one of a large group of animals, with two parts to their bodies and eight legs. Class: Arachnida.

spike *noun* **1.** a tall pointed flower head (**inflorescence**) in which small flowers without stalks grow from a central flower stem **2.** a pointed end of a pole or piece of metal

spiked /spaɪkt/ *adjective* referring to a farm implement with spikes, e.g. a spiked chain harrow

spikelet /ˈspaɪklət/ *noun* part of the flower head of plants such as grass, attached to the main stem without a stalk

spike tooth harrow *noun* a tractor-trailed implement consisting of a simple frame with tines attached where the frame members cross

spinach *noun* an annual plant (*Spinacia oleracea*) grown for its succulent green leaves and eaten as a vegetable

spinach beet *noun* a plant similar to sugar beet, but grown for its leaves which are cooked in the same way as spinach

spine *noun* a pointed structure that is either a modified leaf, as in cacti, or part of a leaf or leaf base

spinner /ˈspɪnə/ *noun* a device used for harvesting potatoes. The potatoes are left on the surface of the soil for picking later.

spinney /ˈspɪni/ *noun* a small wood with undergrowth

spinous /ˈspaɪnəs/ *adjective* referring to a plant with spines

spirochaete /ˈspaɪərəʊkiːt/ *noun* a bacterium with a spiral shape

'But there is a difference between bacteria normally associated with foot-rot and a type seen in the new strain. A spirochaete, a bacterium that can penetrate the skin surface and is more usually linked with digital dermatitis in cattle, has been identified.' [*Farmers Weekly*]

spit *noun* the depth of soil that is dug with a spade

splay leg *noun* a disorder in piglets, which are born unable to stand properly

spoil *verb* (*of food*) to rot or decay

spoilage /ˈspɔɪlɪdʒ/ *noun* the process of food becoming inedible, especially because of poor storage conditions

spore /spɔː/ *noun* the microscopic reproductive body of fungi, bacteria and some non-flowering plants such as ferns

COMMENT: Spores are produced by plants such as ferns or by algae and fungi. They are microscopic and float in the air or water until they find a resting place where they can germinate.

sporicidal /ˌspɔːrɪˈsaɪd(ə)l/ *adjective* able to kill spores

sporicide /ˈspɔːrɪsaɪd/ *noun* a substance that kills spores

Sporozoa /ˌspɔːrəˈzəʊə/ *noun* a type of parasitic Protozoa which includes Plasmodium, the cause of malaria

spot price *noun* the market price for produce or livestock at a specific time

'If most of the feeds needed can be bought forward during dips in the market, the end result is a high quality, highly competitive diet and significant feed cost savings compared to spot prices.' [*Farmers Guardian*]

spp. *abbreviation* species (*plural*) (NOTE: The singular is **sp.**)

spraing /spreɪŋ/ *noun* a disease of potatoes spread by nematodes in the soil

spray *noun* 1. a mass of tiny drops of liquid 2. special liquid for spraying onto a plant to prevent insect infestation or disease ■ *verb* 1. to send out a liquid in a mass of tiny drops ○ *They sprayed the room with disinfectant.* 2. to send out a special liquid onto a plant to prevent insect infestation or disease ○ *Apple trees are sprayed twice a year to kill aphids.*

spraybar /ˈspreɪbɑː/ *noun* an attachment consisting of a horizontal tube with nozzles or jets, used for spraying over a wide area

sprayer /ˈspreɪə/ *noun* a machine which forces a liquid through a nozzle under pressure, used to distribute liquids such as herbicides, fungicides, insecticides and fertilisers

spray irrigation *noun* a system of irrigation using sprinklers which are located along a boom. Some booms rotate and can distribute water over a large circular area.

spray lines *plural noun* a method of distributing irrigation water using flexible hose, mainly used for horticultural crops

spread *verb* to put something such as manure, fertiliser or mulch on an area of ground

spreader /ˈspredə/ *noun* 1. a device used for spreading, e.g. one for spreading granules of fertiliser evenly over a lawn 2. an agent added to an insect spray in order to make sure that the foliage is covered uniformly

sprig *noun* a small shoot or twig

spring *noun* 1. a place where water comes naturally out of the ground 2. the season of the year following winter and before summer, when days become longer and the weather progressively warmer 3. a metal device which, when under tension, tries to resume its previous position □ **spring-tined harrow**, **harrow with spring-loaded tines** a cultivator which has

tines of spring steel which vibrate in the soil. This gives fast seedbed preparation.

springer /ˈsprɪŋə/ *noun* a cow almost ready to calve. Also called **down-calver**

springtail /ˈsprɪŋteɪl/ *noun* a primitive wingless insect very common in soils, where they may do damage to fine roots

spring wheat *noun* wheat which is sown in spring and harvested towards the end of the summer

sprinkle *verb* to water something with drops in a light shower

sprinkler *noun* a hose which sends out a shower of drops

sprout *noun* a little shoot growing out from a plant, with a stem and small leaves ■ *verb* (*of a plant*) to send out new growth

spruce *noun* a temperate softwood coniferous tree. Genus: *Picea.*

spruce-larch adelgid /ˌspruːs lɑːtʃ ə ˈdelɡɪd/ *noun* a relative of the aphid, which may cause serious damage on spruce grown for Christmas trees (*Adelges viridis*)

SPS *abbreviation* Single Payment Scheme

spur *noun* 1. a ridge of land that descends towards a valley floor from higher land above 2. a tubular projection from a flower sepal or petal often containing nectar 3. a short leafy branch of a tree with a cluster of flowers or fruits

spurrey /ˈspʌri/ *noun* a common weed on arable land. Also called **corn spurrey**

square ploughing *noun* a method of ploughing suitable for large areas. A piece of land is ploughed in the centre of a field and then the field is ploughed in a clockwise direction starting from this central point.

squash *noun* a plant of the *Cucurbitaceae* family, e.g. a marrow or cucumber

squirrel *noun* a medium-sized rodent living in trees (NOTE: Squirrels are harmless as far as crops are concerned. There are two types of squirrel in the UK: the grey squirrel (*Sciurus carolinensis*) and the red squirrel (*Sciurus vulgaris*).)

SRPBA *abbreviation* Scottish Rural Property and Business Association

SSSI *abbreviation* Site of Special Scientific Interest

stabilisation lagoon /ˌsteɪbɪlaɪ ˈzeɪʃ(ə)n ləˌɡuːn/ *noun* 1. a pond used for storing liquid waste 2. a pond used for purifying sewage by allowing sunlight to fall on a mixture of sewage and water

stabilise /'steɪbəlaɪz/, **stabilize** *verb* to take measures to prevent soil being eroded, especially from a hillside

stabiliser /'steɪbɪlaɪzə/, **stabilizer** *noun* an artificial substance added to processed food such as sauces containing water and fat to stop the mixture from changing. Also called **stabilising agent** (NOTE: In the EU, emulsifiers and stabilisers have E numbers E322 to E495.)

stability *noun* the state of being stable and not changing

stable *adjective* **1.** steady and not easily moved ○ *a stable surface* **2.** not changing ○ *In parts of Southeast Asia, temperatures remain stable for most of the year.* ■ *noun* a building in which horses are kept

stable fly *noun* a fly which is like the house fly, but with a distinct proboscis which can pierce the skin. It breeds in stable manure and is a serious pest to animals as the bites cause irritation.

stack *noun* a pile of sheaves of grain, hay or straw. Stacks can be round or square.

stag *noun* **1.** a male deer **2.** the male of various animals castrated after maturity **3.** a male turkey

stages of growth *plural noun* same as **growth stages**

staggers /'stægəz/ *noun* a condition of animals in which they stagger about, as in looping-ill and swayback disease. Grass staggers in cattle is caused by hypomagnesaemia.

stake *noun* a thick wooden post, to which a tree or shrub is attached to keep it upright ○ *A newly planted tree needs to have a stake.* ■ *verb* to attach a plant to a stake

stakeholder /'steɪkhəʊldə/ *noun* a person who has an interest in something such as a new environmental policy and is personally affected by how successful it is and how difficult it is to put in place

'Officials told a recent stakeholder meeting that this was one of the conditions the EU had imposed, following discussions with DARD in December over its original Nitrates Action Programme.' [*Farmers Weekly*]

stale seedbed *noun* a method of killing weeds by using a contact herbicide just before drilling

stalk *noun* **1.** the main stem of a plant which holds the plant upright **2.** a subsidiary stem of a plant, branching out from the main stem or attaching a leaf, flower or fruit

stalk rot *noun* a disease of maize caused by a fungus

stall *noun* a single compartment in a stable or cowshed, where an animal can stand or lie down

stallion *noun* an uncastrated full-grown male horse, especially one kept for breeding

stamen /'steɪmən/ *noun* a male part of a flower consisting of a stalk (**filament**) bearing a container (**anther**) in which pollen is produced

stance *noun* ♦ **feed stance**

stand *noun* a group of plants or trees growing together ○ *a stand of conifers*

standard *noun* **1.** something which has been agreed on and is used to measure other things by ○ *set higher standards for water purity* **2.** a plant grown on a single long stem that is kept from forming branches except at the top **3.** a type of fruit tree or rose tree where the stem is about two metres high, on top of which the head is developed □ **standard apple**, **standard pear** an apple or pear tree grown as a standard **4.** a large tree in a woodland ○ *a coppice with standards* ■ *adjective* usual or officially accepted ○ *standard procedures*

COMMENT: In UK orchards, standard trees have been replaced by bush varieties which are easier and cheaper to prune, spray and harvest.

Standard Gross Margin *noun* a measure of the business size of a farm, calculated by looking at the different types of enterprises on the farm and how much each contributes to the overall profit made. Abbr **SGM**

standard man day *noun* eight hours of work, used as a measure for calculating labour costs on a farm. Abbr **SMD**

COMMENT: One milking cow is calculated as equal to ten SMDs, and one hectare of barley or wheat is equal to 50 SMDs; if a farm has less than 275 SMDs it is not counted as a full-time farm.

standing *adjective* growing upright

standing crop *noun* a crop such as wheat which is still growing in a field

standstill *noun* the keeping of animals in the same place for 6 days to prevent the spread of disease

staphylococcal /ˌstæfɪlə'kɒk(ə)l/ *adjective* caused by staphylococci

COMMENT: Staphylococcal infections are treated with antibiotics such as penicillin, or broad-spectrum antibiotics such as tetracycline.

staphylococcal mastitis *noun* a condition of cows caused by several types of staphylococci especially when accompanied by stress resulting from liver fluke or cold conditions. Milk becomes watery and the cow has a high temperature.

staphylococcus /ˌstæfɪləˈkɒkəs/ *noun* a bacterium that causes boils and food poisoning. Genus: *Staphylococcus*. (NOTE: The plural is **staphylococci**.)

staple *noun* the length and fineness of fibres such as wool or cotton, used in determining quality

staple commodity *noun* a basic food or raw material

staple length *noun* the length of the wool fibre

starch *noun* a substance composed of chains of glucose units, found in green plants

COMMENT: Starch is the usual form in which carbohydrate is present in food, especially in bread, rice and potatoes, and it is broken down by the digestive process into forms of sugar. Carbohydrate is not stored in the bodies of animals in the form of starch, but as glycogen.

starchy *adjective* containing a lot of starch ○ *Potatoes are a starchy food.*

starter *noun* a culture of bacteria, used to inoculate animals or to start growth in milk used in cheese production

State Veterinary Service *noun* a nationwide service based in Worcester, with 24 regional offices in the UK, set up by the government to deliver its policies on the health and welfare of livestock. Abbr **SVS**

station *noun* a very large farm, specialising in raising sheep or cattle

Statutory Management Requirement *noun* a set of environmental and animal welfare requirements, which farmers must meet to receive a Single Farm Payment. Abbr **SMR**

steam up *verb* to feed a cow before it calves, to prepare it for the next lactation

stecklings /ˈsteklɪŋz/ *plural noun* young sugar beet plants grown in seedbeds in summer, to be transplanted in the autumn or following spring

steer *noun* a castrated male bovine over one year old. Also called **bullock**

steerage hoe *noun* a hoe mounted behind a tractor and steered by the driver to avoid crop damage

stell /stel/ *noun* a stone shelter for sheep and cattle in upland areas

stem *noun* 1. the main stalk of a plant that holds it upright 2. a subsidiary plant stalk, branching out from the main stalk or attaching a leaf, flower or fruit

stem canker *noun* a fungal disease affecting many types of plant including oilseed rape

stem eelworm *noun* a pest affecting cereals, in particular oats. The plant stem swells and is prevented from growing and producing any ears.

stemmed /stemd/ *adjective* with a stem ○ *a short-stemmed variety of rose*

stem rot *noun* a disease caused by deficiency of nutrients

stem rust *noun* a disease of wheat, infecting the stem

steppe /step/ *noun* a wide grassy plain with no trees, especially in Europe and Asia (NOTE: The North American equivalent of a steppe is a **prairie**.)

sterile *adjective* 1. free from microorganisms 2. infertile or not able to produce offspring

sterilisation /ˌsterɪlaɪˈzeɪʃ(ə)n/, **sterilization** *noun* 1. the action of making something free from microorganisms 2. the action of making an organism unable to produce offspring

sterilise /ˈsterɪlaɪz/, **sterilize** *verb* 1. to make something sterile by killing the microorganisms in it or on it ○ *The soil needs to be sterilised before being used for intensive greenhouse cultivation.* 2. to make an organism unable to have offspring (NOTE: This may be done by various means including drugs, surgery or irradiation.)

sterilised milk *noun* milk prepared for human consumption by heating in sealed airtight containers to kill all bacteria. See Comment at **milk**

sterility /stəˈrɪlɪti/ *noun* 1. the state of being free from microorganisms 2. the inability to produce offspring

stewardship *noun* the protection of the environment for the benefit of future generations of human beings by developing appropriate institutions and strategies

sticker *noun* a substance added to a fungicide or bactericide preparation to help it to stick to the sprayed surface

stigma *noun* the part of a flower's female reproductive organ that receives the pollen grains (NOTE: It is generally located at the tip of the **style**.)

stile /staɪl/ *noun* a set of steps arranged so that people can climb over a wall or fence

stillbirth /ˈstɪlbɜːθ/ *noun* the birth of a dead animal or abortion at a late stage of pregnancy

stillborn /ˈstɪlbɔːn/ *adjective* referring to an animal which is born dead

Stilton /ˈstɪltən/ *noun* English blue cheese

stipula /ˈstɪpjʊlə/ *noun* a newly sprouted feather

stipule /ˈstɪpjuːl/ *noun* one of the pair of wing-like growths at the base of a leaf stalk

stirk /stɜːk/ *noun* a Scottish term for cattle, both male and female, under two years old

stock *noun* **1.** animals or plants that are derived from a common ancestor **2.** a plant with roots onto which a piece of another plant, the **scion**, is grafted. ◊ **rootstock 3.** a supply of something available for future use ○ *Stocks of herring are being decimated by overfishing.* ■ *verb* **1.** to provide a supply of something for future use ○ *a well-stocked garden* ○ *We stocked the ponds with a rare breed of fish.* **2.** to introduce livestock into an area or into a farm

stock breeder *noun* a farmer who specialises in breeding livestock

stock bull *noun* a bull kept for breeding purposes in a pedigree herd

stock farming *noun* the rearing of livestock for sale

stocking density *noun* the number of animals kept on a specific area of land (NOTE: Animal welfare codes lay down rules for the maximum stocking density allowed to ensure that the health and welfare of the animals or birds is good.)

stocking rate *noun* a measure of the carrying capacity of an area in terms of the number of livestock in it at a given time, e.g. the number of animals per hectare

stockman /ˈstɒkmən/ *noun* a farm worker who looks after animals, especially cattle (NOTE: Animal welfare codes lay down rules about how well-trained stockmen should be in order to ensure the good health and welfare of the animals or birds in their care.)

stockproof /ˈstɒkpruːf/ *adjective* referring to a fence which livestock cannot get through

stocky *adjective* referring to an animal with short strong legs

stolon /ˈstəʊlɒn/ *noun* a stem that grows along the ground and gives rise to a new plant when it roots

'White clover's beneficial effect may be related to the large population of microbes associated with its root system and the force the roots exert on the soil in anchoring the plant's stolons to the surface.' [*Dairy Farmer*]

stoma /ˈstəʊmə/ *noun* a pore in a plant, especially in the leaves, through which carbon dioxide is taken in and oxygen is sent out. Each stoma in a leaf is surrounded by a pair of guard cells, which close the stomata if the plant needs to conserve water. (NOTE: The plural is **stomata**.)

stomach digestion *noun* the part of the digestive process which takes place in the animal's stomach

stone *noun* **1.** a single small piece of rock **2.** a hard endocarp that surrounds a seed in a fruit such as a cherry

stone-ground flour *noun* a type of flour made by grinding with millstones

Stoneleigh /ˈstəʊnli/ *noun* the home of the National Agricultural Centre and proposed site for the National Museum of Food and Farming

stone trap *noun* in a combine harvester, a trough with a trap door, which prevents stones passing into the concave

stook /stʊk/ *noun* several (usually twelve) corn sheaves gathered together in a field to form a small pyramid. Also called **shock**

stop *verb* to remove the growing tip of a shoot, to encourage lateral growths

storage drying *noun* a method of drying bales of hay by blowing air through them (NOTE: There are several methods of storage drying. In a building with airtight sides, air is forced up through ventilation holes in the floor. In open barns, radial drying or a centre duct system is used.)

store *noun* **1.** a supply of something kept for future use **2.** □ **store cattle**, **stores**, **store lambs** cattle or lambs bred or bought for fattening. The animals are usually reared on one farm and then sold on to dealers or other farmers. ■ *verb* to keep something until it is needed ○ *Whales store energy as blubber under the skin.*

stover /ˈstəʊvə/ *noun* an inferior type of fodder

straight *noun* an animal foodstuff composed of one type of food

straight fertiliser *noun* a fertiliser that supplies only one nutrient such as nitrogen. Compare **compound fertiliser, mixed fertiliser**

strain *noun* a group within a species with distinct characteristics ○ *They have developed a new strain of virus-resistant rice.*

strake /streɪk/ *noun* an attachment bolted onto the rear wheels of a tractor to improve wheel grip. The strake has spikes which can be extended beyond the tyre and which dig into the soil.

strangles /ˈstræŋɡəlz/ *noun* a disease of mangolds and sugar beet. It occurs in fairly large seedlings after singling. The stem is severely damaged.

Strategy for Sustainable Farming and Food *noun* a scheme administered by Defra which promotes sustainable, ecologically sound farming methods, and provides funding for relevant training and modernisation

straw *noun* **1.** the dry stems and leaves of crops such as wheat and oilseed rape left after the grains have been removed **2.** grass which is mowed after flowering. Compare **hay**

COMMENT: Straw can be ploughed back into the soil. It is often mixed with animal dung to make manure. Non-agricultural uses are varied and include thatching, making paper and making bricks. It can be compressed into bundles to act as fuel and in this way can be used for heating farms and small local industrial buildings.

strawberry *noun* a soft fruit of the *Fragaria* species, used as a dessert fruit, but also preserved as jam

strawberry foot rot *noun* a bacterial disease affecting sheep, causing ulcers

straw burning *noun* a cheap method of disposal of straw, which helps to control diseases

straw chopper *noun* a device fitted to the back of a combine which chops straw into short lengths and drops it on the stubble. Chopped straw is easier to plough in.

straw spreader *noun* a device attached to the back of a combine when the straw is not wanted. The straw is spread over the ground and then ploughed in.

straw walker *noun* the part of a combine harvester where straw is carried away from the threshed grain after it has been separated from the stalks

strength *noun* the ability of wheat flour to produce a yeasted dough capable of retaining carbon dioxide bubbles until the proteins in the bubble walls become relatively rigid, which happens at about 75°C. The milling quality of wheat is measured by the Hagberg test.

streptococcal /ˌstreptəˈkɒk(ə)l/ *adjective* referring to an infection caused by streptococci

streptococcus /ˌstreptəˈkɒkəs/ *noun* a bacterium belonging to a genus that grows in long chains and causes diseases such as strangles and mastitis (NOTE: The plural is **streptococci.**)

streptomycin /ˌstreptəˈmaɪsɪn/ *noun* an antibiotic used against many types of infection, especially streptococcal ones

stress *noun* **1.** a condition where an outside influence changes the composition or functioning of something ○ *Plants in dry environments experience stress due to lack of water.* **2.** a state of anxiety or strain that can affect an animal's health ■ *verb* to worry an animal so that it becomes ill

strike *noun* the infestation of the flesh of sheep by the larvae of blowflies. It causes extreme irritation and death can occur in a short time. ■ *verb* to take root

string bean *noun* same as **French bean**

strip *noun* a long narrow piece, usually of the same width from end to end ○ *a strip of paper* ○ *a strip of land* ■ *verb* to remove a covering from something ○ *Spraying with defoliant strips the leaves off all plants.*

strip cropping *noun* a method of farming in which long thin pieces of land across the contours are planted with different crops in order to reduce soil erosion

strip cultivation *noun* a method of communal farming in which each family has a long thin piece or several long thin pieces of land to cultivate

strip cup *noun* a container into which the first drops of milk are drawn by hand from the teats of the cow before the milking machine is attached to the udder

stripe rust *noun* same as **yellow rust**

strip farming *noun* a method of farming where strips of land across the contours are planted with different crops

strip grazing *noun* a system of grazing which allows animals access to a small part of the field. The rest of the field is protected by a temporary fence, usually electric.

stripper-header *noun* a machine which harvests a crop such as linseed and strips off the seedheads

strippings /ˈstrɪpɪŋz/ *noun* the last drops of milk from a cow's teats at the end of milking session

strobilurin /strɒˈbɪljʊrɪn/ *noun* one of a group of translaminar and protectant fungicides. Strobilurins are used on a wide range of crops.

struck *noun* an acute disease of sheep which is a form of entero-toxaemia. It affects sheep which are one to two years old and is very localised. In Britain it occurs only in the Romney Marsh and in some Welsh valleys.

stubble *noun* the short stems left in the ground after a crop such as wheat or oilseed rape has been cut

stubble burning *noun* formerly, a method of removing dry stubble by burning it before ploughing. Stubble burning was banned under the Crop Residues (Burning) Regulations of 1993.

COMMENT: Stubble burning has the advantage of removing weed seeds and creating a certain amount of natural fertiliser which can be ploughed into the soil. The disadvantage is that it pollutes the atmosphere with smoke, reducing visibility on roads and releasing large amounts of carbon dioxide. This, together with the possible danger that the fire may get out of control, killing small animals and burning trees and crops, means that it is not recommended as a means of dealing with the stalks of harvested plants.

stubble cleaning *noun* working the stubble after harvest, using ploughs, cultivators, and harrows to free the weeds from the soil

stubble turnips *plural noun* quick-growing types of turnip sown into cereal stubble and grown as catch crops

stud *noun* **1.** same as **stud farm 2.** US a male horse kept for breeding **3.** metal nail with a head projecting above the surface

studded roller feed drill *noun* a type of external force feed seed drill in which fluted rollers are replaced by rolls with studs or pegs. It is suitable for drilling most types of seed.

stud farm *noun* a farm where horses are kept for breeding

stud fees *plural noun* money paid to the owner of a stud animal for servicing a female

stump *noun* ♦ **tree stump**

stunt *verb* to reduce the growth of something ○ *The poor soil stunts the growth of the trees.*

sty *noun* a structure for housing pigs, usually with a small run next to it. Also called **pigsty**

style *noun* the elongated structure that carries the stigma at its tip in many flowers

sub- *prefix* **1.** less important than **2.** lower than

subculture *noun* a culture of microorganisms or cells that is grown from another culture ■ *verb* to grow a culture of microorganisms or cells from another culture

subdominant /sʌbˈdɒmɪnənt/ *adjective* (*of a species*) being not as important as the dominant species

subfertility /ˈsʌbfɜːˈtɪlɪti/ *noun* a situation where an animal is less fertile than expected

'When purchasing bulls, also consider fertility. Subfertility in bulls can lead to a 5–10% decrease in pregnancy rate, when compared with a fully fertile bull.' [*Farmers Weekly*]

suboestrus /sʌbˈiːstrəs/ *noun* a situation where a female animal comes on heat, but does not show any of the usual signs

subsidy *noun* money given by a government or organisation to help an industry, charity or other organisation ○ *The reform will result in subsidies for farming being replaced by payments for caring for the environment.*

subsistence /səbˈsɪstəns/ *noun* the condition of managing to live on the smallest amount of resources including food needed to stay alive

subsistence farming, subsistence agriculture *noun* the activity of growing just enough crops to feed the farmer's family and having none left to sell

subsoil /ˈsʌbsɔɪl/ *noun* a layer of soil under the topsoil (NOTE: The subsoil contains little organic matter but chemical substances from the topsoil leach into it.)

subsoiler /'sʌbsɔɪlə/ *noun* a heavy cultivator consisting of a strong frame with long tines attached to it. It is used to break up compacted soil to allow free passage of air and water, a process called 'subsoiling'.

substandard *adjective* not up to standard quality ○ *A load of substandard potatoes was rejected by the buyer.*

substratum /'sʌbstrɑːtəm/ *noun* a layer of rock beneath the topsoil and subsoil (NOTE: The plural is **substrata**.)

subtropical /sʌb'trɒpɪk(ə)l/ *adjective* referring to the subtropics ○ *The islands enjoy a subtropical climate.* ○ *Subtropical plants grow on the sheltered parts of the coast.*

subtropics /sʌb'trɒpɪks/ *plural noun* an area between the tropics and the temperate zone

succession *noun* a series of stages, one after the other, by which a group of organisms living in a community reaches a stable state or **climax**

successional cropping *noun* **1.** the growing of several crops one after the other during the same growing season **2.** the process of sowing a crop such as lettuce over a long period, so that harvesting takes place over a similarly long period

succulent /'sʌkjʊlənt/ *noun* a plant that has fleshy leaves or stems in which it stores water, e.g. a cactus

succulent foods *plural noun* feedingstuffs which contain a lot of water; they are palatable and filling, and usually have a laxative effect. Most root crops, e.g. swedes and turnips, are succulents.

sucker *noun* a shoot which develops from the roots or lower part of a stem

suckle /'sʌk(ə)l/ *verb* to feed with milk from the udder

suckler /'sʌklə/ *noun* a calf or other young animal which is suckling

suckler cow *noun* a cow which rears its own calf and is later used for beef production

Suckler Cow Premium Scheme *noun* until 2005, a subsidy on female cattle forming part of a suckler breeding herd used for rearing calves for meat production (NOTE: Now superseded by the **Single Payment Scheme**.)

suckler herd *noun* a herd of beef cattle, where each dam suckles its own calf or calves

sucklers *plural noun* flowers of clover

suckling *noun* the act of nursing a calf at the udder □ **single suckling**, **double sucking**, **multiple suckling** nursing one calf, two or several calves

suckling pig *noun* an unweaned piglet

sucrose /'suːkrəʊs/ *noun* a sugar that is abundant in many plants, which consists of one molecule of glucose joined to one of fructose

Suffolk /'sʌfək/, **Suffolk Down** *noun* a breed of sheep developed from crosses between the now extinct Norfolk Horn ewes and the Southdown ram. It is a large quick-growing animal with a close short fleece and a black face which has no wool on it. Suffolk crosses perform well under a broad range of farming systems, being equally effective for over-winter storing and for intensive early lamb production.

Suffolk Punch *noun* a heavy draught horse, coloured chestnut. It is shorter and more stocky than the shire and lacks feathers on the fetlocks.

suffrutescent /sə,fruː'tesənt/, **suffruticose** /sə'fruːtɪkəʊz/ *adjective* referring to a perennial plant that is woody at the base of the stem and does not die down to ground level in winter

sugar *noun* **1.** same as **sucrose 2.** any chemical of the saccharide group

sugar beet *noun* a specialised type of beet grown for the high sugar content of its roots. It is cultivated in temperate regions, and in Britain is an especially important crop in East Anglia. The crowns and leaves of the crop are used for feedingstuff, as is also the residue after the sugar content has been extracted from the roots.

sugar beet harvester *noun* a machine for harvesting sugar beet, which may be trailed or self-propelled. The machine cuts off the beet tops, lifts the root, cleans off the soil and conveys the beet to a hopper which is then emptied by a second elevator onto a trailer.

sugar beet topper *noun* an attachment to a sugar beet harvester which collects the sugar beet tops. Some have choppers and blower units, which chop up the tops and then blow them into a trailer.

sugar cane *noun* a large perennial grass, whose stems contain a sweet sap

COMMENT: Sugar cane is rich in sucrose which is extracted and used for making sugar. Cane sugar is now one of the most scientifically produced tropical products, although cutting is still often done by hand. Cane is grown in many tropical and subtropical regions, in particular in the Caribbean. The principal sugar producers are Cuba, India, Brazil, China, Puerto Rico and Hawaii for cane sugar, and Russia, the Ukraine, France and Germany for beet sugar. Rum is a by-product of sugar cane.

suitcase farmer *noun US* a farmer who lives some distance from his or her holding, i.e. more than 30 miles

sulphate *noun* a salt of sulphuric acid and a metal

sulphate of ammonia *noun* same as **ammonium sulphate**

sulphate of potash *noun* same as **potassium sulphate**

sulphur /'sʌlfə/ *noun* a yellow non-metallic chemical element that is essential to biological life. It is used in the manufacture of sulphuric acid and in the vulcanisation of rubber. (NOTE: The usual and recommended scientific spelling of sulphur and derivatives such as sulphate, sulphide and sulphonate is with an -f-, though the spelling with -ph- is still common in general usage.)

COMMENT: In the United Kingdom, the removal of sulphur from the atmosphere means that some crops such as oilseed rape are deficient and sulphur needs to be added to fertilisers.

sulphur deficiency *noun* a lack of sulphur in the soil, leading to deficiency in plants

sulphuric acid /sʌl,fjʊərɪk 'æsɪd/ *noun* a strong acid that exists as a colourless oily corrosive liquid and is made by reacting sulphur trioxide with water. It is used in batteries and in the manufacture of fertilisers, explosives, detergents, dyes and many other chemicals. Formula: H_2SO_4.

summer *noun* the season following spring and before autumn, when the weather is warmest, the sun is highest in the sky and most plants flower and set seed

summer feeding *noun* the feeding of cattle on permanent pastures in the summer months

summer mastitis *noun* an infection of the udder thought to be spread by biting flies. Cows become very ill, lameness may occur and milk is watery and later bloody.

sunflower *noun* an important oilseed crop grown in temperate areas

COMMENT: The oil extracted from the seeds is used for cooking and for margarine production. The residual cake after pressing is a high-protein livestock feed, and the whole plant can be fed to cattle. It is also useful as a green manure plant. Birds can cause serious damage to sunflower crops by feeding on the ripening seeds. The main producing countries are Russia, the Ukraine, Argentina and Romania.

superlevy /'suːpəlevi/ *noun* same as **supplementary levy**

superovulation /,suːpər,ɒvjʊ'leɪʃ(ə)n/ *noun* the process in animal production of injecting hormones to increase the number of eggs released by the ovaries

superphosphate /,suːpə'fɒsfeɪt/ *noun* a chemical compound formed from calcium phosphate and sulphuric acid, used as a fertiliser

supplement *noun* something added in order to make something more complete ○ *vitamin supplements* ■ *verb* /'sʌplɪmənt/ to add to something in order to make it more complete

supplementary levy *noun* in the EU, a payment introduced to penalise milk production over the quota level

supplementary ration *noun* a type of concentrate fed to livestock to supplement feeds of hay and roots

supply chain *noun* the sequence of farmers, processors, distributors and retailers who are involved in taking a product from the farm to the consumer

support *verb* 1. to hold the weight of something ○ *to support the saplings with stakes* 2. to provide what is necessary for an activity or way of life ○ *These wetlands support a natural community of plants, animals and birds.*

support buying *noun* same as **intervention buying**

support energy *noun* the total energy expenditure necessary for the production of plant and animal agricultural foodstuffs

support price *noun* the price at which the EU will buy farm produce which farmers cannot sell, in order to store it. Also called **intervention price**

surface drainage *noun* the removal of surplus water from an area of land by means of ditches and channels

surface-rooting *adjective* referring to a plant whose roots are shallow in the soil. Compare **deep-rooted**

surface runoff *noun* a flow of rainwater, melted snow or excess fertiliser from the surface of land into streams and rivers

surface water *noun* water that flows across the surface of the soil as a stream after rain and drains into rivers rather than seeping into the soil itself. Compare **ground water**

surgeon *noun* a person who has qualified in the treatment of disease by cutting out the diseased part

surplus *adjective* more than is needed ○ *Surplus water will flow away in storm drains.* ■ *noun* **1.** something that is more than is needed ○ *produced a surplus of wheat* **2.** extra produce, produce which is more than is needed

Sussex /'sʌsɪks/ *noun* a beef breed of cattle, similar to the North Devon, that is hardy and adaptable. Dark cherry red in colour, they were originally used as draught animals in preference to draught horses.

sustainability /sə,steɪnə'bɪlɪti/ *noun* the ability of a process or human activity to meet present needs but maintain natural resources and leave the environment in good order for future generations

sustainable agriculture *noun* environmentally friendly methods of farming that allow the production of crops or livestock without damage to the ecosystem

sustainable consumption and production *noun* the idea that agricultural production should not cause environmental damage, exploit workers or use up natural resources that cannot be replaced. Abbr **SCP**

sustainable development *noun* development that balances the satisfaction of people's immediate interests and the protection of future generations' interests

Sustainable Development Commission *noun* an independent body which advises the UK government on ethical considerations in environmental, social and economic development

sustainable energy *noun* energy produced from renewable resources that does not deplete natural resources

Sustainable Farming and Food Strategy *noun* a strategy produced by Defra to support farming and food industries in working towards practices that will lead to a better environment and healthy and prosperous communities

sustainable food chain *noun* a food chain from producer to consumer which is environmentally responsible and sustainable at all stages

sustainable production processes *noun* agricultural production methods which do not damage or deplete natural resources

sustainable products *plural noun* products which are created and supplied using sustainable methods

sustainable society *noun* a society which exists without depleting the natural resources of its habitat

sustainable tourism *noun* the management of tourist activities to ensure minimum disruption of local infrastructure and environment

'CCW is looking into the idea in partnership with the Wales Tourist Board and others, with the aim of helping sustainable tourism by developing resources for visitors and local people.' [*Farmers Guardian*]

sustainable yield *noun* the greatest productivity that can be derived from a renewable resource without depleting the supply in a specific area

SVD *abbreviation* swine vesicular disease

SVS *abbreviation* State Veterinary Service

Swaledale /'sweɪldeɪl/ *noun* a very hardy breed of sheep, with distinctive twisting horns and a black face with a white nose, which originated in the North Pennines of Yorkshire. The fleece has an outer layer of long coarse wool and an inner layer of fine dense wool. The Swaledale ewe is the mother of the popular lowland 'mule' ewe when mated to the Blue-faced (Hexham) Leicester ram.

swamp *noun* an area of permanently wet land and the plants that grow on it

swampland /'swɒmplænd/ *noun* an area of land covered with swamp

swampy /'swɒmpi/ *adjective* referring to land that is permanently wet

sward /swɔːd/ *noun* a cover of grasses and clovers which makes a pasture

sward height record pad *noun* a notebook in which the height of a sward is recorded

swardsman /'swɔːdzmən/ *noun* a farm worker who looks after or grows pasture

swarm *noun* a large number of insects such as bees or locusts flying as a group ■ *verb* (*of insects*) to fly in a large group

swath /swɒθ/ *noun* **1.** a row of grass or other plants lying on the ground after being cut **2.** a row of potatoes which have been lifted and left lying on the ground

swath turner *noun* a haymaking machine used to move individual swaths sideways and turn them over at the same time, so making the drying process faster. It is also used in wet conditions to scatter a swath to dry it more quickly.

swayback disease /'sweɪbæk dɪˌziːz/ *noun* an often fatal disease of lambs caused by copper deficiency in the ewe's diet. Lambs become unsteady and unable to walk. The disease is often a problem when there has been no snow during the winter.

swede *noun* a vegetable (*Brassica rutabaga*) with a swollen root. An important forage crop, it is grown for feeding sheep and cattle, either in the field or as winter feed for housed livestock. Swedes have a slightly higher feeding value and keep better than turnips, so they are often lifted and clamped.

Swedish Red and White /ˌswiːdɪʃ red ən 'waɪt/ *noun* a dual-purpose breed of cattle found in Central and Southern Sweden. The animals are cherry red in colour with white markings.

sweeper bull /'swiːpə bʊl/ *noun* a bull used to serve cows that have not been artificially inseminated

sweet corn *noun* a type of maize in which the grains contain a large amount of sugar rather than starch. It is grown for human consumption. Also called **corn on the cob**

sweetener *noun* an artificial substance such as saccharin added to food to make it sweet

sweet potato *noun* a starchy root crop grown in tropical and subtropical regions

COMMENT: The sweet potato is valuable as famine food in parts of Africa and South America. The main producing countries are Indonesia, Vietnam and Japan. In the Southern USA, the tubers are called 'yams'. The plant has no connection with the ordinary potato.

swell *verb* to grow fat ○ *The rain will help the grapes to swell.*

swidden farming /'swɪd(ə)n ˌfɑːmɪŋ/ *noun* same as **slash and burn agriculture**

swill *noun* waste food from kitchens, formerly used for pig feeding but banned after the foot-and-mouth outbreak of 2001

swine *noun* a collective term for pigs

swine erysipelas *noun* an infectious disease of pigs caused by bacteria. Symptoms include inflammation and skin pustules. The red marks on the skin are diamond-shaped, from which the disease gets its common name of 'diamonds'. It occurs especially in hot muggy weather and in its acute form can be fatal.

swine fever *noun* a notifiable disease of pigs. Its symptoms are fever, loss of appetite and general weakness, and it can be fatal. The disease was eradicated in Britain, but some further cases have occurred. Also called **pig typhoid**

swine vesicular disease *noun* a virus infection similar to foot and mouth disease, but which only attacks pigs. It is a notifiable disease and infected animals are compulsorily slaughtered and movement licences apply.

swinging drawbar *noun* a moveable drawbar which is used to pull heavy trailed equipment

swollen shoot *noun* a viral disease affecting the cocoa tree (NOTE: **swelled – swollen**)

SWT *abbreviation* Scottish Wildlife Trust

sycamore *noun* a large hardwood tree of the maple family. Latin name: *Acer pseudoplatanus*.

symbiont /'sɪmbaɪɒnt/ *noun* one of the set of organisms living in symbiosis with each other. Compare **commensal**

symbiosis /ˌsɪmbaɪ'əʊsɪs/ *noun* a condition where two or more unrelated organisms exist together enabling both to survive. Also called **mutualism**

'The bacteria, a group of 80 different strains that live in symbiosis, are incorporated into the slurry and combat the anaerobic bacteria normally found in slurry pits and which are thought to be responsible for gas emissions and unpleasant odours.' [*Farmers Guardian*]

symbiotic /ˌsɪmbaɪ'ɒtɪk/ *adjective* referring to symbiosis ○ *The rainforest has evolved symbiotic mechanisms to recycle minerals.*

symbiotically /ˌsɪmbaɪˈɒtɪkli/ *adverb* in symbiosis ○ *Colonies of shellfish have parasites that live symbiotically with them.*

symptom *noun* a change in the functioning or appearance of an organism, which shows that a disease or disorder is present

symptomatic *adjective* referring to a medical symptom ○ *The rash is symptomatic of measles.*

syndicate *noun* a group of people or companies working together to make money

syndrome *noun* a group of symptoms and other changes in an organism's functions which, when taken together, show that a particular disease or disorder is present

synthetic insecticide *noun* an insecticide that is made artificially from chemicals that do not occur naturally in plants

system *noun* an arrangement of things or phenomena that act together ○ *a weather system*

systematic *adjective* organised in a planned way

systematic ploughing *noun* the process of ploughing a field by sections, each of which is called a land

systemic /sɪˈstiːmɪk/ *adjective* affecting a whole organism

systemic compound *noun* a compound fertiliser which reaches all parts of a plant

systemic fungicide *noun* a fungicide that is absorbed into a plant's sap system through its leaves or roots and protects the plant from infection by fungi without killing the plant itself. ◊ **systemic pesticide**

systemic herbicide *noun* a herbicide that is absorbed into a plant's sap system through its leaves or roots and is transported through the plant to kill the roots. Also called **systemic weedkiller**

systemic insecticide *noun* an insecticide which is taken up by a plant and enters the sap stream so that biting insects take the insecticide when they suck the sap

systemic pesticide *noun* a pesticide that is absorbed into a plant's sap system through its leaves or roots and protects the plant from pests without killing the plant itself

systemic weedkiller *noun* same as **systemic herbicide**

T

T *abbreviation* tonne

tabanidae /təˈbænɪdaɪ/, **tabanids** *plural noun* a family of horse flies with strong antennae, which are often large and fly fast. Most females suck blood and attack large mammals such as cattle with their blade-like jaws.

Tabanus /təˈbeɪnəs/ *noun* any of the tabanidae

table bird *noun* a poultry bird reared for meat

table chicken *noun* a chicken raised for eating rather than producing eggs

tack *noun* **1.** harness equipment **2.** □ **on tack** taken from or by one farmer to graze on another farmer's fields ○ *He takes 360 store lambs on tack in the autumn to remove surplus grass.*

tackroom /ˈtækrʊm/ *noun* a room for storing harness equipment. The tack should be kept in damp-free conditions.

Taenia /ˈtiːniə/ *noun* a genus of tapeworm

tag *noun* a label attached to an animal to identify it

tailbiting /ˈteɪlbaɪtɪŋ/ *noun* a form of behaviour, especially associated with pigs, in which an animal bites the tail of another animal. The cause is not known but could be due to bad housing.

tail corn, tailings /ˈteɪlɪŋz/ *noun* grains of corn of inferior size

tailpiece /ˈteɪlpiːs/ *noun* an extension of the mouldboard of a plough which helps to press down the furrow slice

taint /teɪnt/ *verb* to give an unpleasant taste to food

take *verb* to grow successfully

take-all *noun* a disease of wheat and barley, causing black discoloration at the base of the stem, premature ripening and white ears containing little or no grain

(NOTE: This is also called 'whiteheads', although whiteheads can be caused by other diseases.)

tall fescue *noun* a very hardy perennial grass often used for winter grazing in hilly or less fertile areas

tallow /ˈtæləʊ/ *noun* a cattle by-product produced by rendering down all the inedible waste, used in the manufacture of soap and formerly incorporated into animal feeds

Tamworth /ˈtæmwɜːθ/ *noun* a breed of pig, red-gold in colour, which makes the animal almost immune to sunburn

COMMENT: Tamworths are widely exported to the USA and Australia because of their ability to stand hot sunshine. They are hardy and can thrive on the roughest land.

tan *verb* to convert animal skins to leather

tandem parlour *noun* a milking parlour where the cows stand in line with their sides to the milker

tank *noun* **1.** a large container for storing fluid ○ *water tank* ○ *fuel tank* **2.** a large container for liquid, part of a spraying machine

tankard *noun* a variety of mangel

tanker *noun* a truck used to carry liquids such as petrol or milk

tank mix *noun* the process of mixing several pesticides into one mixture to be used as a spray

'Removal of the majority of green material will reduce the risk of blight infection following flailing although advice is to include an appropriate fungicide in the tank mix to protect from tuber blight.' [*Farmers Guardian*]

tannin /ˈtænɪn/, **tannic acid** /ˈtænɪk ˈæsɪd/ *noun* **1.** a brownish or yellowish compound formed in leaves and bark

(NOTE: Tannins in forage plants, especially peas and beans, prevent nutrients from the plants being fully absorbed by the grazing livestock.) **2.** a substance capable of changing animal hides to leather, which is resistant to decay, by precipitating the gelatin in the hides as an insoluble compound. All tannins are obtained from plants, including tea, and the bark and galls of oak.

tanning /'tænɪŋ/ noun the process of converting animal skins to leather

COMMENT: Hides are soaked in a mixture of chromium salts and after a certain period of soaking they become leather.

tap noun a pipe with a handle that can be turned to make a liquid or gas come out of a container ■ verb to remove or drain liquid from something

tapeworm /'teɪpwɜːm/ noun a parasitic worm with a long flattened segmented body that lives mainly in the guts of vertebrate animals, including humans. Class: Cestoda.

COMMENT: Tapeworms enter the intestine when a person eats raw fish or meat. The worms attach themselves with hooks to the side of the intestine and grow longer by adding sections to their bodies. Tapeworm larvae do not normally develop in humans, with the exception of the pork and dog tapeworms, *Taenia solium*. Tapeworms seldom need treatment in livestock, although sheepdogs should be wormed regularly.

taproot /'tæp,ruːt/ noun the thick main root of a plant which grows straight down into the soil (NOTE: A taproot system has a main root with smaller roots branching off it, as opposed to a fibrous root system which has no main root.)

tare /teə/ noun **1.** same as **vetch 2.** the weight of a container or of the packaging in which goods are wrapped

Tarentaise /,tærən'teɪz/ noun a breed of dairy cattle from the Savoie region of France. The animals are yellowish fawn in colour, with black muzzle, ears and tail.

target price noun the wholesale price within the EU that market management is intended to achieve for certain products such as wheat. It is linked to the support price.

COMMENT: Target prices are set in terms of fixed agricultural units of account, which are converted into different national currencies using adjusted exchange rates known as 'green rates' (in the UK, the 'green pound'). A system of levies on non-EU agricultural imports is used to protect target prices when they are set above the general level of world prices. In addition, the EU has established an internal price support system based on a set of intervention prices set slightly below the target price. If the level of supply is in excess of what is needed to clear the market at the target price, the excess supply is bought by the Community at the intervention price, thereby preventing overproduction from depressing the common price level.

tariff noun a duty or duties levied by a government on imported or sometimes exported goods

tar oil noun a winter wash used to control aphids and scale insects on fruit trees

tarragon /'tærəgən/ noun an aromatic plant (*Artemisia dracunculus*) of which the leaves are used for seasoning

tassel noun **1.** a male flower of the maize plant **2.** an appendage of hair hanging from the neck of male turkeys ■ verb to produce a tassel ○ *The crop tasselled early.*

tattoo noun a mark made on an animal's body to identify it

TBC abbreviation total bacterial count

tea noun the dried leaves of one or more shrubs of the Camellia family, of which the commonest is *Camelia sinensis*

COMMENT: Tea is one of the hardiest of all subtropical plants. The plants are pruned to form low bushes and to encourage the production of leaves and help plucking. The principal tea-producing countries are India, Bangladesh, Sri Lanka, China, Indonesia and Japan.

teaser ram noun a vasectomised ram, used to stimulate ewes by 'non-fertile' mating prior to the introduction of fertile rams

teat /tiːt/ noun a nipple on an udder. In cattle there are four quarters to the udder, each drained by a teat.

teat chaps plural noun sores on the teat, probably due to abrasions caused by the milking machine

teat cup noun a tube forming part of a milking machine, which fits over the teat of the cow

teat dipping noun a measure for control of mastitis in cattle. The teats are dipped in a cup containing an iodophor disinfectant.

Technical Committee noun a committee appointed by the UK government in 1965, which reviewed the welfare of animals kept under intensive livestock husbandry systems

technology *noun* **1.** the use of scientific knowledge to develop machines and techniques for use in industry **2.** machines and techniques developed using scientific knowledge ○ *The technology has to be related to user requirements.*

tedder /'tedə/ *noun* a machine used to lift and loosen a swath, enabling air to circulate through the cut crop

COMMENT: Tedders may be mounted or trailed, and some are of the rotary type with spring tines which rotate on a vertical axis and move the crop sideways. Others, overshot hay tedders, have a rotor with a series of spring tines which lift the crop over the rotor and back to the ground.

tedding /'tedɪŋ/ *noun* the process of spreading by lifting the swaths of new-mown grass in haymaking, so as to expose more grass to the sun and air and make it dry more quickly

Teeswater /'tiːzwɔːtə/ *noun* a breed of longwool sheep with a dark muzzle. It is used to provide rams for cross-breeding with Swaledale ewes to produce the hybrid Masham.

teg /teg/ *noun* a sheep in its second year

Telemark /'telɪmɑːk/ *noun* a Norwegian breed of dairy cattle. The animals are red with white patches.

tel quel /ˌtel 'kwel/ *adjective* referring to the weight of any type of sugar in tonnes

temporary grassland *noun* arable land sown to ley for a limited period

tenant *noun* a person who pays rent for the use of a farm and land owned by a landlord

Tenant Farmers Association *noun* a group formed to represent the interests of tenant farmers on a national scale. Abbr **TFA**

tender *adjective* **1.** soft or susceptible to damage **2.** referring to a plant which cannot tolerate frost

tenderise /'tendəraɪz/, **tenderize** *verb* to make meat tender by keeping it for a certain time in cold conditions, by applying substances such as papain, by injecting with enzymes, etc.

tenderometer /ˌtendə'rɒmɪtə/ *noun* a device used for testing vining peas to see how firm they are, so allowing harvesting to take place at the right time

tendon *noun* a strip of connective tissue that attaches a muscle to a bone

tendril /'tendrəl/ *noun* a stem, leaf or petiole of a plant modified into a thin touch-sensitive organ that coils around objects, providing support for climbing plants

terminal *adjective* referring to a shoot or bud at the end of a shoot

terminal sire *noun* a sire used in cross-breeding, whose progeny will possess a high rate of growth and good carcass quality, but will not be suitable for breeding themselves

terrace *noun* a flat strip of land across a sloping hillside, lying level along the contours ■ *verb* to build terraces on a mountainside ○ *The hills are covered with terraced rice fields.*

COMMENT: Terracing is widely used to create small flat fields on steeply sloping land, so as to bring more land into productive use, and also to prevent soil erosion.

terrace cultivation *noun* hill slopes cut to form terraced fields which rise in steps one above the other and are cultivated, often with the aid of irrigation

terrain *noun* the ground or an area of land in terms of its physical surface features ○ *mountainous terrain*

terra rossa *noun* a red soil that develops over limestone, found in Spain, Southern France and Southern Italy

terrier *noun* a record of land held and its occupation and use

tertiary *adjective* coming after two other things. ◊ **primary**, **secondary**

Teschen disease /'teʃən dɪˌziːz/ *noun* a virus disease of pigs caused by an enterovirus. It is a notifiable disease which causes fever, paralysis and often death.

testa /'testə/ *noun* the tough protective skin around a seed which protects the embryo inside. Also called **seed coat**

testicle /'testɪk(ə)l/ *noun* same as **testis**

testis /'testɪs/ *noun* one of two male sex glands in the scrotum, producing sperm (NOTE: In mammals, including humans, the paired testes also secrete sex hormones. The plural of **testis** is **testes**.)

tetanus *noun* an infection caused by *Clostridium tetani* in the soil, which affects the spinal cord and causes spasms which occur first in the jaw

COMMENT: Livestock, especially horses, can be affected by the disease. People who work on the land or with soil, such as farm workers or construction workers, should be immunised against tetanus.

tether *noun* a rope, chain or halter, used to tie up animals ■ *verb* to tie up an animal with a rope or chain, so that it cannot move away

tetracycline /ˌtetrə'saɪkliːn/ *noun* an antibiotic used against various bacterial diseases

tetraploid /'tetrəplɔɪd/ *noun* forms of grass and clover with larger seeds and a larger plant than ordinary grass and clover. Deep green in colour, they are lower in dry matter, palatable and digestible. There are also tetraploidal varieties of wheat. (NOTE: The number of chromosomes has been doubled from the normal diploid state in ryegrasses.)

Texan angora goat *noun* a breed of goat with very fine hair, imported from America

Texel /'teksel/ *noun* a breed of sheep from the North of Holland, used to cross-breed as a flock sire

TFA *abbreviation* Tenant Farmers Association

TGE *abbreviation* transmissible gastroenteritis

t/ha *abbreviation* tonnes per hectare

thatch /θætʃ/ *verb* to cover a roof with reeds, straw, grass or other plant material ○ *Reeds provide the longest-lasting material for thatching.* ■ *noun* a covering for a roof made of reeds, straw, grass or other plant material ○ *birds nesting in the old thatch*

theaves /θiːvz/ *plural noun* female sheep between the first and second shearing

theoretical field capacity *noun* the rate of work that would be achieved if a machine were performing its function at its full-rated forward speed for 100% of the time

therophyte /'θerəʊfaɪt/ *noun* an annual plant that completes its life cycle rapidly in favourable conditions, growing from a seed and dying within one season and then surviving the unfavourable season in the form of seeds (NOTE: Many desert plants and plants growing on cultivated land are therophytes.)

thiabendazole /ˌθaɪə'bendəzəʊl/ *noun* a substance used to worm cattle

thiamine /'θaɪəmɪn/, **thiamin** *noun* vitamin B₁, found in yeast, cereals, liver and pork

thicket *noun* a wood of saplings and bushes growing close together

thin *verb* to remove a number of small plants from a crop, so allowing the remaining plants to grow more strongly

thinnings /'θɪnɪŋz/ *noun* small plants removed to let others grow more strongly

thistle *noun* a perennial weed (*Cirsium arvense, Cirsium vulgare*) with spiny or prickly leaves, which grows as an erect plant and has large purple or white flower heads

thorax /'θɔːræks/ *noun* **1.** a cavity in the top part of the body of an animal above the abdomen, containing the diaphragm, heart and lungs, all surrounded by the rib cage **2.** the middle section of the body of an insect, between the head and the abdomen

thorn *noun* **1.** a sharp woody point on plant stems or branches **2.** a plant or tree that has sharp woody points on its stems or branches

thornless /'θɔːnləs/ *adjective* referring to a plant without thorns ○ *He developed a new thornless variety of blackberry.*

thorny /'θɔːni/ *adjective* with thorns

thoroughbred *adjective, noun* (referring to) a horse that is bred for particular characteristics, in particular, a horse bred for racing

thousand grain weight *noun* the weight of a thousand grains, used as an indicator of grain quality

thousand-headed kale, thousand-head kale *noun* a variety of kale grown for feeding to livestock, usually in the winter months. It has many branches and small leaves. The dwarf thousand-head produces a large number of new shoots during the winter.

threadworm /'θredwɜːm/ *noun* a thin parasitic worm which infests the large intestine. Genus: *Enterobius*.

three-point linkage *noun* a method of coupling implements to a tractor. Automatic couplers for three-point linkage permit implements to be attached rapidly and safely.

three-times-a-day *noun* a milking system in which cows are milked three times a day. Using this system can increase milk yields.

thresh /θreʃ/ *verb* to separate grains from stalks and the seedheads of plants

threshing machine *noun* a machine formerly used to thresh cereals, now replaced by the combine harvester

threshold price *noun* in the EU, the lowest price at which farm produce imported into the EU can be sold. This is the price in the home market below which the government or its agencies must buy all the produce offered by producers for sale at that price.

thrift /θrɪft/ *noun* good health in an animal

'PMWS is a complex disease. The devastating impact seen in newly affected herds is well recognised. But it may be difficult to distinguish from other causes of ill thrift if present at a low level.' [*Farmers Weekly*]

thrifty /'θrɪfti/ *adjective* referring to an animal which is developing well

thrips /θrɪps/ *noun* an insect that is a pest of vegetables. Thrips operate on the underside of leaves, leaving spots of sap or other liquid which are red or blackish-brown. Typical examples are the onion thrips, grain thrips and pea thrips.

thrive *verb* (*of an animal or plant*) to develop and grow strongly ○ *These plants thrive in very cold environments.*

throw *verb* to give birth to young

thrunter /'θrʌntə/ *noun* a three-year-old ewe

thrush *noun* a disease affecting the frog of a horse's hoof

Thuya /'θuːjə/ *noun* the Latin name for cedar

thyme *noun* a common aromatic Mediterranean plant (*Thymus*) used for flavouring soups, stuffings and sauces

tick *noun* a tiny parasite that sucks blood from the skin. Order: Acarida.

tick bean *noun* a small bean (*Vicia faba L*), usually used for feeding to horses and other animals

tick-borne fever *noun* an infectious disease transmitted by bites from ticks. In cattle, the disease causes loss of milk yield and a lower resistance to other diseases. In sheep it causes fever, listlessness and loss of weight. Abortions may occur as a result of tick-borne fever.

tick pyaemia /'tɪk paɪˌiːmiə/ *noun* a disease affecting young lambs resulting in limb joint and internal abscesses

tied cottage *noun* a house which can be occupied by the tenant as long as the tenant remains an employee of the landlord

tier *noun* a range of things placed one row above another, e.g. the arrangement of cages in a battery

tilapia /tɪ'læpiə/ *noun* a tropical white fish, suitable for growing in fish farms

tile draining *noun* a means of draining land using underground drains made of clay, plastic or concrete. Special machines called 'tile-laying machines' are available for this work.

till *verb* to prepare the soil, especially by digging or ploughing, to make it ready for the cultivation of crops

tillage /'tɪlɪdʒ/ *noun* the activity of preparing the soil for cultivation

tiller /'tɪlə/ *noun* a shoot of a grass or cereal plant, which forms at ground level in the angle between a leaf and the main shoot. True stems are only produced from the tillers at a later stage in the plant's development.

tillering /'tɪlərɪŋ/ *noun* the process of developing several seedheads in a plant of wheat, barley or oats (NOTE: Tillering leads to the production of a heavier yield, and can be induced by rolling the young crop in the spring when it begins to grow. Tillering is used to compensate for poor establishment. It is also important in grasses.)

tilth /tɪlθ/ *noun* a good light crumbling soil prepared to be suitable for growing plants ○ *Work the soil into a fine tilth before sowing seeds.*

timber *noun* trees which have been or are to be cut down and made into logs

timberline /'tɪmbəlaɪn/ *noun US* same as **treeline**

Timothy /'tɪməθi/ *noun* a palatable tufted perennial grass (*Phleum pratense*). It grows on a wide range of soils and is winter hardy. It is used in grazing mixtures and as a hay plant in conjunction with ryegrass.

tine /taɪn/ *noun* **1.** a pointed spike on a cultivator or harrow. Types of tine include rigid, spring-loaded and spring. **2.** a sharp prong of a fork or rake

tined /'taɪnd/ *adjective* with tines ○ *a spring-tined harrow*

tine harrows *plural noun* sets of curved tines sometimes used when the soil surface is caked or compacted. A tine harrow will break up the soil to depth of several inches.

Tipulidae /tɪ'pjuːlɪdi/ *noun* a family of insects including crane flies and their larvae, leatherjackets, which destroy plant roots

Tir Cymen *noun* an agri-environmental scheme for Wales, closed to new entrants since 1998

Tir Gofal *noun* an agri-environmental scheme for Wales, established in 1999

tissue culture *noun* **1.** plant or animal tissue grown in a culture medium **2.** a method of plant propagation which reproduces clones of the original plant on media containing plant hormones

title *noun* **1.** the right to hold goods or property **2.** a document proving a right to hold property

title deeds *plural noun* documents showing who is the owner of a property

TMR *abbreviation* total mixed ration

toadstool *noun* a fungus that resembles an edible mushroom, but which may be poisonous

tobacco mosaic virus *noun* a virus affecting both tobacco plants and tomatoes

tocopherol /tɒ'kɒfərɒl/ *noun* one of a group of fat-soluble chemicals that make up vitamin E (NOTE: It is particularly abundant in vegetable oils and leafy green vegetables.)

toe-in *noun* the shorter distance between the bases of the front wheels of a tractor, compared to their tops. Toe-in improves steering performance and reduces wear on the front tyres.

Toggenburg /'tɒgənbɜːg/ *noun* a small Swiss breed of goat, which is pale brown in colour with white markings on face, legs and rump. In Britain it has been developed into a larger, darker animal, which is a good milker with a long lactation period.

tolerance *noun* the ability of an organism to accept something, or not to react to something ○ *plants with frost tolerance*

tolerant *adjective* not reacting adversely to something ○ *a salt-tolerant plant*

tolerant variety *noun* a variety of crop which has been developed to withstand a disease or attacks by certain pests

tolerate *verb* not to react adversely to something

toleration /ˌtɒlə'reɪʃ(ə)n/ *noun* the ability to tolerate something, or the act of tolerating something ○ *poor toleration of high temperatures*

tomato *noun* an important food crop that produces a round fruit with a bright red skin and pulpy flesh with many seeds. The ripe fruit are used in salads and many cooked dishes, and also pressed to make juice and sauces. Large quantities are also canned.

tom turkey *noun* a male turkey

ton *noun* **1.** a unit of measurement of weight, equal to 1016kg. Also called **long ton 2.** *US* a unit of measurement of weight, equal to 907kg. Also called **short ton**

tonne *noun* a unit of measurement of weight, equal to 1000kg. Also called **metric ton**

top dressing *noun* a fertiliser applied to a growing crop

topknot /'tɒpnɒt/ *noun* a tuft of hair on the top of an animal's head, found in certain breeds of cattle and sheep

top link sensor *noun* the mechanism by which most draught controls sense the draught on a tractor implement. It uses the top link of the three-point linkage.

topper /'tɒpə/ *noun* a machine used to cut the tops off sugar beet. ◊ **pasture topper**

topper unit *noun* a unit forming part of a two- or three-stage system, with a chopper and blower unit

topping *noun* the process of cutting the leaves and stems from the sugar beet root. It must be done accurately, as overtopping reduces yield.

tops *plural noun* leaves and stems of plants such as sugar beet which are cut off and used as fodder for cattle and sheep or made into silage

top-saving attachment *noun* an attachment to a topper unit which collects the tops of sugar beet after they have been cut off

topsoil /'tɒpsɔɪl/ *noun* the top layer of soil, often containing organic material, from which chemical substances are washed by water into the subsoil below

total bacterial count *noun* a system of calculating the strength of an infection by counting the number of bacteria present in a sample quantity of liquid taken from the animal. Abbr **TBC**

total mixed ration *noun* a winter feed for livestock involving mixing of concentrates with roughage and allowing the animals free feeding of the mix. Abbr **TMR**

Toulouse /'tuːluːz/ *noun* a medium-large grey and white breed of goose, which originates in France

tow *verb* to pull another wheeled vehicle or implement

towbar /'təʊbɑː/ *noun* a strong bar at the back of a car or tractor, to which another vehicle can be attached to be pulled along

tower silo *noun* a tall circular tower used for storing silage

Townshend, Viscount (1674–1738) an 18th-century Norfolk landowner. Nicknamed 'Turnip' Townshend, he did much to make the Norfolk four-course rotation system popular.

tox- /tɒks/, **toxi-** /tɒksi/ *prefix* poison

toxaemia /tɒk'siːmiə/ *noun* blood poisoning. ◊ **pregnancy toxaemia**

toxic *adjective* referring to a substance that is poisonous or harmful to humans, animals or the environment

toxicity /tɒk'sɪsɪti/ *noun* the degree to which a substance is poisonous or harmful ○ *They were concerned about the high level of toxicity of the fumes.*

toxico- /tɒksɪkəʊ/ *prefix* poison

toxicological /ˌtɒksɪkə'lɒdʒɪk(ə)l/ *adjective* referring to toxicology ○ *Irradiated food presents no toxicological hazard to humans.*

toxicologist /ˌtɒksɪ'kɒlədʒɪst/ *noun* a scientist who specialises in the study of poisons

toxicology /ˌtɒksɪ'kɒlədʒi/ *noun* the scientific study of poisons and their effects on the human body

toxicosis /ˌtɒksɪ'kəʊsɪs/ *noun* poisoning

toxic substance *noun* a substance that is poisonous or harmful to humans, animals or the environment

toxin *noun* a poisonous substance produced by microorganisms. ◊ **mycotoxin**

toxoplasmosis /ˌtɒksəʊplæz'məʊsɪs/ *noun* an infectious disease affecting ewes, which causes pregnant animals to abort

t/pa *abbreviation* tonnes per annum

TPO *abbreviation* tree preservation order

trace *noun* a very small amount of something ○ *There are traces of radioactivity in the sample.*

traceability /ˌtreɪsə'bɪlɪti/ *noun* the concept that each stage in the supply chain from farm to consumer can be traced so that the quality of the food can be guaranteed

'A new survey has found Welsh food is riding high among consumers who are identifying and asking for local produce when eating out – and seemingly do not mind paying that little bit extra. Commissioned by the Welsh Development Agency and carried out by Beaufort Research, the survey also highlights the importance of traceability, with consumers said to be increasingly interested in the origin of their food.' [*Farmers Guardian*]

trace element *noun* a chemical element that is essential to organic growth but only in very small quantities

COMMENT: Plants require traces of copper, iron, manganese and zinc. Humans require the trace elements chromium, cobalt, copper, magnesium, manganese, molybdenum, selenium and zinc.

traces *plural noun* side-straps or chains by which a horse pulls a cart or implement

tracklayer /'træsleɪə/, **tracklaying tractor** *noun* a heavy-duty caterpillar tractor, used mainly for earthmoving and drainage work

tractor /'træktə/ *noun* a heavy vehicle with large wheels used for a range of tasks. On farms it is mainly used for pulling cultivation and spraying equipment.

COMMENT: The general-purpose tractor does most of the work on arable and livestock farms and may have either two- or four-wheel drive. Lighter tractors, usually two-wheel drive models, are used by market gardeners. More powerful four-wheel drive tractors are needed for ploughing and heavy cultivation. The heaviest tractors are tracklayers, or 'crawlers', which are used for very heavy work and on heavy soils. Besides pulling trailed implements such as balers, forage harvesters and drills, the tractor's hydraulic system can be used to raise and lower mounted implements and operate lifting and loading equipment. The hydraulic system provides the power for fertiliser spreaders, hedge trimmers and a variety of other implements. Medium-sized tractors develop 40–60hp, while large models can develop as much as 200hp. Very modern tractors have in-cab computers which can tell the driver how much ground has been covered, and how much fuel has been used, and can even advise on the gear which will give the most economic use of fuel. In the UK, since 1970, all new tractors must be fitted with an approved protective cab or frame to prevent the driver being crushed in accidents where the tractor rolls over.

tractor-mounted loader *noun* a loader which is mounted on a tractor, and not trailed

tractor vaporising oil *noun* a fuel formerly used in many tractors, but now replaced by diesel oil. Abbr **TVO**

Trade Union Sustainable Development Advisory Committee *noun* a forum for consultation on environmental issues between the Government and the main TUC-affiliated trade unions in the UK. Abbr **TUSDAC**

trafficability /ˌtræfɪkəˈbɪlɪti/ *noun* the ability of soil to take machinery or stock without significant soil damage. It is related to the soil water content.

trail *noun* **1.** a path or track ○ *created a new nature trail in the forest* **2.** a mark or scent left by an animal ○ *on the trail of a badger* ■ *verb* to pull another wheeled vehicle or implement

trailed implements *plural noun* implements such as harrows which are pulled behind a tractor. See Comment at **mount**

trailer *noun* a machine used for carrying purposes. Trailers are of the two-wheel or four-wheel types, and are used for carrying cereal and root crops, and for general use on the farm.

trailing *adjective* referring to a plant whose shoots lie on the ground

train *verb* to make plants, especially fruit trees and climbing plants, become a certain shape, by attaching shoots to supports or by pruning

tramline /ˈtræmlaɪn/ *noun* a path left clear for the wheels of tractors to drive over. They are used as guidemarks for spraying and when applying fertiliser so that damage to crops is kept to a minimum.

trans- /trænz/ *prefix* through or across

transgenesis /trænzˈdʒenəsɪs/ *noun* the process of transferring genetic material from one organism to another

transgenic /trænzˈdʒenɪk/ *adjective* **1.** referring to an organism into which genetic material from a different species has been transferred using the techniques of genetic modification **2.** referring to the techniques of transferring genetic material from one organism to another ■ *noun* an organism produced by genetic modification

'Meanwhile in New Zealand the government has given the green light for further research with transgenic cows, despite objections from anti-GM and green groups. The research is part of a long term programme to develop milk capable of treating multiple sclerosis.' [*Dairy Farmer*]

transhumance /trænsˈhjuːməns/ *noun* the practice of moving flocks and herds up to high summer pastures and bringing them down to a valley again in winter

translaminar fungicide /ˌtrænzlæmɪnə ˈfʌŋgɪsaɪd/ *noun* a fungicide that is absorbed into a plant's system through its leaves

translocate /ˌtrænsləˈkeɪt/ *verb* to move substances through the tissues of a plant

translocated herbicide *noun* a herbicide that kills a plant after being absorbed through its leaves

translocation /ˌtrænsləʊˈkeɪʃ(ə)n/ *noun* a movement of substances through the tissues of a plant

'Timing of application is crucial and must be when the weed and grass are growing to ensure good control and minimal check to grass growth. Grass should not be grazed for seven days or cut for 28 days after application to allow translocation to take place.' [*Dairy Farmer*]

transmissible gastro-enteritis *noun* a very infectious disease, which mainly affects very young pigs. Abbr **TGE**

transmissible spongiform encephalopathy *noun* the name given to the group of spongiform encephalopathies which affect animals or humans, including scrapie and BSE. Abbr **TSE**

transmit *verb* to pass on a disease to another animal or plant ○ *Some diseases are transmitted by insects.* (NOTE: **transmitting – transmitted**)

transpiration /ˌtrænspɪˈreɪʃ(ə)n/ *noun* **1.** the loss of water from a plant through its stomata **2.** the removal of moisture from the soil by plant roots, which pass the moisture up the stem to the leaves

COMMENT: Transpiration accounts for a large amount of water vapour in the atmosphere. A tropical rainforest will transpire more water per square kilometre than is evaporated from the same area of sea. Clearance of forest has the effect of reducing transpiration, with an accompanying change in climate: less rain, leading to eventual desertification.

transpire *verb* (*of a plant*) to lose water through stomata ○ *In tropical rainforests, up to 75% of rainfall will evaporate or transpire into the atmosphere.*

transplant *noun* **1.** taking a growing plant from one place and planting it in the soil in another place **2.** a plant taken from one place and planted in the soil in another place ■ *verb* to take a growing plant from one place and plant it in the soil in another place

transplanter /træns'plɑːntə/ *noun* a machine for transplanting seedlings, especially used for planting brassicas

COMMENT: There are two types of transplanter. One is the hand-fed machine, where a worker feeds seedlings into the machine as it passes over the field. The other is the automatic transplanter, where seedlings are raised in special containers and placed in the machine before transplanting starts.

transport *noun* **1.** a system of moving things from one place to another ○ *road and rail transport* ○ *an integrated transport policy* Also called **transportation 2.** the activity of moving something from one place to another (NOTE: Animal welfare codes lay down rules on how animals and birds should be treated during transport in order to ensure that their health and welfare is good.)

tray *noun* a flat shallow container, usually made of plastic, in which seeds can be sown in a greenhouse

tree *noun* a plant typically with one main woody stem that may grow to a great height

tree farming *noun* the growing of trees for commercial purposes

COMMENT: Schemes have been started to grow trees on surplus agricultural land to provide fuel for wood-burning power stations. Farmers will be paid to plant genetically engineered poplars and willows. 90,000 tonnes of dry wood are expected each year. Grants are available to cover some of the costs of establishing short rotation coppice as an energy crop.

treeline /'triːlaɪn/ *noun* **1.** a line at a specific altitude, above which trees will not grow ○ *The slopes above the treeline were covered with boulders, rocks and pebbles.* **2.** a line in the northern or southern hemi-sphere, north or south of which trees will not grow

tree nursery *noun* a place where trees are grown from seed until they are large enough to be planted out

tree preservation order *noun* an order from a local government department that prevents a tree from being cut down. Abbr **TPO**

tree ring *noun* same as **annual ring**

tree stump *noun* a short section of the trunk of a tree left in the ground with the roots after a tree has been cut down

tree surgeon *noun* a person who specialises in the treatment of diseased or old trees, by cutting or lopping branches

trefoil /'triːfɔɪl/ *noun* a leguminous plant, the thin wiry form of a small-flowered yellow clover, sometimes grown in pasture mixtures. It is a useful catch crop, and thrives in marshy acid soils.

trematode /'tremətəʊd/ *noun* a fluke, a parasitic flatworm

trembles *noun* same as **louping-ill**

trench *noun* a long narrow hole in the ground

trenching /'trenʃɪŋ/ *noun* a method of double digging which loosens the soil to a depth of two feet, twice as deep as in plain digging

trial *noun* a test carried out to see if something works well

triazine /'traɪəziːn/ *noun* one of a group of soil-acting herbicides, e.g. Atrazine and Simazine. Maize is tolerant to these substances. (NOTE: Triazines will no longer be approved for use in the UK after December 2007.)

triazole /'traɪəzɒl/ *noun* one of a group of systemic, protectant and curative fungicides, e.g. tebuconazole or myclobutanil. Triazoles can be used on a wide range of crops.

trichinosis /ˌtrɪkɪ'nəʊsɪs/, **trichiniasis** /ˌtrɪkɪ'naɪəsɪs/ *noun* a disease caused by infestation of the intestine by larvae of roundworms or nematodes, which pass round the body in the bloodstream and settle in muscles. Pigs are usually infected after eating raw swill.

trichlorophenoxyacetic acid /ˌtraɪklɔːrəʊfenˌɒksɪəˌsetɪk 'æsɪd/ *noun* a herbicide which forms dioxin as a by-product during the manufacturing process and is effective against woody shrubs

(NOTE: This herbicide is no longer approved for use in the UK.)

Trichomonas /ˌtrɪkəˈməʊnəs/ *noun* a species of long thin parasite which infests the intestines. *Trichomonas foetus* is a cause of infertility in cattle.

trickle irrigation, trickle system *noun* same as **drip irrigation**

trifolium /traɪˈfəʊliəm/ *noun* the crimson clover (*Trifolium incarnatum*), a plant which does best on calcareous loams and is grown after cereals as a catch crop. It is planted in mixed herbage as a winter annual for forage, particularly for sheep.

trifuralin /traɪˈfluːrəlɪn/ *noun* a commonly used herbicide incorporated into the soil before planting a wide range of crops (NOTE: It is under review for withdrawal from use in the European Union.)

trim *verb* to cut off the end parts of something, e.g. the shoots from a bush or a hedge, usually to give the object a neater shape

trimmer /ˈtrɪmə/ *noun* an implement for trimming hedges

trimmings /ˈtrɪmɪŋz/ *plural noun* small pieces of vegetation which have been cut off a hedge when trimming

trip device *noun* a device used to sense when a person is too close to a hazard and isolate the hazard before contact can occur. Trip devices can take the form of trip bars, as used on some rotating arm bale wrappers.

triple-purpose animal *noun* a breed of animal, usually cattle, which is used for three purposes, i.e. for milk, for meat and as a draught animal

tripoding /ˈtraɪpɒdɪŋ/ *noun* the process of drying hay on a wooden frame in the field. It is rarely practised in the UK, but still common in some parts of Europe. Tripoding is also used as a means of drying out peas.

triticale /ˌtrɪtɪˈkeɪli/ *noun* a new cereal hybrid of wheat and rye. It combines the yield potential of wheat with the winter hardiness and resistance to drought of rye.

COMMENT: Increasingly used in the UK, triticale replaces winter and spring feed barleys. It has a high level of disease resistance and a reduced demand for chemical fertiliser. The name is made up from the Latin words for wheat (Triticum) and rye (Secale).

Triticum /ˈtrɪtɪkəm/ *noun* the Latin name for wheat

trocar /ˈtrəʊkɑː/ *noun* a pointed rod which slides inside a cannula to draw off liquid or to puncture an animal's stomach to let gas escape, such as in the treatment of bloat

Trondheim /ˈtrɒndhaɪm/ ♦ **Blacksided Trondheim**

tropical *adjective* referring to the tropics ○ *The disease is carried by a tropical insect.*

trotter /ˈtrɒtə/ *noun* a foot of a pig or sheep

trough *noun* **1.** a long narrow area of low pressure with cold air in it, leading away from the centre of a depression **2.** a long narrow open wooden or metal container for holding water or feed for livestock

trough space *noun* the length of trough that should be allowed per animal in an enclosure, so that they each have space to feed comfortably

'Troughs should be no more than 500 metres apart, otherwise there is a risk that the distance that the cow is required to travel for a drink will impinge on production. Trough space should be sufficient to allow 15% of the herd to drink at one time.' [*Farmers Guardian*]

truck farming *noun US* a term used to describe intensive vegetable cultivation at a considerable distance from the urban markets where the produce is sold

trug /trʌg/ *noun* a low fruit or garden basket made of willow strips fastened to a strong framework of ash or chestnut

trunk *noun* the main woody stem of a tree

truss /trʌs/ *noun* **1.** a bundle of hay or straw **2.** a compact cluster or flowers or fruit such as tomatoes

tryptophan /ˈtrɪptəfæn/ *noun* an essential amino acid

TSE *abbreviation* transmissible spongiform encephalopathy

tuber /ˈtjuːbə/ *noun* a swollen underground stem or root, which holds nutrients and which has buds from which shoots develop ○ *A potato is the tuber of a potato plant.*

COMMENT: Potatoes, cassavas and sweet potatoes are all tubers from which new shoots develop.

tubercle /ˈtjuːbək(ə)l/ *noun* **1.** a small raised area on part of a plant or an animal **2.** a small tuber

tuberculin /tjʊˈbɜːkjʊlɪn/ *noun* a substance which is derived from the culture of the tuberculosis bacillus and is used to test cattle for the presence of tuberculosis

tuberculin testing *noun* the testing of cattle for the presence of bovine tuberculosis

tuberculosis /tjʊˌbɜːkjʊˈləʊsɪs/ *noun* an infectious disease caused by the tuberculosis bacillus, where infected lumps form in tissue and which affects humans and other animals. Cattle and pigs are more commonly affected than other species. It is a notifiable disease.

COMMENT: Tuberculosis in cattle has been eradicated in the UK, by a policy of tuberculin testing and the slaughter of animals which react to the test. Bovine tuberculosis is a notifiable disease and can affect certain wild animals, in particular badgers. Infected badgers are believed by some people to be able to transmit the disease to cattle though this has not been proved.

tuberous /ˈtjuːbərəs/ *adjective* **1.** like a tuber **2.** referring to a plant that grows from a tuber

Tull, Jethro (1674–1740) an 18th-century gentleman farmer. He invented the mechanical seed drill and the horse-drawn hoe.

tunnel *noun* a long enclosure, covered with a semicircular roof

tunnel cloche *noun* a long continuous covering over rows of plants, usually made of plastic

tunnel drying *noun* a method of storage drying of hay, where the bales are stacked in the form of a tunnel over a central duct through which unheated air is blown

tup /tʌp/ *noun* an uncastrated male sheep ■ *verb* to serve a ewe

turbary /ˈtɜːbəri/ *noun* a place where turf or peat is dug for fuel

turf *noun* a surface earth covered with grass, with its roots matted in the soil

turkey *noun* a large poultry bird raised for meat (NOTE: The adult males are called **cocks** or **toms**, the adult females are **hens**.)

turnip *noun* a brassica plant that has a swollen root, is an important forage crop and is also used as a vegetable

COMMENT: Turnips can be harvested by machine and stored outdoors in clamps. In milder areas they can be left growing in the fields and used when needed. Turnips are often grazed off in the field.

turnip-rooted cabbage *noun* same as **kohlrabi**

turn out *verb* to put animals out to pasture, after they have been kept indoors during the winter ○ *The ewes are turned out in March.*

turn out time, turnout *noun* a season, usually in the spring, when animals which have been kept indoors during the winter are let out to grass

TUSDAC *abbreviation* Trade Union Sustainable Development Advisory Committee

tussock grass *noun* a coarse grass growing in tufts

TVO *abbreviation* tractor vaporising oil

twice-a-day *noun* a milking system in which cows are milked two times a day

twig *noun* a small woody growth from the branch of a tree, bearing leaves, flowers or fruit

twin *noun* one of two babies or animals born at the same time from two ova fertilised at the same time or from one ovum that splits in two □ **twin embryos** embryos of twin young, used in ET

twine /twaɪn/ *noun* a strong string used for binding bales ■ *verb* to coil round a support

twin lamb disease *noun* same as **pregnancy toxaemia**

twinning /ˈtwɪnɪŋ/ *noun* the act of giving birth to twins

twinter /ˈtwɪntə/ *noun* a two-year-old ewe

twist *noun* a disease of cereals and grasses which causes malformation of the leaves and stalks due to the growth of internal fungus. This may prevent the ear emerging from its sheath (*Dilophospora alopecuri*).

twitch *noun* **1.** same as **couch grass 2.** same as **louping-ill**

2,4-D *noun* a herbicide that is absorbed into a plant through its leaves and is especially effective against broadleaved weeds growing in cereals

two-sward system *noun* grazing system where the area being grazed is kept separate from the area being conserved for cutting

two-tooth *noun* a sheep showing two permanent incisors, approximately 18 months old

U

U *symbol* Good (*in the EUROP carcass classification system*)

udder *noun* the mammary gland of an animal, which secretes milk. It takes the form of a bag under the body of the animal with teats from which the milk is sucked.

COMMENT: In the cow there are four glands, each with a teat. In the ewe there are two glands, each with a teat, and in the sow there are two glands per teat and from 12 to 20 teats.

udder oedema *noun* a livestock disease in which the udder becomes swollen, probably due to increased pressure in the milk vein around the time of calving. It can be caused by excess sodium in diet.

ugli /ˈʌgli/ a trademark for a citrus fruit similar to a grapefruit, but of uneven shape

UHT *abbreviation* **1.** ultra heat treated **2.** ultra high temperature

UHT sterilisation *noun* sterilisation of milk at very high temperatures. Milk which has been treated in this way may be stored for periods of up to one year.

UKASTA *abbreviation* United Kingdom Agricultural Supply Trade Association

UKBAP *abbreviation* United Kingdom Biodiversity Action Plan

UKEPRA *abbreviation* United Kingdom Egg Producer Retailer Association

ulcer *noun* an open sore in the skin or in mucous membrane, which is inflamed and difficult to heal

ulcerated /ˈʌlsəreɪtɪd/ *adjective* covered with ulcers

ulceration /ˌʌlsəˈreɪʃ(ə)n/ *noun* **1.** a condition where ulcers develop **2.** the development of an ulcer

Ulmus /ˈʌlməs/ *noun* the Latin name for elm

Ulster White /ˌʌlstə ˈwaɪt/ *noun* a breed of pig popular for bacon production in Northern Ireland. It is quite rare today, having been replaced by the Large White.

ultra heat treated *adjective* (*of milk*) treated by sterilising at temperatures above 135°C, and then put aseptically into containers. Ultra heat treated milk has a much longer shelf-life than normal milk.

ultra high temperature *adjective* referring to something such as milk which has undergone a process of sterilisation at very high temperatures. Abbr **UHT**

ultramicroscopic /ˌʌltrəˌmaɪkrəˈskɒpɪk/ *adjective* too small to be seen with a light microscope

ultrasonics /ˌʌltrəˈsɒnɪks/ *noun* using high-frequency sound waves to tell what is below the skin of a live animal. By using ultrasonics, it is possible to tell the amount of fat layers and the muscle area.

umbel /ˈʌmbəl/ *noun* a flower cluster in which the flowers all rise on stalks from the same point on the plant's stem, e.g. on a carrot or polyanthus

umbellifer /ʌmˈbelɪfə/ *noun* a plant belonging to the Umbelliferae

Umbelliferae /ˌʌmbəˈlɪfəri/ *noun* a family of herbs and shrubs, including important food plants such as carrot, parsnip and celery

umbelliferous /ˌʌmbəˈlɪfərəs/ *adjective* referring to a plant of the Umbelliferae

unavailable water *noun* water in the soil which is held in the smallest soil pores and so is not available for plants

uncastrated /ˌʌnkæˈstreɪtɪd/ *adjective* referring to a male animal which has not been castrated. Also called **entire**

unchitted /ʌnˈtʃɪtɪd/ *adjective* referring to seed which has not been chitted

uncropped land *noun* land on which crops are not currently being grown or have never been grown

uncultivated /ʌnˈkʌltɪveɪtɪd/ *adjective* not cultivated ○ *uncultivated land and semi-natural areas* ○ *The field was left uncultivated over winter to allow ground-nesting birds such as skylarks to nest and rear young.*

under- *prefix* **1.** below or underneath **2.** less than or not as strong

undercoat /ˈʌndəkəʊt/ *noun* a coat of fine hair under the main coat of some animals

underdeveloped countries /ˌʌndədɪveləpt ˈkʌntriz/ *plural noun* countries which have not been industrialised

under-drain *noun* a drain under the surface of the soil, e.g. a mole drain or pipe drain

underfeeding /ˌʌndəˈfiːdɪŋ/ *noun* the action of giving an animal less feed than it needs

'Regular condition scoring has raised awareness of underfeeding of cows in early lactation, says Mr Ward. He also points out that acidosis problems were not helping.' [*Farmers Weekly*]

undergrowth *noun* shrubs and other plants growing under large trees

undershot wheel /ˈʌndəʃɒt ˌwiːl/ *noun* a type of waterwheel where the wheel rests in the flow of water which passes underneath it and makes it turn. Compare **overshot wheel** (NOTE: It is not as efficient as an overshot wheel where the water falls on the wheel from above.)

undersow /ˈʌndəsəʊ/ *verb* to sow a grass mixture after an arable crop has established, so that both develop at the same time. Cereals are most often used as cover crops.

understorey /ˈʌndəstɔːri/ *noun* the lowest layer of small trees and shrubs in a wood, below the canopy

undulant fever /ˈʌndjʊlənt ˌfiːvə/ *noun* same as **brucellosis**

unenclosed land /ˌʌnɪnləʊzd ˈlænd/ *noun* an area of land without any walls or fences round it

UNEP *abbreviation* United Nations Environment Programme

Ungulata /ˌʌŋgjuːˈlɑːtə/ *plural noun* grazing animals which have hoofs, e.g. cattle, sheep, goats and horses

ungulate /ˈʌŋgjʊleɪt/ *adjective* having hoofs (NOTE: Ungulates are divided into two groups, odd-toed such as horses or even-toed such as cows.) ■ *noun* a grazing animal that has hooves, e.g. a horse

unheated /ʌnˈhiːtɪd/ *adjective* referring to something which is not heated

unheated glasshouse, cold glasshouse *noun* a glasshouse with no heating

unimproved /ˌʌnɪmˈpruːvd/ *adjective* referring to land which has not been well looked after, and has not been improved by fertilising and proper husbandry

union *noun* the point of contact between a scion and stock in a grafted fruit tree

unisexual /ˌjuːniˈseksjʊəl/ *adjective* describes a plant which has either male or female flowers, but not both

unit *noun* **1.** a component of something larger **2.** a quantity or amount used as a standard, accepted measurement ○ *The internationally agreed unit of pressure is the millibar.*

unit cost *noun* the cost of one item, calculated as the total product cost divided by the number of units produced

United Kingdom Agricultural Supply Trade Association *noun* former name for **Agricultural Industries Confederation**

United Kingdom Biodiversity Action Plan *noun* the United Kingdom's response to the Convention on Biological Diversity in 1992, which contains plans to conserve threatened species and habitats. Abbr **UKBAP**

United Nations Environment Programme *noun* an organisation which promotes international cooperation in the sustainable use of resources. Abbr **UNEP**

United States Department of Agriculture *noun* a department of the US federal government, which deals with agricultural matters. Abbr **USDA**

unit of account *noun* a currency used for calculating the EU budget and farm prices

unit of fertiliser *noun* 1% of one hundredweight

unnecessary pain, unnecessary distress *noun* any suffering caused to livestock which can and should be avoided, instances of which will cause a farm to fail quality and subsidy-related inspections. Abbr **UPUD**

unpaired /ʌnˈpeəd/ *adjective* referring to a chromosome which is not associated with another chromosome of the same type ○ *an unpaired X chromosome in males*

unpalatable /ʌnˈpælətəb(ə)l/ *adjective* having an unpleasant taste

unpasteurised /ʌnˈpæstʃəraɪzd/, **unpasteurized** *adjective* referring to something such as milk which has not been pasteurised ○ *Unpasteurised milk can carry bacilli.*

unploughed land /ˌʌnplaʊd ˈlænd/ *noun* land which has not been ploughed

unsaturated fat *noun* a fat which does not have a large amount of hydrogen and so can be broken down more easily

unthrifty /ʌnˈθrɪfti/ *adjective* not thriving *or* not growing well

untreated milk /ʌnˈtriːtɪd mɪlk/ *noun* milk which has not been treated and which is sold direct from the farm to the public (NOTE: In the UK this is called 'green top milk' and sales are now banned.)

unweaned /ʌnˈwiːnd/ *adjective* referring to a young animal which has not yet been weaned

upgrade *verb* to make improvements to a herd by repeated crossing with superior males

upland /ˈʌplənd/ *noun* an inland area of high land ○ *The uplands have different ecosystems from the lowlands.* ■ *adjective* referring to an upland ○ *upland farming*

upland crop *noun* a crop that is grown in hilly areas

upland farm *noun* same as **hill farm**

uptake *noun* the taking in of trace elements or nutrients by a plant or animal

'Poor rooting is responsible for poor nutrient uptake and drought stress during the growing season. This phenomenon is not unique to just direct drilled crops, but the system exaggerates the problems which can be diluted by tillage.' [*Arable Farming*]

UPUD *abbreviation* unnecessary pain, unnecessary distress

urban fringe *noun* an area of land use where the urban activities meet the rural, usually a source of conflict between townspeople and farmers

urea /juˈriːə/ *noun* a popular fertiliser used as a top dressing to supply nitrogen. Formula: $CO(NH_2)_2$.

uric acid /ˌjuərɪk ˈæsɪd/ *noun* a chemical compound which is formed from nitrogen in waste products from the body

urine *noun* a liquid secreted as waste from an animal's body

Urticaceae /ˌɜːtɪˈkeɪsiiː/ *noun* a family of shrubs and herbs with stinging leaves such as the common stinging nettle (*Urtica dioica*)

urticaria /ˌɜːtɪˈkeəriə/ *noun* an allergic reaction, e.g. to injections or certain foods, where the skin forms irritating reddish patches

uterus *noun* an organ inside which the eggs or young of animals develop (NOTE: In humans and other mammals it is often called the womb and has strong muscles to push the baby out at birth.)

V

vaccinate *verb* to use a vaccine to give a person immunisation against a specific disease ○ *She was vaccinated against smallpox as a child.*

vaccination /ˌvæksɪˈneɪʃ(ə)n/ *noun* the action of vaccinating someone against a disease. ◊ **immunisation** (NOTE: Originally the words **vaccination** and **vaccine** applied only to smallpox immunisation, but they are now used for immunisation against any disease)

vaccine *noun* a substance which contains the germs of a disease, used to inoculate or vaccinate someone against it ○ *The vet is waiting for a new batch of vaccine to come from the laboratory.* ○ *New vaccines are being developed all the time.*

COMMENT: A vaccine contains the germs of the disease, sometimes alive and sometimes dead, and this is injected into the animal so that its body will develop immunity to the disease. The vaccine contains antigens, and these provoke the body to produce antibodies, some of which remain in the bloodstream for a very long time and react against the same antigens if they enter the body naturally at a later date when the animal is exposed to the disease.

vaccinia /vækˈsɪniə/ *noun* same as **cowpox** (*technical*)

vacuum silage *noun* silage placed in large polythene bags, usually by a baler specially adapted for this purpose. Air is excluded, so preventing the development of moulds and the green crop is conserved in succulent form.

vagal indigestion /ˈveɪg(ə)l ˌɪndɪdʒestʃ(ə)n/ *noun* a disease of livestock due to malfunction of the vagus nerve which controls the activity of the stomach and intestines

valine /ˈveɪliːn/ *noun* an essential amino acid

value added *noun* the difference between the cost of the materials purchased to produce a product and the final selling price of the finished product (NOTE: In agriculture, activities that add value include butchering, milling wheat or turning milk into cheese.)

vanilla *noun* a tropical climbing plant (*Vanilla planiolia*) which produces long pods, used for flavouring in confectionery

variable premium *noun* an extra payment which varies according to production quality

variant *noun* a specimen of a plant or animal that is different from the usual type

variant CJD *noun* a form of Creutzfeldt-Jakob disease which was observed first in the 1980s, especially affecting younger people. Abbr **vCJD**

variegated /ˈveəriəgeɪtɪd/ *adjective* referring to a plant with different-coloured patches

variegation /ˌveərɪˈgeɪʃ(ə)n/ *noun* a phenomenon in some plants where two or more colours occur in patches on the leaves or flowers

varietal /vəˈraɪətəl/ *noun* a variety of cultivated plant such as a grapevine

variety *noun* a named cultivated plant ○ *a new variety of wheat* Also called **cultivar**

vasectomise /vəˈsektəmaɪz/, **vasectomize** *verb* to perform a vasectomy on an animal

vasectomy *noun* an operation to cut the duct which takes sperm from the testicles, so making the animal infertile

vCJD *abbreviation* variant CJD

veal *noun* meat of a young calf fed solely on a milk diet, slaughtered between three and fifteen weeks old

veal crate system *noun* an intensive method of veal production, where calves are kept in crates. It was abandoned in the UK, but is still practised elsewhere.

vector /'vektə/ noun an insect or animal which carries a disease or parasite and can pass it to other organisms ○ *The tsetse fly is a vector of sleeping sickness.*

'Despite their leanings towards infected milk Defra still cannot rule out pigs eating infected wildlife as source of the infection – badgers are well known to be vectors of the disease, as are deer.' [*Farmers Guardian*]

veer *verb* 1. (*of the wind*) to change in a clockwise direction, in the northern hemisphere ○ *Winds veer and increase with height ahead of a warm front.* Opposite **back** 2. to change direction, especially as in an uncontrolled movement ○ *The aircraft veered off the runway into the grass.*

vegetable *noun* a plant grown for food, especially plants grown for leaves, roots or pods or seeds that are usually cooked ○ *Green vegetables are a source of dietary fibre.*

COMMENT: The main vegetable crops being grown on a field scale are broad, green and navy beans, vining and dried peas, Brussels sprouts, cabbages, carrots, cauliflowers, celery, onions, turnips and swedes and potatoes. In 2005, 139,010 hectares of potatoes and 48,043 hectares of other vegetables were cultivated for human consumption in England and Wales.

vegetable oils *plural noun* oils obtained from plants and their seeds, which are low in saturated fats

vegetable protein *noun* protein obtained from cereals, oilseeds, pulses, green vegetables and roots, which provides for the feeding requirements of both humans and livestock

vegetarian *noun* a person who does not eat meat ■ *adjective* referring to vegetarians or their diet ○ *He is on a vegetarian diet.*

vegetation *noun* 1. plants that are growing ○ *The vegetation was destroyed by fire.* ○ *Very little vegetation is found in the Arctic regions.* 2. the set of plants that is found in a particular area ○ *He is studying the vegetation of the island.*

vegetative /'vedʒɪtətɪv/ *adjective* referring to plants ○ *The loss of vegetative cover increases the accumulation of carbon dioxide in the atmosphere.*

vegetative propagation *noun* the artificial reproduction of plants by taking cuttings or by grafting, not by seed

venison *noun* meat from deer

ventilate *verb* to cause air to pass in and out of a place freely

ventilation *noun* the process of air passing in and out of a place freely ○ *A constant supply of air for ventilation purposes is always available from the air conditioning system.*

ventilator /'ventɪleɪtə/ *noun* a device that causes fresh air to pass into a room or building

ventricle /'ventrɪk(ə)l/ *noun* 1. a chamber of the heart that receives blood from the atria and pumps it to the arteries 2. one of the cavities of the vertebrate brain that connects with the others and contains cerebrospinal fluid

verandah /və'rændə/ *noun* a type of housing for poultry or pigs with a slatted or wire floor, through which the droppings fall

verge *noun* 1. the edge or boundary of something 2. an area of grass and other plants at the side of a road ○ *Roadside verges, especially motorway verges, offer security from human disturbance, and the wildlife quickly adapts to the noise and wind generated by passing vehicles.*

vermicide /'vɜːmɪsaɪd/ *noun* a substance that kills worms

vermiculite /vɜː'mɪkjuːlaɪt/ *noun* a substance that is a form of silica processed into small pieces. It is used instead of soil in horticulture because it retains moisture.

vermifuge /'vɜːmɪfjuːdʒ/ *noun* a substance used to get rid of parasitic worms in the intestines of livestock

vermin *noun* 1. an organism that is regarded as a pest ○ *Vermin such as rats are often carriers of disease.* ◊ **pest** (NOTE: The word **vermin** is usually treated as plural.) 2. insects such as lice which live on other animals as parasites

vernal /'vɜːn(ə)l/ *adjective* referring to the spring

vernalisation /ˌvɜːnəlaɪ'zeɪʃ(ə)n/, **vernalization** *noun* 1. a requirement by some plants for a period of cold in order to develop normally 2. the technique of making a seed germinate early by refrigerating it for a time

verroa /və'rəʊə/ *noun* a disease which affects bees

vertebrate /'vɜːtɪbrət/ *noun* an animal that has a backbone ■ *adjective* referring to animals that have a backbone ► compare (all senses) **invertebrate**

vertical-looking radar *noun* radar equipment used for analysis of features such as insect populations and movement. Abbr **VLR**

verticillium wilt /ˌvɜːtɪ'sɪliəm ˌwɪlt/ *noun* a plant disease caused by a fungus, which makes leaves become yellow and wilt. It is a notifiable disease of hops, and also a serious disease in lucerne and clovers.

vet *noun* same as **veterinary surgeon**

vetch /vetʃ/ *noun* a leguminous plant (*Vicia sativa*). Vetches can be sown with oats as an arable silage. Also called **tare**

veterinarian /ˌvet(ə)rɪ'neəriən/ *noun* US same as **veterinary surgeon**

veterinary *adjective* referring to the care of sick animals

Veterinary Investigation Diagnosis Analysis *noun* a method of data recording and retrieval for veterinary diagnostic laboratories. Abbr **VIDA**

Veterinary Laboratories Agency *noun* an executive agency of Defra which diagnoses, tracks and researches disease in animals. Abbr **VLA**

Veterinary Medicines Directorate *noun* an executive agency of Defra which regulates the development and use of veterinary medicines. Abbr **VMD**

veterinary science *noun* the scientific study of diseases of animals and their treatment

veterinary surgeon *noun* a person who is qualified to give medical treatment to animals

Veterinary Surveillance Strategy *noun* a 10-year initiative by Defra to monitor animal diseases in farms, so that the information can be used to plan future health and welfare practices

V-graft *noun* a method of grafting, where the stem of the stock is trimmed to a point, and the stem of the cutting is split to allow it to be fitted over the point of the stock

VI *abbreviation* Voluntary Initiative

vibriosis /ˌvɪbri'əʊsɪs/ *noun* a venereal disease in cattle which leads to a high incidence of infertility and abortion. It can be prevented by vaccination.

vice *noun* a bad habit in an animal, e.g. the habit of biting other animals' tails

Vicia /'vɪsiə/ *noun* the Latin name for beans such as broad beans

Victoria /vɪk'tɔːriə/ *noun* ♦ **plum**

VIDA *abbreviation* Veterinary Investigation Diagnosis Analysis

vigorous *adjective* growing strongly ○ *Plants put out vigorous shoots in a warm damp atmosphere.*

'Tight grazing is the cheapest and most effective way to encourage tillering, particularly during spells of rapid growth. Keep grazing down to 4–5cm to encourage tillers and give grass a competitive advantage over less vigorous grasses and weeds.' [*Farmers Guardian*]

vigour *noun* strength and energy (NOTE: The US spelling is **vigor**.)

vine *noun* **1.** a plant that supports itself by climbing up something or creeping along a surface **2.** a flexible stem of a vine plant **3.** same as **grapevine**

vine crops *plural noun* crops (*Cucurbitaceae*) such as cucumber, marrow, gourds and melons, which are annuals and produce long trailing shoots and heavy fleshy fruit

viner /'vaɪnə/ *noun* a machine for harvesting vining peas

vineyard *noun* a plantation of grapevines

vining /'vaɪnɪŋ/ *noun* the harvesting of peas for processing

vining peas *plural noun* peas used for canning or freezing

violet root rot *noun* a common disease of sugar beet in which a violet-coloured fungus (*Helicobasidium purpureum*) grows on the surface of the root. It lowers sugar content of the plant.

viral *adjective* referring to or caused by a virus ○ *a viral disease*

viral strike *noun* any apparently new virus disease, borne by wind or vectors, which travels through a wide area causing devastating effects for a time, especially in large livestock units

virgin *adjective* in its natural state, untouched by humans ○ *Virgin rainforest was being cleared at the rate of 1000 hectares per month.*

virgin land *noun* land which has never been cultivated

virus *noun* a microorganism consisting of a nucleic acid surrounded by a protein coat which can only develop in other cells, and often destroys them

COMMENT: Viruses produce disease in man, animals and plants. Many common diseases such as measles or the common cold are caused by viruses; viral diseases cannot be treated with antibiotics, which only destroy bacteria. Viruses can be transmitted from one animal to another, reproducing the same disease. Insects, particularly aphids, transmit certain virus diseases in plants.

virus pneumonia *noun* ♦ **enzootic pneumonia**

virus yellows *noun* a disease of sugar beet and mangolds as a result of which the leaves turn yellow and the sugar content is greatly reduced. Crops are most at risk when virus-carrying aphids infest the plants at the two-leaf stage.

viscera /'vɪsərə/ *plural noun* the internal organs, in particular the intestines and other contents of the abdomen

vitamin *noun* a substance not produced in the body, but found in most foods, and needed for good health

COMMENT: Vitamins are a group of chemical compounds found in a variety of foodstuffs, which are necessary for the healthy regulation of physical processes in an animal's body. Vitamin deficiencies in animals can cause serious health problems and, once identified, can be cured by the use of mineral supplements either as an individual simple substance or by complex mixtures added to normal rations and supplied according to the animals' needs.

vitamin A *noun* a vitamin which is soluble in fat and can be synthesised in the body from precursors, but is mainly found in food such as liver, vegetables, eggs and cod liver oil. Also called **retinol**

COMMENT: Lack of Vitamin A affects the body's growth and resistance to disease and can cause night blindness. The primary source of this vitamin is the green plant. It is of great importance for dairy cows: lack of the vitamin leads to retardation of growth in young stock and in adult animals appears to lower their resistance to infectious diseases.

vitamin B$_1$ *noun* a vitamin found in yeast, liver, cereals and pork. Also called **thiamine**

vitamin B$_2$ *noun* a vitamin found in eggs, liver, green vegetables, milk and yeast. Also called **riboflavin**

vitamin B$_6$ *noun* a vitamin found in meat, cereals and molasses. Also called **pyridoxine**

vitamin B$_{12}$ *noun* a water-soluble vitamin found especially in liver, milk and eggs but not in vegetables, and important for blood formation, nerve function, and growth. Also called **cyanocobalamin** (NOTE: A deficiency of B$_{12}$ causes pernicious anaemia.)

vitamin B complex *noun* a group of vitamins which are soluble in water, including folic acid, pyridoxine and riboflavine

vitamin C *noun* a vitamin which is soluble in water and is found in fresh fruit, especially oranges and lemons, raw vegetables and liver. Also called **ascorbic acid** (NOTE: Lack of vitamin C can cause anaemia and scurvy.)

vitamin D *noun* a vitamin which is soluble in fat, and is found in butter, eggs and fish (NOTE: It is also produced by the skin when exposed to sunlight. Vitamin D helps in the formation of bones, and lack of it causes rickets in children.)

vitamin E *noun* a vitamin found in vegetables, vegetable oils, eggs and wholemeal bread

vitamin K *noun* a vitamin found in green vegetables such as spinach and cabbage, which helps the clotting of blood and is needed to activate prothrombin

viticulture /'vɪtɪkʌltʃə/ *noun* the cultivation of grapes

viviparous /vɪ'vɪpərəs/ *adjective* **1.** referring to an animal such as a mammal or some fish that give birth to live young. Compare **oviparous 2.** reproducing by buds that form plantlets while still attached to the parent plant or by seeds that germinate within a fruit

VLA *abbreviation* Veterinary Laboratories Agency

VLR *abbreviation* vertical-looking radar

VMD *abbreviation* Veterinary Medicines Directorate

VOC *abbreviation* volatile organic compound

volatile oils *plural noun* concentrated oils from a scented plant used in cosmetics or as antiseptics

volatile organic compound *noun* an organic compound which evaporates at a relatively low temperature. Abbr **VOC** (NOTE: Volatile organic compounds such as ethylene, propylene, benzene and styrene contribute to air pollution.)

Voluntary Initiative *noun* a five year programme of measures aimed at minimising the environmental impact of crop protection products. It was introduced in 2003 under agreement between the agriculture industry and the government to prevent the introduction of a pesticide tax. Abbr **VI**

voluntary restraint agreement *noun* an agreement by which farmers agree not to spray in windy conditions. Such agreements are not legally binding. Abbr **VRA**

volunteer, volunteer plant *noun* a plant that has grown by natural propagation, as opposed to having been planted ○ *Volunteer cereals are a problem in establishing oilseed rape.*

vomiting and wasting disease *noun* a disease of piglets, symptoms of which include vomiting and loss of appetite

VRA *abbreviation* voluntary restraint agreement

Vulgare /vʌlˈgɑːrɪs/, **Vulgaris** *adjective* a Latin word meaning 'common', often used in plant names

W

wages *plural noun* money paid to an employee for work done

wall barley grass *noun* a weed (*Hordeum murinum*) found in grassland

walnut *noun* a hardwood tree of the genus *Juglans*, with edible nuts. The timber is used in furniture making.

WAOS *abbreviation* Welsh Agricultural Organisation Society

warble fly *noun* a parasitic fly whose larvae infest cattle. Infestation by warble fly is a notifiable disease.

COMMENT: Eggs of warble flies are laid on the legs or bellies of cattle. On hatching, the maggots burrow into the skin and cause swellings on the back of the animal. When adult, the maggots leave the body through the skin and fall to the ground to pupate. They cause severe irritation, loss of condition and in young animals may cause death.

warbles /ˈwɔːb(ə)lz/ *plural noun* swellings on the backs of cattle caused by the warble fly

ware growers *plural noun* farmers who grow potatoes for consumption, not for seed

ware potatoes *plural noun* potatoes grown for human consumption, as opposed to those grown for seed

COMMENT: Ware potatoes are used in many ways: for crisps, chips, canning and dehydration. Good quality ware potatoes should not be damaged or diseased, should not be green in colour, and should be between 40 and 80mm in size.

warfarin /ˈwɔːf(ə)rɪn/ *noun* a substance used to poison rats, to which many rats in some areas are now resistant

warping /ˈwɔːpɪŋ/ *noun* a farming practice which permits a river to flood low-lying land to cover it with silt in which crops will be grown

wart *noun* a small often infectious growth, caused by a virus, that appears on the skin of an animal, or a similar growth on a plant

wart disease *noun* a notifiable disease of potatoes, in which warts appear on the surface of the tubers, and develop into large eruptions which may become larger than the potatoes themselves

WASK *abbreviation* Welfare of Animals (Slaughter or Killing) Regulations 1995

waste *noun* material that is thrown away by people or is an unwanted by-product of a process ○ *household waste* ○ *industrial waste* ■ *adjective* without a specific use and unwanted ○ *Waste products are dumped in the sea.* ○ *Waste matter is excreted by the body in the faeces or urine.* ■ *verb* to use more of something than is needed

'There are estimated to be 300 000 tonnes of non-natural wastes produced on agricultural holdings in England and Wales each year. These include a wide range of materials such as waste packaging, silage plastics, metal, tyres, oils and animal health products. (Agricultural Waste: Opportunities for Change. Information from the Agricultural Waste Stakeholders' Forum 2003)'

wasteland *noun* an area of land that is no longer used for agriculture or for any other purpose ○ *Overgrazing has produced wastelands in Central Africa.*

waste lime *noun* lime obtained from industrial concerns after it has been used as a purifying material

water *noun* a liquid which forms rain, rivers, lakes and the sea and which makes up a large part of the bodies of organisms. Formula: H_2O. ■ *verb* to give water to a plant

COMMENT: Water is essential to plant and animal life. Water pollution can take many forms: the most common are discharges from industrial processes, household sewage and the runoff of chemicals used in agriculture.

water abstraction *noun* the diversion or removal of water from any surface or underground source for some purpose, such as for irrigation

Water Act 1989 *noun* an Act of Parliament which made it an offence to cause a discharge of poisonous, noxious or polluting matter or solid matter to any controlled water under the responsibility of the National Rivers Authority. Controls are also in force to ensure that silage, slurry and fuel oil installations are of adequate standard.

water balance *noun* **1.** a state in which the water lost in an area by evaporation or by runoff is replaced by water received in the form of rain **2.** a state in which the water lost by the body in urine and perspiration or by other physiological processes is balanced by water absorbed from food and drink

waterbowl /'wɔːtəbəʊl/ *noun* a container for water in a stable or loose-box

water buffalo *noun* a large buffalo with a grey-black coat and long backward-sloping horns (*Bubalus bubalis*), which is kept for its meat and used as a draught animal especially in Asia

water catchment *noun* the act of rainwater being collected in a place, whether naturally (in a surface pool) or deliberately (using a water catchment system)

watercourse /'wɔːtəkɔːs/ *noun* a stream, river, canal or other flow of water

waterfowl /'wɔːtəfaʊl/ *plural noun* birds which spend much of their time on water, e.g. ducks

Water Framework Directive *noun* a basis for future policy decisions in the European Union, setting objectives for water use and management and waste water disposal. Abbr **WFD**

waterlogged *adjective* referring to soil that is saturated with water and so cannot keep oxygen between its particles (NOTE: Most plants cannot grow in waterlogged soil.)

water management *noun* the careful and appropriate use of water

water meadow *noun* a grassy field near a river, which is often flooded

water melon *noun* a plant of the genus *Citrullus vulgaris* with large green fruit with watery pink flesh

water meter *noun* a device that records the amount of water that passes through a pipe, e.g. to monitor the water intake of animals

water mill *noun* a mill which is driven by the power of a stream of water which turns a large wheel

watershed *noun* a natural dividing line between the sources of river systems, dividing one catchment area from another

water-soluble *adjective* able to dissolve in water

water table *noun* the area below the soil surface at which the ground is saturated with water

waterwheel /'wɔːtəwiːl/ *noun* a wheel with wooden steps or buckets that is turned by the flow of water against it and itself turns machinery such as a mill wheel or an electric generator. ◊ **overshot wheel**, **undershot wheel**

COMMENT: Waterwheels have two purposes. When placed over a moving stream, the wheel is turned by the pressure of the moving current and so drives machinery, as in a watermill. In the second case, the wheel is also placed over a moving stream, but the water collects in the buckets on the wheel, which, as it turns, raises the full buckets and tips the water into an irrigation channel.

watery mouth *noun* a disease affecting new-born lambs

WATO *abbreviation* Welfare of Animals (Transport) Order 1997

wattle /'wɒt(ə)l/ *noun* **1.** rods and twigs woven together to make a type of fence **2.** a piece of fleshy skin hanging down below the throat of birds such as the turkey

wean *verb* to remove a young animal from the milk source of its mother. Weaning is common at 5 weeks.

weaner /'wiːnə/ *noun* a young animal which has been weaned, especially a young pig

weather *noun* daily atmospheric conditions such as sunshine, wind and precipitation in an area ■ *verb* to change the state of soil or rock through the action of natural agents such as rain, sun, frost or wind or by artificially produced pollutants

weathering *noun* the alteration of the state of soil or rock through the action of natural agents such as rain, sun, frost or wind or by artificially produced pollutants

weatings /'wiːtɪŋz/ *plural noun* a by-product of milling wheat, made up of brans of various particle sizes and varying amounts of attached endosperm, which is used as a feedingstuff. Also called **wheatings**

web conveyor *noun* a machine used to move material along a moving web; found on all types of harvesters and some processing machines

wedge *noun* ♦ **Dorset wedge silage**

weed *noun* a plant that grows where it is not wanted, e.g. a poppy in a wheat field (NOTE: Some weeds are cultivated plants, for example oilseed rape growing in hedgerows.)

COMMENT: Weeds compete with crops for nutrients and water; the presence of weeds can lower the quality of a crop and often make it more difficult to harvest. Some weeds may taint milk when eaten by cows and some are poisonous and can affect livestock. Weeds also harbour pests and diseases which can spread to crops. Chemical control of weeds is an additional cost, but weeds can be controlled by good rotations and tillage treatment.

weed beet *noun* a type of beet which is regarded as a weed because it produces seeds as opposed to roots which can be harvested. Weed beet affects sugar beet crops and can harbour rhizomania. It is controlled by limiting bolters and so preventing cross-pollination. The most effective control is by hand-pulling bolters.

weedkiller /'wiːdkɪlə/ *noun* same as **herbicide**

weevil /'wiːv(ə)l/ *noun* a kind of beetle which feeds on grain, nuts, fruit and leaves. The larvae of grain beetles feed on the stored grain where they also pupate.

Weil's disease /'weɪlz dɪˌziːz/ *noun* a sometimes fatal disease of humans caused by Leptospira bacteria, caught from the urine of infected cattle or rats

welfare *noun* the fact of being happy, healthy and well-looked-after

welfare code *noun* an official set of rules for making sure that animals are healthy and happy in a particular situation, e.g. in quarantine, or when being transported

welfare legislation *noun* a law or set of laws that makes it illegal to cause harm or distress to animals while caring for them

Welfare of Animals (Transport) Order 1997 *noun* a piece of legislation which sets out the minimum standard of welfare for animal in transport, including guidelines on vehicle condition, journey times and necessary documentation. Abbr **WATO** (NOTE: New EU legislation on animal welfare in transport is due come into force in 2007.)

Welfare of Animals (Slaughter or Killing) Regulations 1995 *noun* a piece of legislation which sets out rules under which animals should be slaughtered or killed, including making it an offence to cause unnecessary pain or distress during the slaughter process. Abbr **WASK** (NOTE: The original regulations from 1995 have been regularly updated, including updates in 1999, 2000 and 2003.)

Welfare of Farmed Animals (England) Regulations 2000 *noun* a piece of legislation which sets out guidelines for the humane treatment of animals on farms, including separate schedules on the treatment specific types of animals such as laying hens, cattle or pigs (NOTE: This legislation also requires that anyone attending to farm animals should have full knowledge of the relevant animal welfare code for that type of animal. Similar pieces of legislation are also in place for Scotland, Wales and Northern Ireland.)

well *noun* a hole dug in the ground to the level of the water table, from which water can be removed by a pump or bucket

Welsh *adjective* referring to a breed of pig, white in colour, with lop ears. It is one of the older breeds of British pig.

Welsh black *noun* a hardy dual-purpose breed of cattle formed when the northern Anglesey strain was bred with the Castlemartin strain. Welsh blacks produce a reasonable milk yield and very lean meat.

Welsh half bred *noun* a cross between a border Leicester ram and a Welsh mountain ewe

Welsh mountain *noun* a hardy breed of sheep, well adapted to wet conditions. The animals are small with white faces and very fine fleece, and only the rams have horns.

Welsh mule *noun* a cross between a Blue-faced Leicester and a ewe of one of the Welsh mountain breeds

Wensleydale /ˈwenslideɪl/ *noun* **1.** a longwool breed of sheep. The animals are large and polled, and the skin of the face, legs and ears is blue. Wensleydales are now rare, but are still found in Yorkshire. **2.** a type of hard white cheese

Wessex Saddleback *noun* one of two saddleback breeds now joined with the Essex Saddleback to give the British Saddleback, a dual-purpose breed of pig, now rare

Western *noun* same as **Wiltshire horn**

Westerwold ryegrass /ˌwestəwəʊld ˈraɪɡrɑːs/ *noun* an annual type of ryegrass, which is a fast-growing summer crop

wet-feeding *noun* a method of feeding livestock such as pigs in which the animal has access to dry feedingstuffs and water at the same time

wether /ˈweðə/ *noun* a castrated male sheep

wetlands /ˈwetlændz/ *noun* an area of land which is often covered by water or which is very marshy

wet mash *noun* mash feed mixed with water

wet pluck *noun* the process of removing the feathers when the carcass is wet. This is easier than dry plucking, but may harm the skin.

WFA *abbreviation* Whole Farm Approach

WFD *abbreviation* Water Framework Directive

WFU *abbreviation* Women's Food and Farming Union

wheat *noun* a cereal crop grown in temperate regions. Genus: *Triticum*. (NOTE: Wheat is one of the major arable crops.)

COMMENT: The two main species of wheat grown are *Triticum aestivum* which is grown for bread flour, and some varieties of which produce the most suitable flour for cakes and biscuits, and *Triticum durum* which is grown for pasta. Cereal drilling usually takes place in the UK between September and April. Winter wheat usually yields higher quantities than spring wheat and is harvested before it. Spring wheat varieties are grown in those areas with more extreme climates as they are the quick-maturing varieties. Spring wheat is grown in the prairie provinces of Canada, in the Dakotas and Montana in the USA, and in the more northerly parts of the steppe wheat belt in Russia. Winter varieties account for about three-quarters of the total output of wheat in the USA, and over 80% of exports. The state of Kansas grows 30% of the hard red winter wheat grown in the USA.

wheat blossom midge *noun* a pest that affects wheat

wheat bulb fly *noun* a fly whose larvae feed on the roots of wheat. The central shoot turns yellow and dies.

wheatfeed /ˈwiːtfiːd/ *noun* same as **wheat offals**

wheatgerm /ˈwiːtdʒɜːm/ *noun* the central part of the wheat seed, which contains valuable nutrients

wheatings /ˈwiːtɪŋz/ *plural noun* another spelling of **weatings**

wheatmeal /ˈwiːtmiːl/ *noun* brown flour with a large amount of bran, but not as much as is in wholemeal

wheat offals *noun* the embryo and seed coat of the wheat grain, used as animal feed

whey /weɪ/ *noun* a residue from milk after the casein and most of the fat have been removed. Whey is used as pig feed.

whip *noun* a short stick with a lash attached, used to control horses

whip and tongue cutting *noun* a form of graft in which the stock and scion are cut diagonally to form large open surfaces with a small notch in each. The surfaces are bound together tightly with twine.

whipworm /ˈwɪpwɜːm/ *noun* a variety of worm affecting pigs, especially weaners

Whitbred shorthorn /ˌwɪtbred ˈʃɔːthɔːn/ *noun* a breed of white beef cattle

white bird's-eye *noun* same as **chickweed**

white clover *noun* a type of perennial clover (*Trifolium repens*). There are several varieties including the large-leaved variety suitable for silage or hay and the small-leaved variety which is quick to establish and keeps out weeds and other grasses.

white corpuscle *noun* a blood cell which does not contain haemoglobin

White-faced Woodland *noun* a large hill breed of sheep, with white face and legs and pinkish nostrils. The ram has heavy twisted horns. Found mainly in the South Pennines, it has been crossed with other hill breeds to give them its size and vigour. Also called **Penistone**

whiteheads /ˈwaɪthedz/ *noun* same as **take-all**

White Leghorn *noun* a laying breed of poultry

white lupin *noun* a new strain of lupin (*Lupinus albus*) that is able to withstand cold. Seeds are 40% protein and at least 12% edible oil.

white mulberry *noun* a tree grown for its leaves, on which silkworms feed

white mustard *noun* a crop grown to increase the organic content of the soil by using it as a green manure

White Park *noun* a rare breed of cattle, white in colour with either black or red muzzle, eyelids, ears and feet. It is one of the most ancient breeds of British cattle.

White Plymouth Rock *noun* a large heavy breed of table poultry

white rot *noun* a fungal disease of onions and leeks. The leaves turn yellow and a white mass appears on the bulb.

whites *noun* same as **metritis**

white scour *noun* a disease affecting young calves

White Wyandotte /ˌwaɪt ˈwaɪəndɒt/ *noun* a dual-purpose breed of poultry

whole crops *plural noun* crops used for silage which do not need wilting

Whole Farm Approach *noun* an official system of communication between the Government and farmers which is more streamlined and aims to avoid duplicated information and 'red tape'. Abbr **WFA**

wholefood *noun* food such as brown rice or wholemeal flour that has not been processed and so contains the vitamins, minerals and fibre that are removed by processing

wholegrain /ˈhəʊlɡreɪn/ *noun* a cereal grain containing the whole of the original seed, including the bran

wholemeal *noun* flour that contains a large proportion of the original wheat seed, including the bran

wholesale *adjective, adverb* selling in bulk to shops, who then sell in smaller quantities to individual buyers

wholesale seed merchant *noun* a merchant who sells seed in bulk

wild *adjective* not domesticated

wild boar *noun* a species of feral pig, common in parts of Europe, but extinct in the UK. Wild boars are preserved for hunting, but are now bred on farms. Their meat is dark, with very little fat, and is of high value.

wild chamomile *noun* same as **mayweed**

wild crop *noun* a crop which is harvested by man, but not cultivated, e.g. wild berries or herbs

wildlife *noun* wild animals of all types, including birds, reptiles and fish ○ *Plantations of conifers are poorer for wildlife than mixed or deciduous woodlands.* ○ *The effects of the open-cast mining scheme would be disastrous on wildlife, particularly on moorland birds.*

wildlife reserve *noun* an area where animals and their environment are protected

wild oats *plural noun* several species of annual weeds, including (*Avena fatua*) and (*Avena ludoviciana*), found among cereal crops, and now largely controlled by selective herbicides, although manual weeding or roguing is also used

wild onion *noun* a perennial weed affecting cereal crops, beans and rape. Also called **crow garlic**

wild radish *noun* same as **runch**

wild white clover *noun* a variety of small-leaved white clove which is slow to get established, but is an essential part of a long ley. It is drought resistant and very productive.

William /ˈwɪljəm/ *noun* ♦ **pear**

willow *noun* a temperate hardwood tree that often grows near water. Genus: *Salix*. (NOTE: Willow is sometimes grown as a crop and is coppiced or pollarded to produce biomass for fuel.)

wilt *noun* **1.** the drooping of plants particularly young stems, leaves and flowers, as a result of a lack of water, too much heat or disease **2.** one of a group of plant diseases that cause drooping and shrivelling of leaves (NOTE: It is caused by fungi, bacteria, or viruses that block the plant's water-carrying vessels.) ■ *verb* to droop or shrivel through lack of water or great heat, or to cause or allow plants to lose firmness in their stems and leaves so they are suitable for use as silage

wilting *noun* limpness found when plant tissues do not contain enough water. ◊ **permanent wilting point**

Wiltshire cure /ˈwɪltʃə ˌkjʊə/ *noun* a special method of mild curing and smoking sides of bacon over wood fires

Wiltshire horn *noun* a distinctive white-faced breed of sheep, with curled horns. It grows a coat of thick matted hair and is found in the Midlands and Anglesey. It is a hardy breed, producing rapid-growing lambs. Also called **Western**

wind *noun* air which moves in the lower atmosphere, or a stream of air ○ *The weather station has instruments to measure the speed of the wind.*

windbreak /'wɪndbreɪk/ *noun* a hedge or line of trees, planted to give protection from the wind to land with growing crops

wind chill factor *noun* a way of calculating the risk of exposure in cold weather by adding the speed of the wind to the number of degrees of temperature below zero

windmill /'wɪndmɪl/ *noun* **1.** a construction with sails which are turned by the wind, providing the power to drive a machine. ◊ **panemone 2.** same as **wind turbine**

COMMENT: Windmills were originally built to grind corn or to pump water from marshes. Large modern windmills are used to harness the wind to produce electricity.

windrow /'wɪndrəʊ/ *noun* a row of the cut stalks of a crop, gathered together and laid on the ground to be dried by the wind

windrowed /'wɪndrəʊd/ *adjective* referring to a crop which has been lifted and left in a swath

windrower /'wɪndrəʊə/ *noun* a machine which lifts a crop such as potatoes and leaves it in a swath on the surface of the soil

windrow pick-up *noun* a pickup mechanism which lifts a crop into a harvester

wind turbine *noun* a turbine driven by wind

'According to Mr Schultze, there are over 16,500 wind turbines in German fields, though new developments were slowing due to resistance from environmental groups. Farmers were therefore turning increasingly to biogas plants, producing electricity and heat for their own businesses and for local communities.' [*Farmers Weekly*]

wing *noun* **1.** one of the feather-covered limbs of a bird or membrane-covered limbs of a bat that are used for flying **2.** an outgrowth on a seed case of seeds such as sycamore dispersed by wind **3.** the lower part of the ploughshare behind the point

winnow /'wɪnəʊ/ *verb* to separate grain from chaff. Originally this was done by throwing the grain and chaff up into the air, the lighter chaff being blown away by the wind.

winter *noun* the season of the year, following autumn and before spring, when the weather is coldest, the days are short, most plants do not flower or produce new shoots and some animals hibernate ■ *verb* to spend the winter in a place

winter burn *noun* leaf burn in winter

winter feeding *noun* a system of feeding livestock during the winter months, giving them feeds of hay, silage and concentrates

winter greens *plural noun* hardy varieties of Brassica which are grown for use during the winter

winter hardy *adjective* able to survive through the winter outside

wintering grounds *plural noun* area where birds come each year to spend the winter

winter kill *noun* the death of plants in winter

winter wash *noun* an egg-killing spray applied to fruit trees in the dormant winter period. Tar oil is the commonest winter spray.

winter wheat *noun* wheat of a variety sown in the autumn or early winter months and harvested early the following summer

wireweed /'waɪəwiːd/ *noun* a common name for 'knotgrass' (*Agriotes lineatus*)

wireworm /'waɪəwɜːm/ *noun* the shiny thin hard-bodied larvae of the click beetle, which feeds on the roots of cereals and other plants

wither *verb* (*of plants, leaves, flowers*) to shrivel and die

withers /'wɪðəz/ *noun* a ridge between the shoulder blades of an animal

withstand *verb* not to be affected by ○ *some plants can withstand very low temperatures*

wolds /wəʊldz/ *plural noun* areas of low chalk or limestone hills. Wolds are characterised by having few hedges and no surface water.

Women's Food and Farming Union *noun* an association for women in farming, focusing on sustainability and environmental considerations. Abbr **WFFU**

wood *noun* **1.** a large number of trees growing together **2.** a hard tissue which forms the main stem and branches of a tree **3.** a construction material that comes from trees

wood alcohol *noun* same as **methanol**

wood ash *noun* ash from burnt wood, a source of potash

wooden tongue *noun* same as **actino-bacillosis**

woodfuel /ˈwʊdfjuːəl/ *noun* wood which is used as fuel

woodland /ˈwʊdlənd/ *noun* an area in which the main vegetation is trees with some spaces between them

Woodland Grant Scheme *noun* an agri-environmental scheme aimed at ensuring good management of forests and woodland

woodlot /ˈwʊdlɒt/ *noun* a small area of land planted with trees

woody /ˈwʊdi/ *adjective* referring to plant tissue which is like wood *or* which is becoming wood

wool *noun* soft curly hair, the coat of the domesticated sheep. Wool is also produced by goats and rabbits.

wool ball *noun* a mass of wool found in the first or fourth stomachs of lambs. Small amounts of wool swallowed by the lamb collect to form a ball which can increase in size until it blocks the stomach and causes death.

wool fat *noun* lanolin, a fat which covers the fibres of sheep's wool

woolsorter's disease /ˈwʊlsɔːtəz dɪˌziːz/ *noun* same as **anthrax**

work *verb* to cultivate land

workability /ˌwɜːkəˈbɪlɪti/ *noun* the ability of soil to be cultivated. It is an interaction between climatic conditions and the physical condition of the soil.

work days *plural noun* the number of days when land can be worked with acceptable risk of damage to soil structure during the main activities of tillage, drilling and harvesting. Heavy soils have fewer work days than light soils. In general, good ground conditions exist when the soil is below field capacity.

World Food Programme *noun* part of the Food and Agriculture Organization of the United Nations. The programme is intended to give international aid in the form of food from countries with food surpluses.

World Trade Organization *noun* an international organisation set up with the aim of reducing restrictions in trade between countries (replacing GATT). Abbr **WTO**

World Wide Fund for Nature *noun* full form of **WWF**

worm *noun* **1.** an invertebrate animal with a soft body and no limbs, e.g. a nematode or flatworm **2.** an invertebrate animal with a long thin body and no legs that lives in large numbers in the soil. Also called **earthworm** ■ *verb* to treat an animal in order to remove parasitic worms from its intestines

worm cast *noun* waste earth rejected by an earthworm

wormer /ˈwɜːmə/ *noun* a substance used to worm animals such as cattle

worms *plural noun* a condition in which an animal is infested with parasitic worms which can cause disease

COMMENT: Parasitic worms infest most animals, but especially cattle and sheep, and can be removed with anthelmintics. Wormed cattle may give higher yields of milk than untreated animals, but tests are not conclusive. Various substances are used in worming, such as thiabendazole or fenbendazole.

worrying *noun* the chasing of sheep and other livestock, by dogs which are not controlled by their owners

WTO *abbreviation* World Trade Organization

WWF *noun* an international organisation, set up in 1961, to protect endangered species of wildlife and their habitats, and now also involved with projects to control pollution and promote policies of sustainable development. Full form **World Wide Fund for Nature**

Wyandotte /ˈwaɪəndɒt/ *noun* ♦ **White Wyandotte**

wych elm *noun* ♦ **elm**

XYZ

X chromosome *noun* a chromosome that determines sex. ◊ **Y chromosome**

xeno- /ˈzenəʊ/ *prefix* different

xenobiotics /ˌzenəʊbaɪˈɒtɪks/ *plural noun* chemical compounds that are foreign to an organism

xero- /ˈzɪərəʊ/ *prefix* dry

xeromorphic /ˌzɪərəʊˈmɔːfɪk/ *adjective* referring to a plant which can prevent water loss from its stems during hot weather

xerophilous /zɪˈrɒfɪləs/ *adjective* referring to a plant which lives in very dry conditions

xerophyte /ˈzɪərəfaɪt/ *noun* a plant which is adapted to living in very dry conditions

xerosere /ˈzɪərəʊsɪə/ *noun* a succession of communities growing in very dry conditions

xylem /ˈzaɪləm/ *noun* the tissue in a plant which transports water and dissolved minerals from the roots to the rest of the plant. Compare **phloem**

yard *noun* **1.** a unit of length in the US and British Imperial Systems equal to 3 ft or 0.9144 m. Abbr **yd 2.** an open space in a farm, surrounded on three sides by barns, stables and farm buildings

yard and parlour *noun* a system of housing dairy cattle in yards and bringing them through a parlour for milking

yarr /jɑː/ *noun* same as **corn spurrey**

yarrow /ˈjærəʊ/ *noun* a common weed (*Achillea millifolium*) which can cause taints in milk. Also called **milfoil**

Y chromosome *noun* a chromosome that determines sex, carried by males and shorter than an X chromosome. ◊ **X chromosome** (NOTE: A male usually has an XY pair of chromosomes.)

yd *abbreviation* yard

yearling /ˈjɪəlɪŋ/ *noun* an animal aged between one and two years

yeast *noun* a single-celled fungus that is used in the fermentation of alcohol and in making bread

yellow cereal fly *noun* a pest which affects early-sown wheat crops and causes the death of the plant's central shoot

yellow dwarf virus *noun* a fungal disease affecting barley, and also wheat and grass. The leaves turn red and yellow and yields are reduced. The disease is carried by wingless aphids and the common name for it is 'BYDV'.

yellowing /ˈjeləʊɪŋ/ *noun* **1.** a condition where the leaves of plants turn yellow, caused by lack of light **2.** a sign of disease or of nutrient deficiency

COMMENT: Yellow diseases are often caused by viruses, but may also be caused by bacteria and fungi. Yellowing is a common symptom when there is a deficiency of elements which are important to chlorophyll production, such as iron and magnesium.

yellow rattle *noun* an annual weed (*Rhinanthus minor*) found in grasslands

yellow rust *noun* a fungal disease (*Puccinia striiformis*) of cereals, mainly affecting wheat and barley. Yellow pustules form on leaves, stems and ears. Also called **stripe rust**

yellows /ˈjeləʊz/ *noun* **1.** a general term for any plant disease in which the leaves become yellow **2.** jaundice in animals

yelt /jelt/ *noun* a young female pig

yew /juː/ *noun* a coniferous tree or large shrub (*Taxus baccata*)

COMMENT: All varieties of the British yew tree are poisonous and stock which eat the leaves or berries suffer vomiting and in severe cases die.

yew poisoning *noun* poisoning through eating yew berries or leaves

YFC *abbreviation* Young Farmers' Club

yield *noun* the quantity of a crop or a product produced from a plant or from an area of land ○ *The usual yield is 8 tonnes per hectare.* ○ *The green revolution increased rice yields in parts of Asia.* ■ *verb* to produce a quantity of a crop or a product ○ *The rice can yield up to 2 tonnes per hectare.* ○ *The oil deposits may yield 100000 barrels a month.*

yoghurt, yogurt, yoghourt *noun* soured milk in which fermentation is accelerated by the introduction of specific bacterial microorganisms

yoke *noun* **1.** a wooden crosspiece fastened over the necks of two oxen **2.** a pair of oxen

yolk *noun* **1.** the yellow central part of an egg **2.** greasy material present in sheep wool

Yorkshire /ˈjɔːkʃə/ *noun* a breed of large white pig, similar to the Large White. This name is not much used in the UK.

Yorkshire fog *noun* a weed grass *(Holcus lanatus)* able to grow under poor conditions. It is unpalatable and of little value.

Young Farmers' Club *noun* a social organisation for young farmers. Abbr **YFC**

Zadoks scale *noun* a scale used to show the growth stages of a plant from germination to ripening

Zea /ziə/ *noun* the Latin name for maize or corn

zebu /ˈziːbuː/ *noun* a humped cattle of the tropics; a domesticated Asiatic cattle breed with a pronounced shoulder hump and prominent dewlap. In the USA, it is called a 'Brahman'.

zero grazing *noun* the practice of harvesting forage crops and taking the green material to feed housed livestock

zero tillage *noun* a technique using herbicides instead of tilling the soil before sowing an arable crop by direct drilling

zigzag harrow *noun* a light harrow used for final seedbed work, and also for covering sown seeds. The frames are zigzag in shape, with short tines bolted to them.

zinc *noun* a white metallic trace element, essential to biological life. It is used in alloys and as a protective coating for steel.

COMMENT: Zinc deficiency in plants prevents the expansion of leaves and internodes; in animals zinc forms part of certain enzyme systems and is present in crystallised insulin.

zoo- /zəʊə, zuːə/ *prefix* animal

Zoonoses Order /ˌzəʊəˈnəʊsiːz ˌɔːdə/ *noun* an order under which the presence of conditions such as salmonellosis and brucellosis, which affect both animals and humans, must be notified

zoonosis /ˌzəʊəˈnəʊsɪs/ *noun* a disease that a human can catch from an animal, e.g. tuberculosis and disorders caused by Salmonella bacteria

'The USDA considers TB to be a serious zoonosis which is dangerous to humans. It is seen as a grave threat to the cattle industry that will debilitate cattle and cause exports to be stopped.' [*Farmers Guardian*]

zoophyte /ˈzəʊəfaɪt/ *noun* an animal that looks like a plant, e.g. a sea anemone

zootechnology /ˈzuːəʊtekˌnɒlədʒi/ *noun* the use of modern technological advances in animal breeding to increase quality and production

zucchini /zʊˈkiːni/ *noun* an Italian or American name for courgettes, the fruit of the marrow at a very immature stage in its development, cut when between 10 and 20cm long. Zucchinis may be green or yellow in colour.

zygote /ˈzaɪɡəʊt/ *noun* a fertilised ovum, the first stage of development of an embryo

SUPPLEMENTS

Weights and Measures
Agricultural Land Classification
Nitrogen Fertilisers
Gestation Periods
Oestrous Cycles
Zadoks Scale
World Commodity Markets

Weights and Measures

Conversion tables: Metric - Imperial

Length:
1 millimetre (mm)	=	0.0394 in.
1 centimetre (cm)	=	0.3937 in.
1 decimetre (dm)	=	3.9370 in.
1 metre (m)	=	1.0936 yds
1 kilometre (km)	=	0.6214 mile

Weight:
1 milligram (mg)	=	0.0154 grain
1 gram (gm)	=	0.0353 oz
1 kilogram (kg)	=	2.2046 lb
1 tonne (t)	=	0.9842 ton

Area:
1 cm²	=	0.1550 sq. in.
1 m²	=	1.1960 sq. yds
1 are (a)	=	119.60 sq. yds
1 hectare (ha)	=	2.4711 acres
1 km²	=	0.3861 sq. mile

Capacity: 1 cm³
	=	0.0610 cu. in.
1 m³	=	1.3080 cu. yds
1 litre	=	0.2200 gallon
1 hectolitre	=	2.7497 bushels

Conversion tables: Imperial - Metric

Length:
1 inch (in.)	=	2.54 cm
1 foot (ft)	=	0.3048 m
1 yard (yd)	=	0.9144 m
1 rod	=	4.0292 m
1 chain	=	20.117 m
1 furlong	=	201.17 m
1 mile	=	1.6093 km

Weights and Measures *continued*

Conversion tables: Imperial - Metric

Weight:	1 ounce (oz)	=	28.350 g
	1 pound (lb)	=	0.4536 kg
	1 stone (st)	=	6.3503 kg
	1 hundredweight (cwt)	=	50.802 kg
	1 ton	=	1.0161 tonnes

Area:	1 sq. in.	=	6.4516 cm²
	1 sq ft	=	0.0929 m²
	1 sq. yd	=	0.8361 m²
	1 acre	=	4046.9 m²
	1 sq. mile	=	259.0 ha

Capacity:	1 cu. in.	=	16.837 cm³
	1 cu. ft	=	0.0273 m³
	1 cu. yd	=	0.7646 m³
	1 pint	=	0.5683 l
	1 quart	=	1.1365 l
	1 gallon	=	4.5461 l
	1 bushel	=	36.369 l
	1 fluid ounce (fl. oz.)	=	28.413 cm³

US measures

Dry measures	1 US pint	=	0.9689 UK pint
	1 US bushel	=	0.9689 UK bushel

Liquid measures	1 US fl. oz.	=	1.0408 UK fl. oz.
	1 US pint	=	0.8327 UK pint
	1 US gallon	=	0.8327 UK gallon

Agricultural Land Classification

Grade 1	high quality, suitable for intensive arable farming (roots, cereals, vegetables) and horticulture
Grade 2	medium to high quality, suitable for arable farming and intensive grazing (dairy cattle)
Grade 3a/b	medium quality, suitable for hardwood forestry and rotational cropping
Grade 4	medium to poor quality, suitable for rough grazing (beef cattle and sheep), softwood forestry and limited cereal cropping
Grade 5	poor quality, very rough with rocky outcrops, suitable for grazing (hardy breeds) and very limited cropping

Nitrogen fertilisers

		Nitrogen content (approximate)
Chemical:	Ammonium nitrate	33.5%
	Ammonium sulphate	20.5%
	Anhydrous ammonia (liquid)	82%
	Aqua ammonia (liquid)	25%
	Calcium nitrate	15.5%
	Diammonium phosphate	18%
	Low-pressure solution (liquid)	39%
	No pressure solutions (liquid)	30%
	Urea	45%
Organic:	Alfalfa meal	2.5%
	Bloodmeal	13%
	Cattle manure	0.5%
	Chicken manure	0.9%
	Cottonseed meal	6%
	Fish emulsion	5%
	Fish scrap	9%
	Hoof and horn	14%
	Sheep manure	0.9%

Gestation periods

	months	*days*
Cow	9	283-284
Ewe	4-5	144-150
Sow	3	114
Mare	11	340
Goat	4-5	144-150

Oestrous cycles

	length of oestrus	*interval between*
Cow	18 hours	21 days
Ewe	36 hours	17 days
Sow	45 hours	21 days
Mare	120 hours	21 days
Goat	40 hours	20 days

Zadoks Scale

0	**Germination**
00	dry seed
05	root emerges
07	shoot emerges

1	**Seedling**
11	first leaf unfolds
12	second leaf unfolds
13	third leaf unfolds

2	**Tillering**
21	main shoot, one side shoot
22	main shoot, two side shoots

3	**Stem**
31	one node
32	two nodes

4	**Boots**
43	boots visible
49	awns visible

5	**Inflorescence**
52	¼ of inflorescence visible
54	½ of inflorescence visible
56	¾ of inflorescence visible
58	inflorescence complete

6	**Anthesis**
65	½ of anthesis complete
69	anthesis complete

7	**Milk development**
75	medium milk

8	**Dough development**
85	soft dough

9	**Seed ripening**
91	kernel hard
93	kernel loose
95	dormant seed
96	viable seed

World Commodity Markets

Argentina	Bolsa de Cereales, Buenos Aires	grains
Australia	Sydney Futures Exchange	wool, cattle, electricity
Austria	Wiener Börse	raw skins and hides, leather, driving belts, technical leather products, timber
Brazil	Bolsa de Mercadorias & Futuros	gold, coffee, alcohol, sugar, cotton, cattle, soybean, corn
Canada	Winnipeg Commodity Exchange	canola, canola meal, feed wheat, flaxseed, feed barley
China	Shanghai Gold Exchange	gold
	Shanghai Futures Exchange	copper cathode, aluminium ingot
France	MATIF (Marché a Terme International de France)	European rapeseed futures, milling wheat futures, corn futures, sunflower seeds
Germany	Südwestdeutscher Warenbörsen (Mannheimer Produktenbörse, Stuttgarter Waren- und Produktenbörse, Frankfurter Getreide- und Produktenbörse, Wormser Getreide- und Produktenbörse)	grain, fodder, oilseed, eggs roughage, potatoes, fuel oil
	Warenterminbörse, Hanover	potatoes, hogs, wheat, rapeseed, heating oil, recyclable paper
	Bremer Baumwollbörse	cotton
Hong Kong	Chinese Gold and Silver Exchange	gold, silver
Hungary	Budapest Commodity Exchange	grain, livestock, financials
India	Tobacco Board, Andhra Pradesh	tobacco
	Coffee Board, Bangalore	coffee
	Central Silk Board, Mumbai	silk
	Tea Board of India, Calcutta	tea
	Cardamom Board, Cochin	cardamom
	Coir Board, Cochin	coir
	Rubber Board, Kerala	rubber
Italy	Borsa Merci Telematicade Mediterraneo	bergamot orange, essential oil of bergamot, tangerine, orange, lemon, mandarin, grapefruit, oil, wine

Japan	Central Japan Commodity Exchange	gasoline, kerosene, eggs, azuki beans, soybeans
	Hokkaido Grain Exchange	corn, soybean, soybean meal, azuki bean, arabica coffee, robusta coffee, raw sugar futures
	Kanmon Commodity Exchange	broiler, corn, soybean, redbean, refined sugar
	Kansai Commodities Exchange	frozen shrimp, coffee, corn, soybeans, azuki beans, raw sugar, raw silk
	Osaka Mercantile Exchange	aluminium, cotton, rubber
	Tokyo Commodity Exchange	aluminium, gold, silver, platinum, palladium, gasoline, kerosene, crude oil, rubber
	Tokyo Grain Exchange	corn, soybean, coffee, raw sugar, redbean
	Tokohashi Dried Cocoon Exchange	silk cocoons
Kenya	Coffee Board of Kenya	coffee
	East African Tea Trade Association	tea
	Kenya Tea Development Authority	tea
Malaysia	Malaysian Rubber Board	rubber
	Malaysia Derivatives Exchange	crude palm oil, interest rate futures, government securities futures
Netherlands	Euronext, Amsterdam	pigs, potatoes
Singapore	Singapore Commodities Exchange	rubber, coffee
U.K.	Liverpool Cotton Association	raw cotton
	Euronext LIFFE (London International Financial Futures and Options Exchange)	cocoa, coffee, sugar, wheat, barley, weather futures
	London Metal Exchange	aluminium, copper
	International Petroleum Exchange	crude oil, gas oil, natural gas
USA	Mid-American Commodity Exchange	gold, silver, platinum
	Kansas City Board of Trade	wheat, natural gas, stock indexes
	New York Board of Trade (NYBOT, parent company of Coffee, Sugar & Cocoa Exchange, New York Cotton Exchange, New York Futures Exchange)	cocoa, coffee, cotton, sugar

World Commodity Markets *continued*

USA cont.

New York Mercantile Exchange (NYMEX)	crude oil, gasoline, heating oil, natural gas,, propane, coal, gold, propane, silver, platinum, palladium, copper, aluminium
Chicago Board of Trade	corn, oats, soya bean oil, wheat, soya beans, rough rice, gold, silver, Treasury bonds, Treasury notes, other interest rates, stock indexes
Chicago Mercantile Exchange	beef, dairy, forest, e-livestock, hogs, crude oil, natural gas, weather futures, chemical futures, foreign currencies
Minneapolis Grain Exchange	spring wheat
BrokerTec Futures Exchange	government securities
Merchants Exchange	barge freight rates, energy products
NASDAQ LIFFE	securities futures
FutureCom	cattle